Panorama

of International Horticultural Expo.

—International Horticultural Exposition 2011, Xi'an

全景世园 ——2011 西安世界园艺博览会

佳图文化 编

中国林业出版社

图书在版编目（ＣＩＰ）数据

全景世园 : 2011 西安世界园艺博览会 / 佳图文化 主编 . -- 北京 : 中国林业出版社 , 2011.7
ISBN 978-7-5038-6241-0

Ⅰ . ①全… Ⅱ . ①佳… Ⅲ . ①园艺－博览会－西安市－ 2011 －图集 Ⅳ . ① S68-282.411

中国版本图书馆 CIP 数据核字 (2011) 第 123755 号

策划：佳图文化
主编：王志 龙志伟

中国林业出版社 • 建筑与家居图书出版中心
责任编辑：李 顺
电话、传真：（010）83223051

出 版：中国林业出版社（100009 北京西城区德内大街刘海胡同 7 号）
印 刷：利丰雅高印刷（深圳）有限公司
发 行：新华书店北京发行所
电 话：（010）83224477
版 次：2011 年 7 月第 1 版
印 次：2011 年 7 月第 1 次
开 本：889mm×1092mm 1 / 12
印 张：35
字 数：200 千字
定 价：398.00 元

PREFACE

International Horticultural Exposition is the highest-level and professional international exhibition, which is also called International Horticultural Festival. It is a big exhibition for gardening and horticultural works as well as exotic flowers and herbs all around the world that enhances the communication between countries. It is also the largest A1-grade international horticultural exposition integrating cultural and scientific achievements together.

In 2011, Xi'an, an origin of China, an ancient capital, the starting point of the Silk Road, and the hometown of Terracotta Warriors, has become the focus of the world. It is the third time that International Horticultural Exposition came into China after 1999 in Kunming and 2006 in Shenyang.

On April 28, Xi'an International Horticultural Exposition held its grand opening in the Chan-Ba Ecological District, Xi'an.

"Every time when I participate in the International Horticultural Expos, my heart beats faster!", many visitors can't help saying. When walking in international horticultural exhibition, cameras and eyes cannot catch all the things that touch the heart, because each artistic decoration, flower and sculpture are the beauty of art which you may neglect in a hurry. With the passion of "discovering beauty and recording beauty", we have taken photos for all the gardens and cooperated with the chief designers and related units to select the best pictures from thousands of photos and publish them. The book has included almost all the contents of Xi'an International Horticultural Exposition, such as all the exhibition parks, artistic landscape decorations, flowers, grass, signboard and indicating system, squares and service facilities. We have displayed all the aspects of Xi'an Horticultural Expo entirely, professionally and vividly by real pictures, technical drawings, brief introduction, and clear typesetting to ensure the readers to experience the "supernatural workmanship" in the expo.

The landmarks of International Horticultural Exposition 2011 Xi'an are Chang'an Tower, Theme Pavilion, Greenhouse and Guangyun Entrance which are designed to be spectacular and magnificent by radical masters. In addition, the special service zones, including Romance by the Ba River, Southeast Asian Street, European Avenue and Ba Liu Yi, have embraced the landscape to create "the fun of landscape and leisure".

In the Ecological District of 4,180,000m^2, there are 109 outdoor exhibition gardens, including Master Garden, Creativity Park, World Garden, Feature Park, Enterprise Park, International Garden, Domestic Garden, etc. Nine masterpieces of world top designers' are gathering in the Master Garden which is extremely rare. The representative gardening works of 43 countries have made a communication on horticulture in the International Garden. The Domestic Garden has included regional and classic gardens of 40 provinces of China. Nine enterprise gardens have appealed the theme of "scientific environmental protection" to the whole world. Five world gardens have shown the charm of classic European gardens to Chinese people. Ten universities have interpreted the multiple meanings of the "Creativity Park". And the Feature Park has created five botanical gardens.

Xi'an International Horticultural Exposition has also exhibited delicate works of art and sculptures from home and abroad. We hope that our choreographed book will bring the beauty of gardening, horticulture, architecture and art to the readers.

Editor

June 2011

前言

世界园艺博览会是最高级别的专业性国际博览会，也叫世界园艺节。它是世界各国园林园艺精品、奇花异草的大联展，是以增进各国的相互交流，集文化成就与科技成果于一体的规模最大的 A1 级世界园艺博览会。

2011，华夏源脉、千年帝都、丝路起点、秦俑故乡——中国西安聚焦世界的目光，这是继 1999 年昆明、2006 年沈阳之后，世界园艺博览会第三次来到中国。

4 月 28 日西安世界园艺博览会在西安浐灞生态区盛大开幕！

“每一次与园艺会相会，都心跳不已！”这是众多游览者情不自禁发出的感叹。游走于世园会上，镜头和肉眼都无法把触动心灵的景物一一揽收，因为即使您在匆忙中略过的那一个小品、一个个花雕都是美的艺术。正源于发现美记录美的情感迸发，特地对所有展园进行了专业拍摄，同时诚邀了主要设计师及相关单位的通力合作，从数千张实景图中进行严谨而专业的筛选才得以顺利出版。全书中包含了本次博览会的所有展园以及园林小品、花草绿化、标识指示系统、广场、配套服务设施等几乎涵盖所有西安世园的内容。我们以最真切的实景与技术工艺图相结合，用简练精辟的描述、开阔明朗的排版，力求最全面、最专业、最生动地展示西安世园的方方面面，让读者一探世园会设计的种种“巧夺天工”。

世园标志性建筑分别是长安塔、创意馆、自然馆和广运门，恢弘宏大是标志性建筑的闪亮之处，当中激进前卫的大师作品极具震撼力。此外还有灞上人家、椰风水岸、欧陆风情以及灞柳驿等特色服务区，这些服务区与景观相互辉映，打造“山水与闲情之乐”。

在总面积 4 180 000 的生态区上，西安世园会的室外展园共 109 个，如大师园、创意园、世界庭院、专类园、企业园、国际园、内地园等。9 个国际顶尖级设计师的作品相聚于大师园，这种现象极为罕见；43 个国家的代表作品进行了国际级园艺的交流；40 个中国不同省份重塑传统地域性的经典；以科学环保为大主题的 9 大企业园呼吁声音传遍各地；5 个世界庭院展现欧罗巴洲的经典庭院，让中国人民一睹西式庭院的风采；10 所高校诠释了庭院“创意”的多重含义；专类园打造了 5 个植物园奇观……

西安世园会还展示来自国内外的精美艺术品、雕塑等，同时更希望通过我们的精心编排能让读者充分领略园林、园艺、建筑、艺术之美。

编者

2011 年 6 月

CONTENTS
目录

Master Garden
大师园

Creativity Park
创意园

International Garden
国际园

Hong Kong Garden, Macau Garden and Taipei Garden
港澳台北园

223

Domestic Garden
内地园

235

绿地集团

Feature Garden

专类园

347

Characteristic Creations

特色创意

363

Master Garden

大师园

In order to ensure the exhibition level of International Horticultural Exposition 2011 Xi' an, nine master exhibitions are displayed for the first time. The Master Garden is located in the International Exhibition, consisting of nine 1, 000m2 gardens with different styles which are strongly originated and culture-integrated.

Nine gardeners from China, U.S., Britain, Germany and Netherlands present their remarkable works. Because the exposition is held in China, the designers blend the Chinese spirit and elements into their works in various patterns. Meanwhile, these works have shown the highest level of each sect in the world.

Wang Xiangrong, the only Chinese designer of the master garden and the designer of Quadrangle Garden, puts that the whole exposition is a garden of greatest creativity. It is a unique case like a splendid fashion show.

Characterized by nature, the nine exhibitions have created landscapes like mountain and water, spaces and caves, presenting the designers' trial and exploration to display western culture by Chinese architecture and expressing the harmonious relationship between human and nature. The master garden would be the permanent sight of Xi' an after the Horticultural Exposition as the memorable gardening work.

为了保证 2011 西安世园会园艺的展出水准，2011 西安世园会第一次设置了 9 个大师展园。大师园位于世园会国际展园区内，由 9 个占地各 1000 平方米左右、风格迥异的园子组成，是原创性极强、中西园林文化融合的佳作。

来自中国、美国、英国、德国、荷兰等国的 9 个园艺大师奉献了自己的精彩作品，由于世园会在中国举办，各国设计师在进行创作时都不约而同地以各种形式将中国精神、中国元素融入到作品中，同时这些作品也将反映出当今国际各个流派的最高水平。

大师园中唯一的中国设计师、四盒园的设计者王向荣认为，大师园是整个世园会最具创造性的花园，属于特殊案例，如同一场华丽的时装秀而绝非普通意义上的服装。

9 个展园以自然为特色，创造了山水、空间、洞穴等多种景观，处处体现出用中国建筑表现西方文化的尝试和探索，以不同形式诠释了人与自然和谐共生的关系。大师园作为具有纪念意义的园林景观作品，在世园会结束后将成为西安永久的景点。

Garden with a Labyrinth in the Mountain

山之径迷园

Chief designer: Benedetta Tagliabue

About the Designer: The registered female architects are less than 10% in the world and Benedetta Tagliabue is one of them. As one of the founders of the new and prominent Spanish architecture firm EMBT, receiver of Spain 1st National Award of Building Economy and of the European ITALSTAD Award, she is also in charge of the designer of the Spanish Pavilion in World Expo 2010 Shanghai.

主要设计师：Benedetta Tagliabue

设计师介绍：Benedetta Tagliabue是全世界只有不到10%的女性注册建筑师之一，是西班牙新锐建筑事务所EMBT创立人之一，曾获西班牙第一届全国建筑经济奖、欧洲ITALSTAD奖等，主持设计了上海世博会西班牙国家馆。

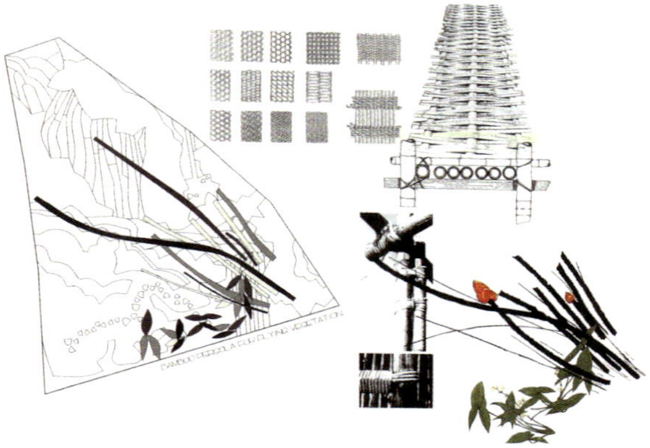

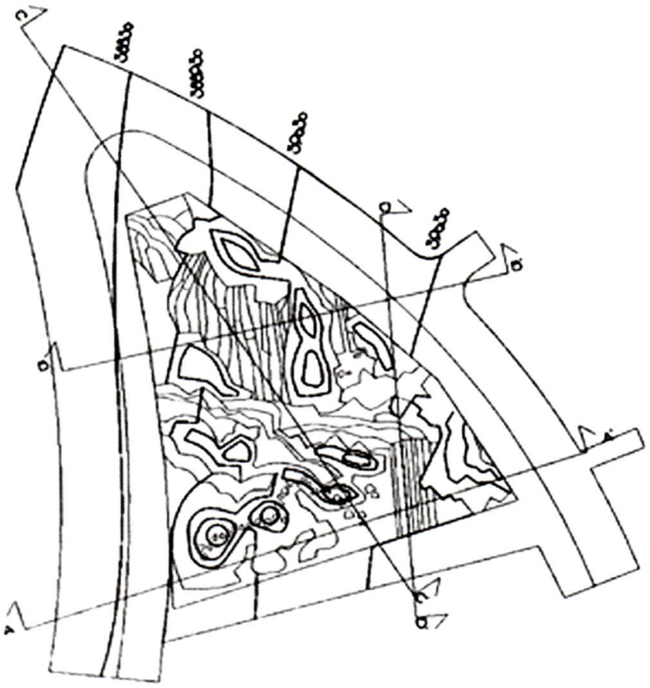

Introduction:

Inspired by Chinese landscape painting, it aims to create a natural manmade landscape in the Garden with Labyrinth of Mountainous Paths. The garden is designed in triangle shape. Trees and other plants are scattered in the hilly land orderly. The materials used in the garden are gotten from nature. The arrangement of the trees has considered the microclimate: pines and cypresses are planted in the higher land, flowering shrubs and herbaceous plants are set in the lower land. The garden is enclosed by rattan fence which are covered by climbing plants to ensure the privacy of the garden.

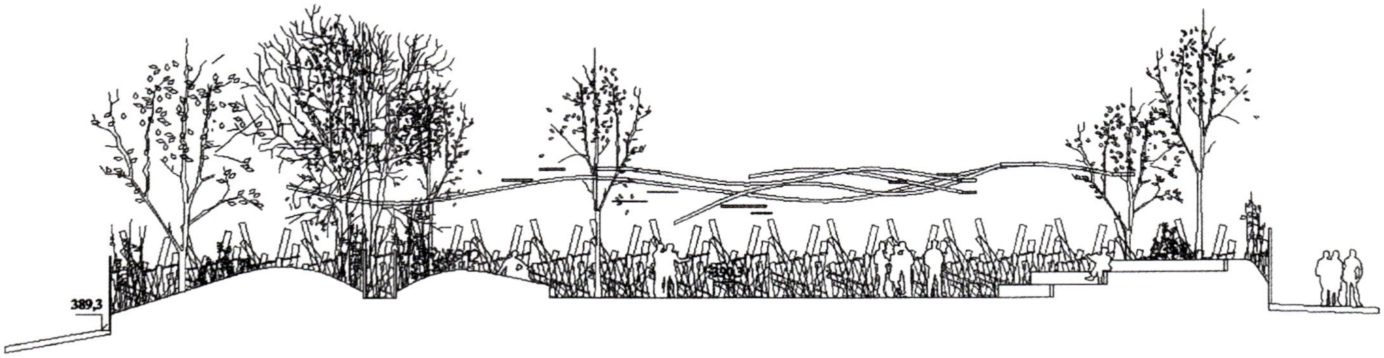
389,3

390,3

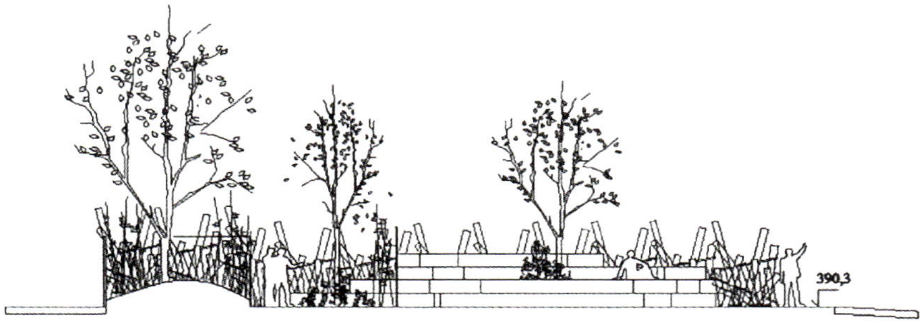
390,3

390,3

展园介绍：

山之迷径园的设计灵感源自中国山水画，致力于构建一处人造景观，使其尽可能表现自然。整个园子呈一个三角形展开，三角形的山体上，树木等植物错落分布，密而不乱，为花园所选的材料全部取自天然。树木的排布考虑到了生长各需的小气候，松柏类种植在花园中地势较高的地方，花灌木和草本植物则种植在地势较低处。整个花园由藤编篱笆围合起来，以确保花园的私密性，并在其上爬满攀援植物。

Passageway Garden

通道园

Chief Designer: Vladimir Sitta

About the Designer: Vladimir Sitta, professor of University of Western Australia and president of TERRAGRAM in Australia. His design style is full of romanticism, he once received Berlin Peter Joseph Ronna National Awards in 1981 and 1986, and Certificate High Distinction of Australia Society of Landscape Architects in the year of 2002.

主要设计师：Vladimir Sitta

设计师介绍：通道园的建筑师 Vladimir Sitta 是西澳大利亚大学教授，澳大利亚 TERRAGRAM 总裁。他的设计风格带有浓重的浪漫主义色彩。曾获得 1981 年、1986 年度柏林彼得约瑟夫伦讷国家奖、2002 年度澳大利亚景观建筑师学会特优奖。

Introduction:

Walls distributed in the garden, they are dotted with some holes, let people to catch a glimpse of indistinct scenery in the front garden. At the end of the garden, there is a moon gate, it is a metaphor of life cycle. Twisty pine seems to challenge the gravity, testifying that it will stay evergreen, this eternity form a contrast with the instant of the plants in the garden. Visitors can sow the seeds of the plants in the gravel ground, so the seeds would taking root and germinating at will in the gravel and present different states. Anytime anywhere, life of plants is seeding, breeding and sprawl.

展园介绍：

这座花园里主要分布着墙体，其间点缀着一些开口，让人们可以一瞥前方花园那若隐若现的景色。花园的尽头有一扇月洞门，这也是对生命周期的隐喻。扭曲成型的松树似乎在挑战地心引力，证明它的万年长青。这种永恒与花园里植被的瞬间形成了对比。在园子里，游客可将植物种子撒播在砂砾地上，种子可在砂砾地上随意扎根和发芽，并呈现不同状态。无论何时何地，植物生命都是播种、繁殖和蔓生。

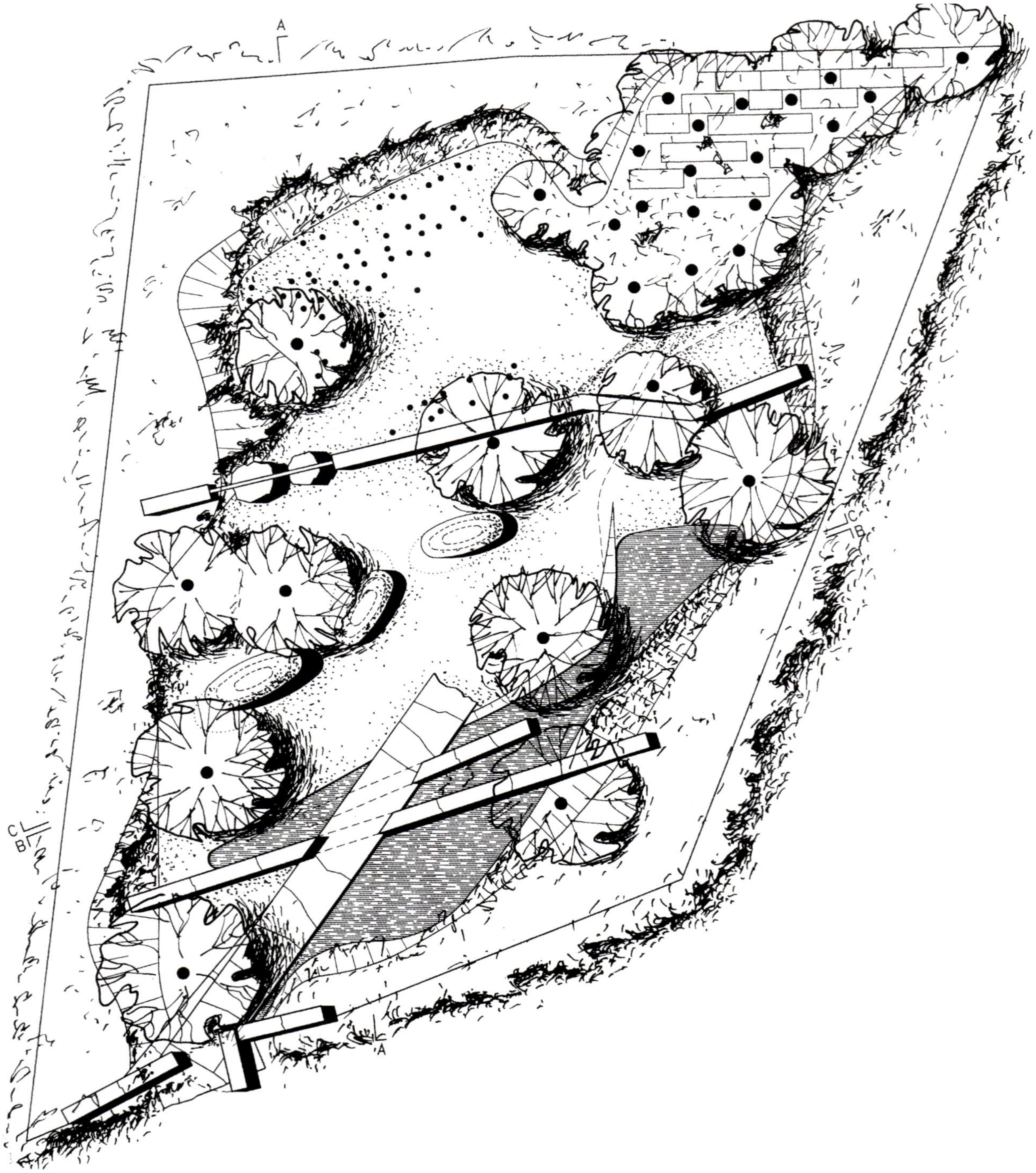

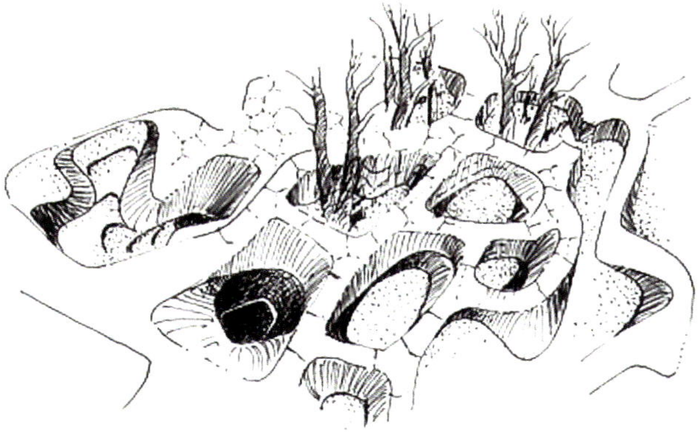
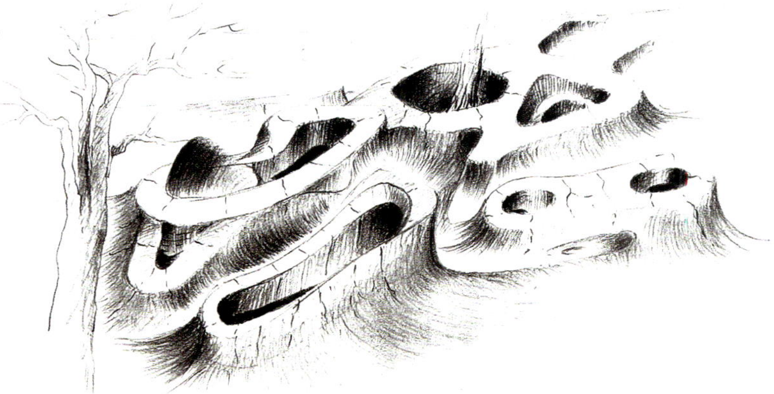
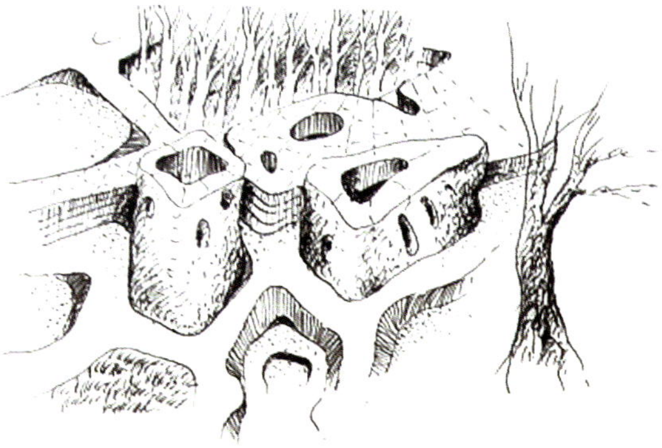

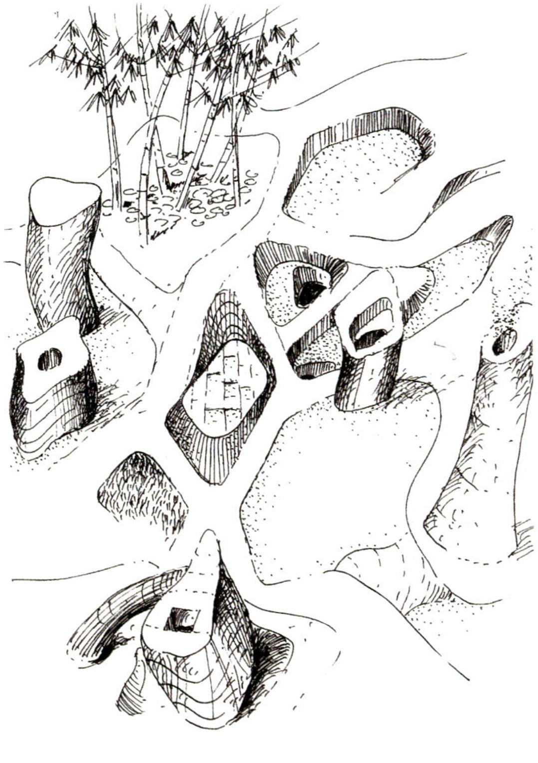

Labyrinth Garden

万桥园

Chief Designer: Adriaan Geuze

Architect Introduction: Adriaan Geuze is a famous Dutch landscape architect, professor for Landscape Architecture of Harvard University and the leading architect of West 8 Urban Design & Landscape Architecture.

主要设计师：Adriaan Geuze

设计师介绍：Adriaan Geuze 是荷兰著名风景园林师，哈佛大学风景园林学教授,West 8 城市规划与景观设计事务所主创设计师。

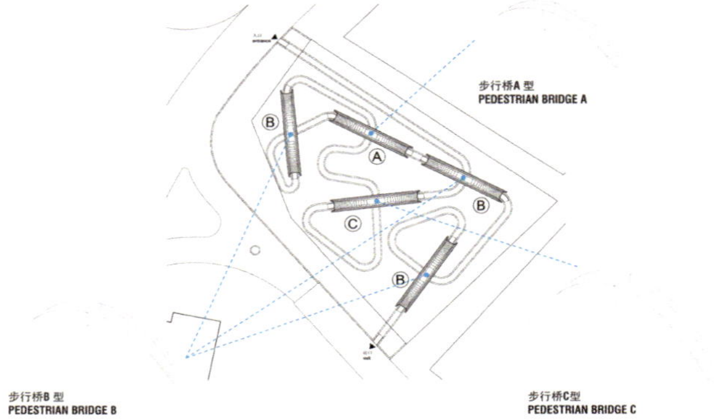

Introduction:

The design concept of Garden of Bridges aims at emulating the impact on vision, size and psychology from natural landscape. The Garden of Bridges is located beside the lake and near the center of the garden area. At the beginning of the design, the architect wanted to create a special exhibit garden according to the geographical advantage. The design concept is simple and powerful: bridges, paths and bamboos.

There is only one entrance and one exit for a gravel path that runs through the whole garden. Getting on the top of the bridge through the narrow and steep stairs, one could overlook not only the bamboo grove underneath but also all the magnificent gardens of the horticultural exhibition with carved beams and painted rafters. Every bridge is selected exquisitely like the layers of the mountains and corresponds with the overall layout of the horticultural exhibition, giving different view and experience to the visitors on the bridges.

In the Garden of Bridges, the modern concrete pouring technique is adopted. There are three types of the five bridges: one is high, three are medium and one is low. In order to control the cost, all bridges are manufactured in the factory by concrete molds. Every bridge is spliced by two identical half-arc concrete molds, becoming the arch bridge. The inner surface of the bridge is kept glossy except for the handrails which help the visitors climb. The external structure is decorated by the concrete turning model of bamboo printing.

The five prefabricated concrete bridges will save more materials and cost than cast-in-situ concrete. The remote industrial manufacturing could strictly control the quality and simplify the uncertain factors in the construction site.

The architect had tried different template texture and finally chose the projection pattern of bamboo leaf. The bamboo leaf is from the nature and one of the common theme of Chinese ink paintings. After the concrete slurry is inserted into the template, the template is withdrawn before the concrete totally curdles. The texture would change influenced by the humidity of the concrete and comes out rich effect. After the module of the bridge is formed, the internal and external surfaces of the concrete would be painted with red epoxy, making the bridge more and more beautiful as time goes by.

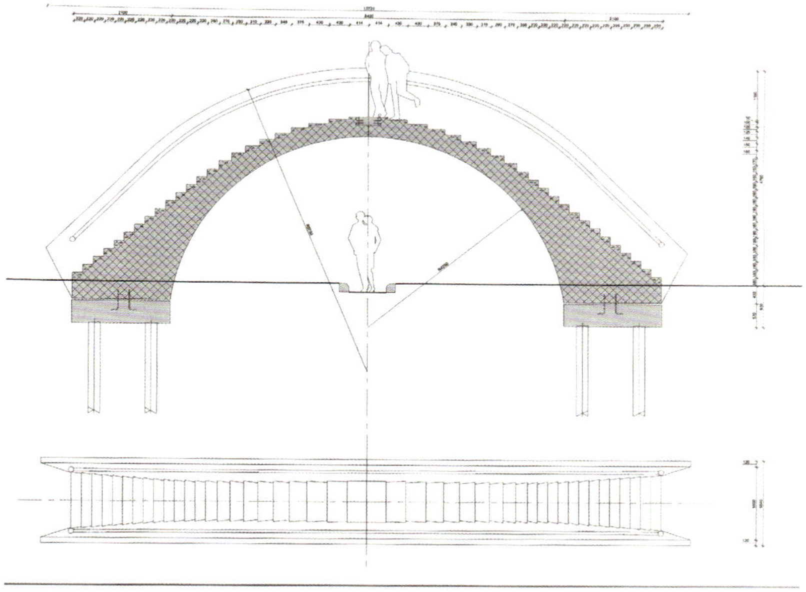

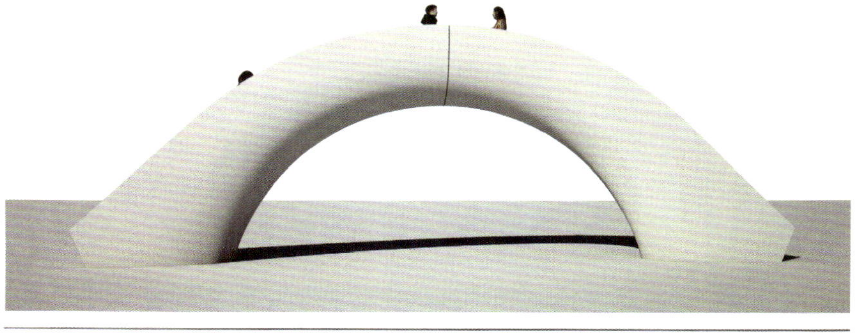

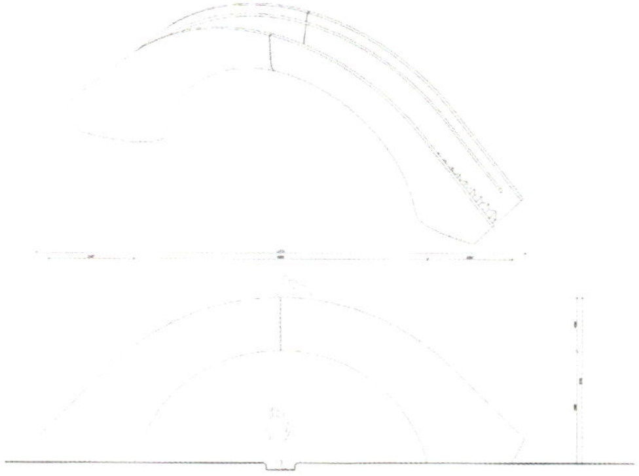

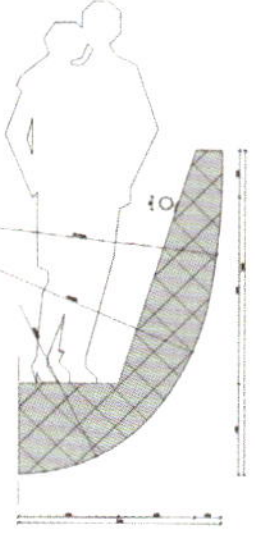

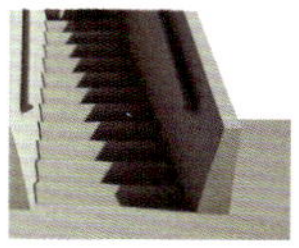

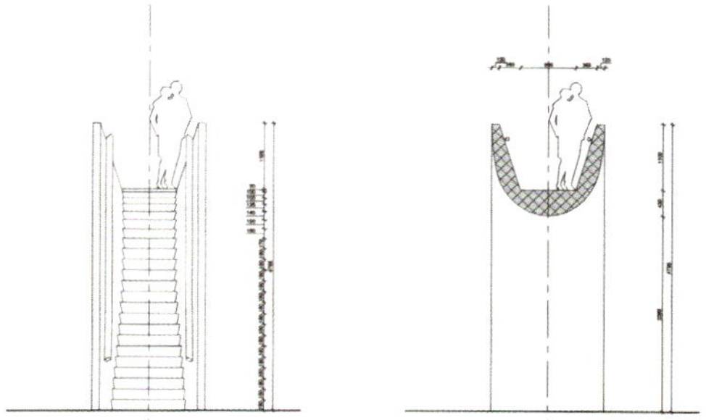

展园介绍：

设计概念万桥园试图效仿的是自然山水带给人的视觉、尺度和心理上的冲击力。万桥园所在区域邻湖且接近园区中心，设计之初，就希望借助地理优势打造一座别样的展园。设计的概念非常简单有力：桥、小径和竹子。

一条砾石小径只设一个入口和一个出口，贯穿全园。沿着狭窄陡峭的楼梯艰难地登上桥顶，一目了然的不仅仅是远在脚下的竹林、周围雕梁画栋精彩绝伦的园林，世园会的山山水水已尽收眼底。每座桥的位置经过精心选择，如桂林的山峰般层层叠叠，亦与世园会的总体布局相呼应，使参观者站在每座桥上都能得到不同的视角和景观体验。

万桥园里，桥的制作工艺都采用了现代的混凝土浇筑技术。万桥园五座桥共分高中低三种类型，一高三中一低。为控制成本，所有的桥都通过混凝土模具在工厂完成。每座桥都是由完全相同的两个半弧混凝土模件现场拼接而成，合在一起就构成拱桥造型。桥的内表面保持光滑，仅留有扶手方便参观者攀爬。外部结构以混凝土翻模竹叶印花装饰。

这五座预制筑模混凝土桥将比现浇混凝土节约材料和成本。异地工业化制造可以严格控制质量，简化现场施工中的不确定因素。

设计师在设计中尝试各种不同的模板纹理，最终选择了竹叶的投影图案。竹叶取材于自然主题，也是中国水墨画最常见的题材之一。模板嵌入混凝土浆表面后，在混凝土尚未完全凝固时取出，形成的纹理受混凝土湿度影响变形，在似与非似间达到丰富而非单一的肌理效果。在桥的模块成型后，将以手工给混凝土内外表面，特别是外部纹理涂以红色环氧树脂。中国红是最佳的选择。由于纹理凹凸不一，岁月和环境将进一步雕蚀这些纹理，使桥随着时间的演进呈现出愈来愈美的状态。

Quadrangle Garden

四盒园

Chief Designer: Wang Xiangrong

About the Designer:Wang Xiangrong is the chief designer of Atelier DYJG, the professor and doctoral supervisor in College of Landscape Architecture, Beijing Forestry University, associate editor of journal— Chinese Landscape Architecture, chief editor of journal— Landscape Architecture, member of Editorial Committee of magazines of Landscape Design and Modern Landscape Architecture, honorary professor of Hong Kong University, guest professor of Southwest University, Nanjing Forestry University and Zhejiang Forestry College.

主要设计师：王向荣

设计师介绍：北京多义景观规划设计事务所主创设计师、北京林业大学园林学院教授、博士生导师，《中国园林》学刊副主编、《风景园林》学刊主编，《景观设计》、《现代园林》等杂志编委，香港大学荣誉教授，西南大学、南京林业大学、浙江林学院等大学客座教授。

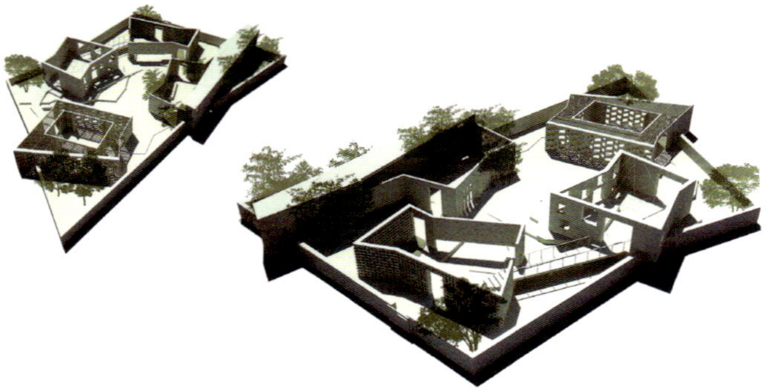

Introduction:

Quadrangle Garden, as one of the master gardens of Horticultural Expo, is named after Chinese homophonic quadrangle courtyard. Its general shape is similar to the layout of quadrangle courtyard, the difference is, the small parks in quadrangle garden are in irregular shapes, it seems to get rid of the traditional constraint of square model to demonstrate its own charm.

In the north of the Garden, there is an open door, which is the entrance. Stepping into the entrance, you will see the Spring Box of the Garden, the box is constructed by white-washed wall and stone, create a spring-like beautiful scenery. Sitting on the bench, you may see the landscape in the main courtyard and the surrounding booming bamboo grove through the windows on the wall.

Summer Box: It is a flower pergola made by wood, crawling with vines, which shaped a nice place for enjoying coolness. Across the corridor, as you were in a farm house of northwest. For the different overlap joints, the pergola reflects a strong light effect and gives you a dramatic transparent sight experience.

Autumn Box: It is built up by stone, its ground is 1m higher than the central patio, there are lots of square holes on the wall, formed frames one after another. The roof of this box is paved with metal net, seen from afar, it looks like a stone garden with a metal cap on the top.

Winter Box: it is built up by black bricks. Due to the different ways of bricklaying, the four faces of this box express different

South elevation

West elevation

transparent degree as well. The interior and exterior of this box are white sandstone ground, just like winter snow.

The ground is in the integrated dynamic line shape, which adds somewhat grace to this garden. Fresh plants, white sandstones and black sandstones adorn inside, just like a Chinese brush drawing. In terms of design, quadrangle garden pursues space artistic conception and poetry, and people' s meditation and experience in the garden, which demonstrated the Oriental' s comprehension to the garden.

展园介绍：

作为世园会大师园之一的“四盒园”，以四合院的谐音来命名。四盒园的大概形状也有点类似四合院的布局，但是不同的是四盒园内的几个小园区都是不规则的形状，似乎是在强力摆脱传统四合院方方正正布局模式的束缚，各自展示着自己的魅力。

四盒园的南边有一扇可以开启的大门，这便是花园的入口。踏进主要入口便是四盒中的“春盒”。由白粉墙和石材建造的“春盒”，营造了春天般的美景。坐在长椅上，透过墙上的门窗，可以看到主庭院的景色以及四周春意盎然的竹丛。

“夏盒”：一个用木头做成的花架屋。在这里，葡萄藤爬满了花架，搭成了一个纳凉的好地方，穿过绿荫的长廊，如同走进一个西北的农家院。由于木头的搭接方式不同，花架具有强烈的光影效果和戏剧性的视线通透体验。

“秋盒”：由石头砌筑而成，其地面比中心庭院高出 1m，墙上有许多正方形的窗洞，形成一个个画框。秋盒的顶面是由金属网铺成的，远远看去就像是石头砌成的园子上扣了一个“金刚顶”。

还有一个由青砖砌筑而成的盒子，那是“冬盒”。由于砖砌的方式不同，盒子的四个面也表现出了不同的通透程度。“冬盒”的里外是白色沙石地面，如同冬雪。

花园的地面统一在整体的流动线形之中，给园子增添了几许柔美之感。葱翠的植物、白色沙石和黑色沙石点缀其中，宛如水墨画一般。四盒园的设计上追求花园的空间意境和诗意，以及人们在花园中的静思与体验，展示了东方人对花园的理解与感悟。

Botanist Garden

植物学家花园

Chief Designer: Bridget Baines, Eelco Hooftman

About the Designer: Bridget Baines and Eelco Hooftman are the leading figures in Edinburgh College of Art, founders the renowned landscape design studio Gross. Max, receivers of international awards such as the First Prize in Competition of Royal Institute of British Architects (RIBA) and the First Prize in Colchester Invitational Tournament.

主要设计师：Bridget Baines, Eelco Hooftman

设计师介绍：Bridget Baines, Eelco Hooftman是英国爱丁堡艺术学院领军人物，著名Gross. Max景观设计事务所主创人，曾获英国皇家建筑师学会竞赛一等奖、科尔切斯特邀请赛一等奖等国际奖项。

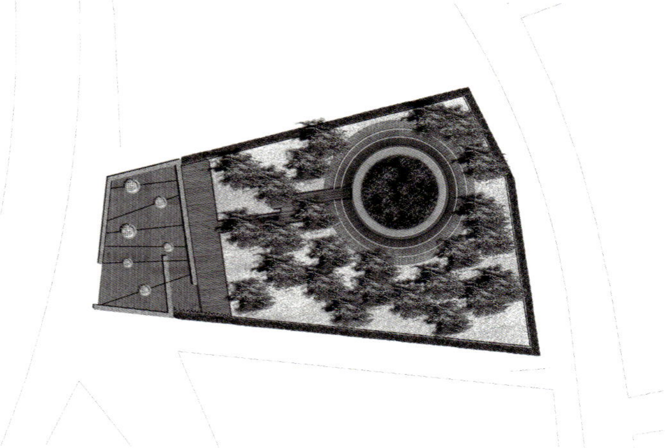

Introduction:

The design idea of "Botanist Garden" is in memory of E. H. Wilson, one of the most famous British botanists. He had investigated in our country several times in the first 11 years of the 20th century.

In order to memorize this botanist who was close to China, the two architects absorbed many Chinese elements. Enlightened by China's Terracotta Army, they use terracotta as the major raw material for the garden pavement and wall construction, combining the local prefabricated bricks and flat tiles. In the research of the project, they visited the north part of Xi'an, Chenlu Ancient Town of Tongchuan City – the famous Town of Pottery, the brick factory of Fuping Town and the International Pottery Museum.

The Botanist Garden contains a series of special space: front yard, outer courtyard and inner room. In the front yard, the transition from outside the horticultural exhibition to the inner garden is made. The floor of the courtyard is covered by the embedded clays and flat tiles, enclosed by the brick wall. The baskets filled with crickets are hung up by the electric wire which passes through the front yard.

The most significant feature of the garden is the plants. There are

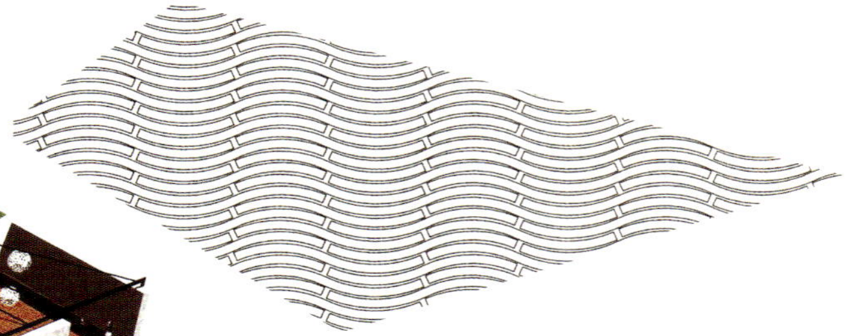

many Metasequoia glyptostroboides in the outer courtyard of the central courtyard. Under these trees there is a large sheet of flowering plants and ferns including Meconopsis punicea and Meconopsis integrifolia.

There are thousands of plants searching from every corner of China by E. H. Wilson in the garden. Metasequoia glyptostroboides was the living fossil of Cupressaceae before and then discovered by a Chinese botanist in the transition zone of Hubei and Sichuan. While the Meconopsis punicea and Meconopsis integrifolia, described as the most beautiful Alps plants, were said to be first found by Wilson in the highland of Tibet. The garden inside the enclosure has the special plant varieties such as Lilium regale, the most famous discovery of Wilson.

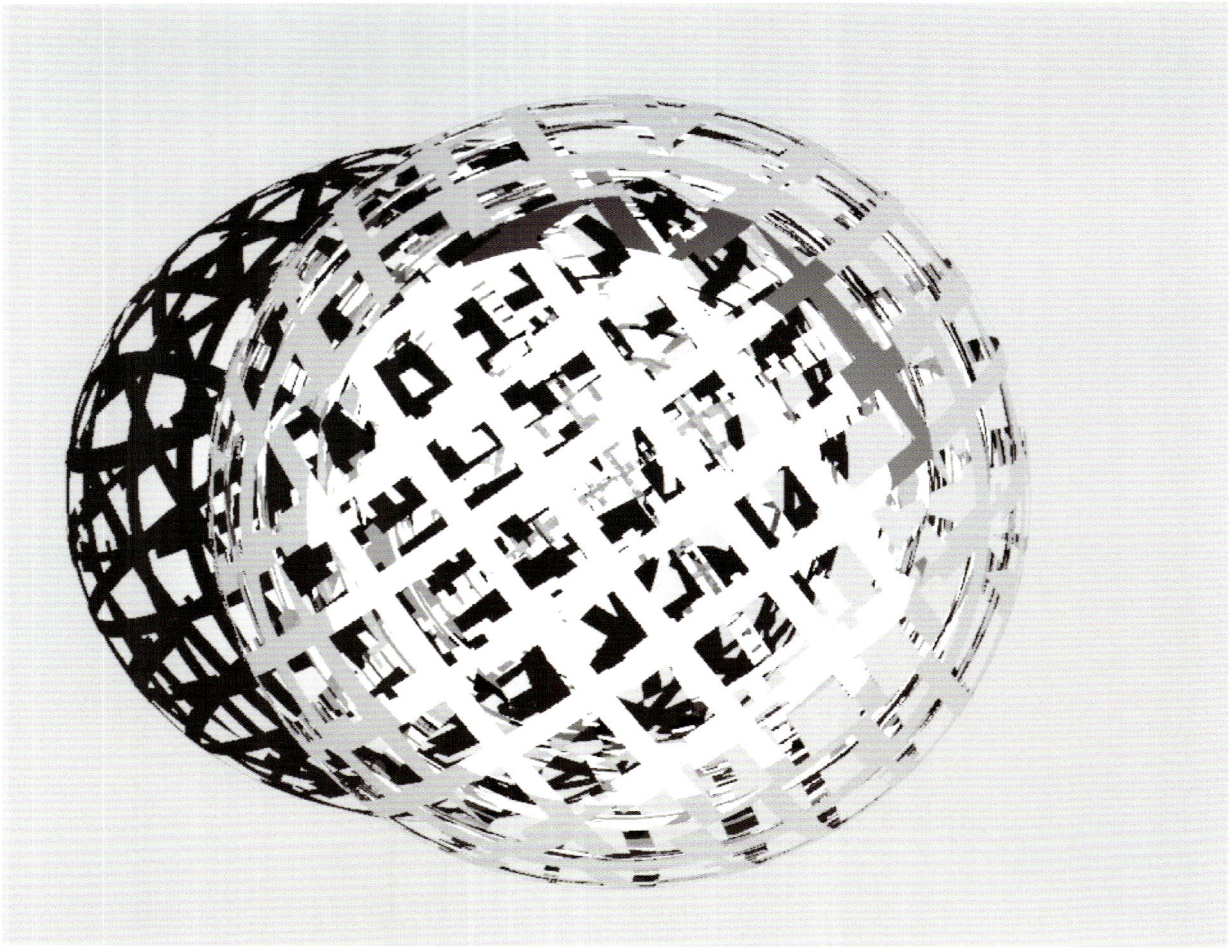

展园介绍：

“植物学家花园”的设计思路是为了缅怀植物学家 E・H・威尔逊。他是大不列颠最著名的植物引种家之一，在 20 世纪的前 11 年中他曾多次到我国考察。

为了缅怀这位与中国有着密切关系的植物学家，两位设计师在设计中吸收了许多中国元素。他们受中国兵马俑的启发，在花园地面铺装和墙体施工中采用赤土作为主要的原料，所选材料是当地预制砖块和平板砖的结合。在项目的调研中，他们参观了位于西安北部、有着“陶器之乡”的铜川陈炉古镇以及富平县的砖厂和国际陶艺博物馆。

“植物学家花园”包括了一系列的独特空间：前庭、外院和内室。前庭营造了从世园会的外部到花园内部空间的过渡。庭院的地面用内嵌的陶土和平板砖进行铺装，四周用砖墙进行围合。电线将装着蟋蟀的竹篮悬挂起来，穿过前庭的空间。

这座园子最大特色在于这里种植的植物。中心庭院的外院种有许多水杉，在这些树林下，铺着一大片开花植物和蕨类植物，有红花绿绒蒿和全缘绿绒蒿。

而在园内的花园，将向游客展示成千上万种由 E・H・威尔逊在中国各地搜集而来的植物。水杉最初被当做一种柏科活化石，随后被一位中国植物学家在湖北和四川的交接带发现，而红花绿绒蒿和全缘绿绒蒿，这两种植物据称都是由威尔逊在西藏的高地中首次发现的，威尔逊将这两种植物描述成为最美丽的阿尔卑斯山植物。围合内的花园里特殊植物品种包括峨眉百合，这正是威尔逊最有名的发现之一。

Loess Garden

黄土园

Chief Designer: Stig L Andersson

Introduction of the Designer: It is mainly designed by Stig L Andersson, famous landscape architect in Denmark, leading figure in landscape design in Scandinavia, and the founder of the renowned landscape design corporation SLA. Stig regards landscape architecture as a process but not a kind of design.

主要设计师：Stig L Andersson

设计师介绍：黄土园由丹麦著名景观设计师，北欧景观设计领军人物，著名的 SLA 景观设计公司主创人 Stig L Andersson 设计。他将风景园林视为一种过程，而不是一种设计。

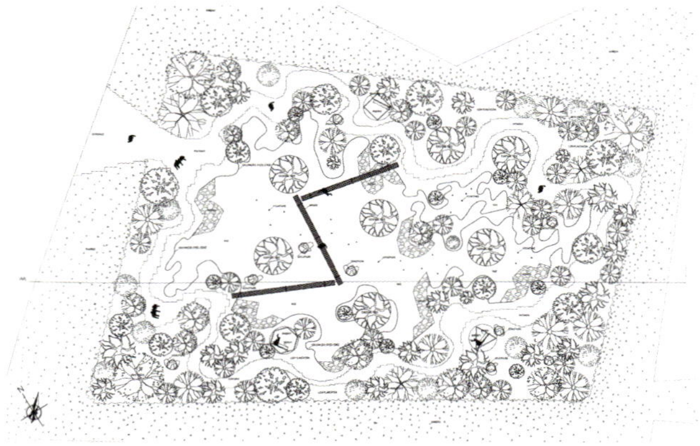

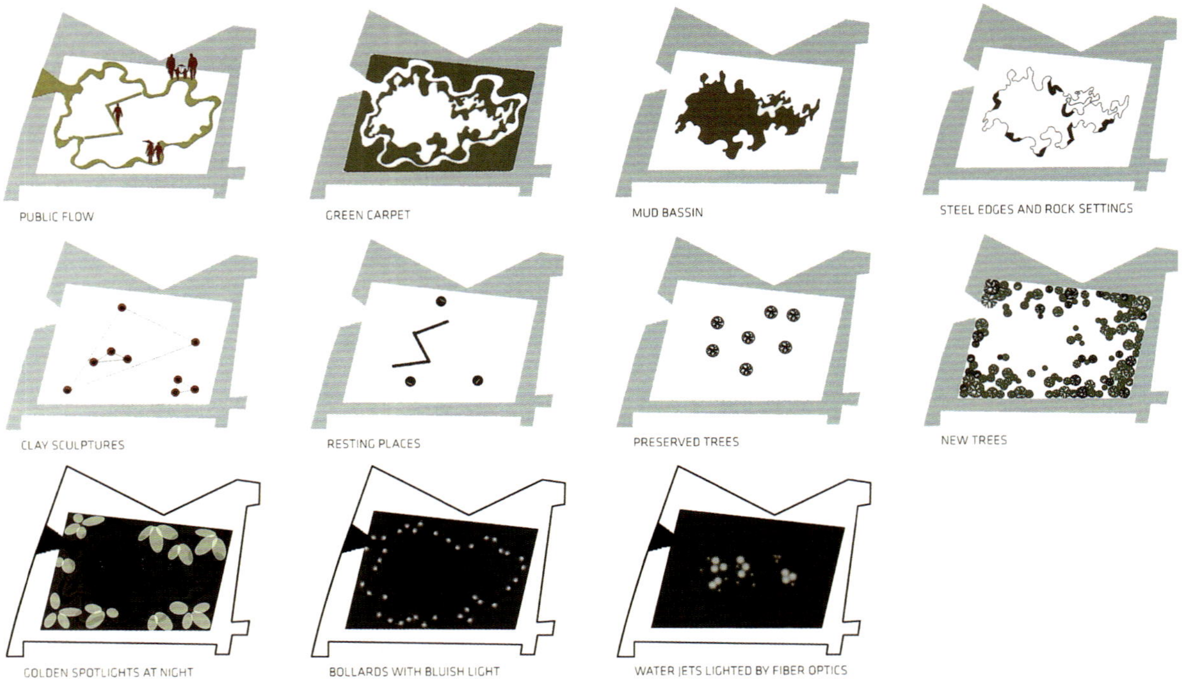

Introduction:

In Loess Garden, all the elements are concerned with "earth". It features clay elements and an irregular flat-bottom water pool. All the loess is collected from the Yellow River. The civilization of "earth" lies in the power of storage. The project has also collected earth from seven rivers in the north of Xi' an and other provinces of China. After dried in the sun, the earth is kept in the local warehouses. The water pool is filled up again by the dry earth.

The visual focus of the garden is the irregular flat-bottom water pool which is 315 m^2 in size and 40cm in depth. When the pool dries up, the clay begins to crack and present a moon-shaped surface. And when the fountain wets the clay, the clay appears again to create marvelous phenomenon. The water of the fountain all collects from rivers and rainwater.

At the bottom of the water pool, there are some inclined sculptures and vertical gold-plating steel edge to show the perfect combination of art and modernity. The stones are local materials and the steel edge is piled up by vertical materials to keep natural appearance. Zigzag form has made the pool lively and mellow.

Nine terra-cotta clay sculptures in the garden are handmade by baking to carry on the traditional handicraft of hundreds of years. Along the small path, three rest stops are designed with brick

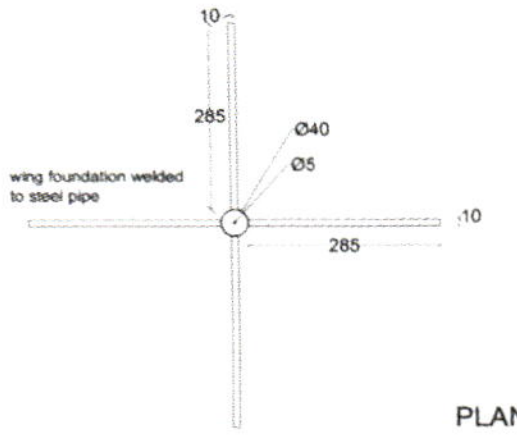

PLAN

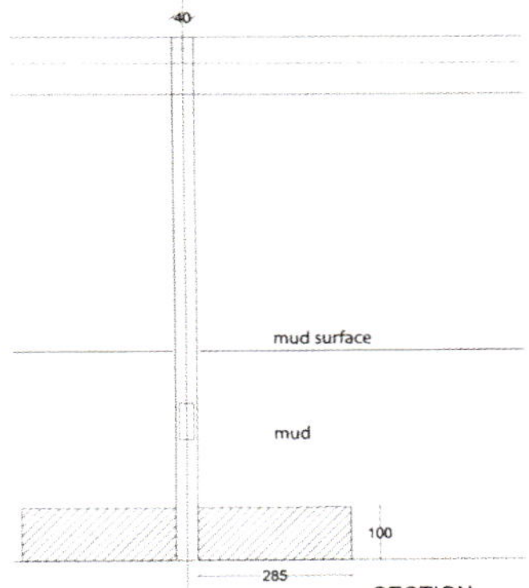

SECTION

FOUNTAIN PRINCIPLE

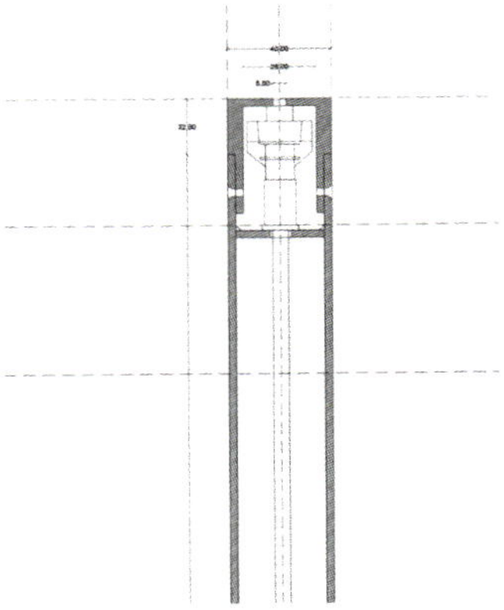

DETAIL SECTION

BRICK PATH PRINCIPLE

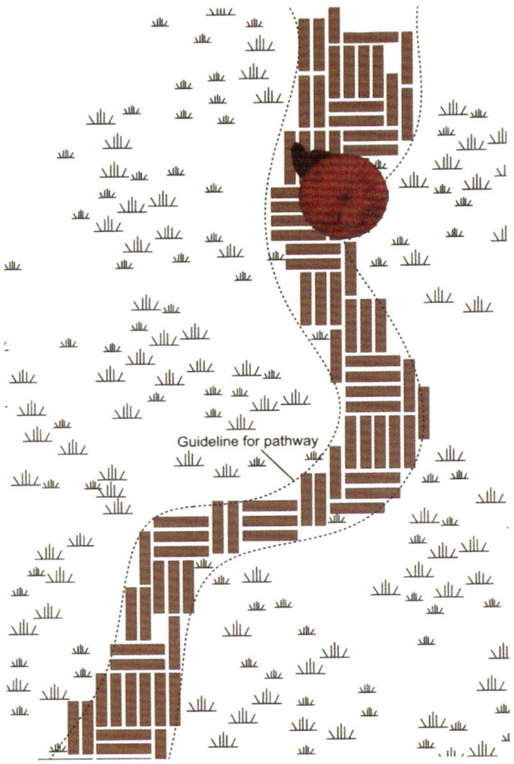

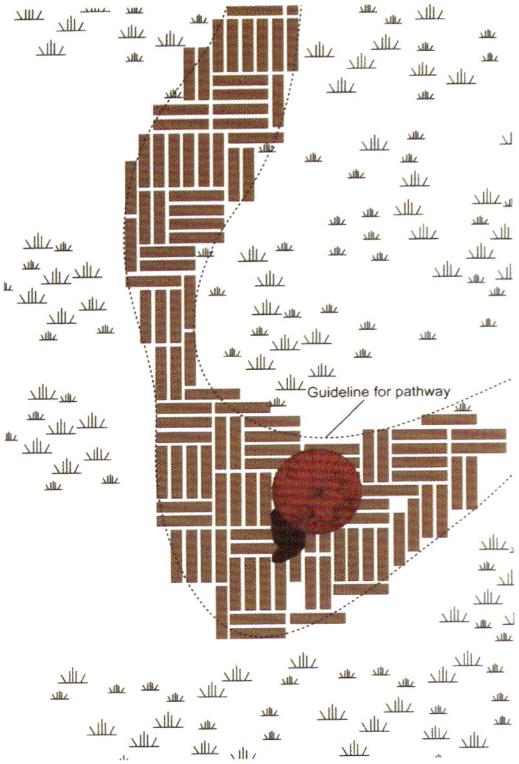

STEEL BRIDGE PRINCIPLE

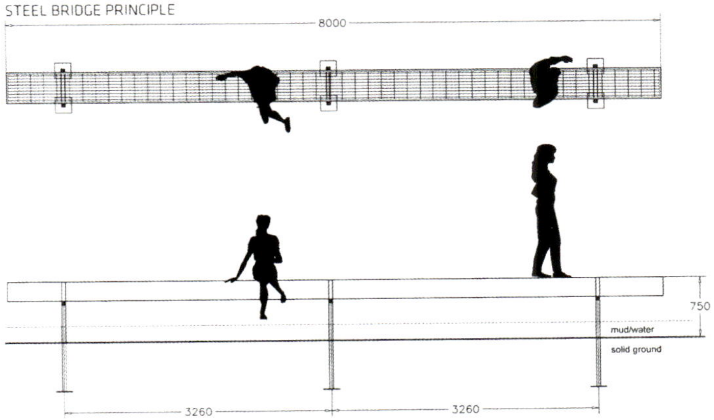

bases, looking like moon holes carved in the traditional garden walls, for passing and light transmitting. The geometrical window holes in the wall have led the sight to the unique landscape. Around the pool, there is a 140m long brick road running through the garden. The road becomes narrow or wide like a zigzag river passing through a series of obstacles and flowing forward. The garden is in rectangle shape, surrounded by luxuriant bamboo forest. Seven retained trees and 83 newly planted trees are standing in the garden. Five extraneous flowering plants are introduced and set in different landscape spots. And radix ophiopogonises are planted on the ground.

There is also a small bridge in the garden. It is made out of shining yellow iron bars with three segments which are connected in different directions to form a big "Z". It has combined hardness with softness perfectly.

Fiber optical technology is applied to lighten Loess Garden at night. Designed with lighting system, the water of the fountain will jump in the white light. Lower pillars along the small path are designed with down lamps which give off azury light to highlight the zigzag path and lighten the trees or plants on ground. The spotlight with filter has projected the head of frog on the mud surface. Golden light has increased the atmosphere of clay to create a natural environment.

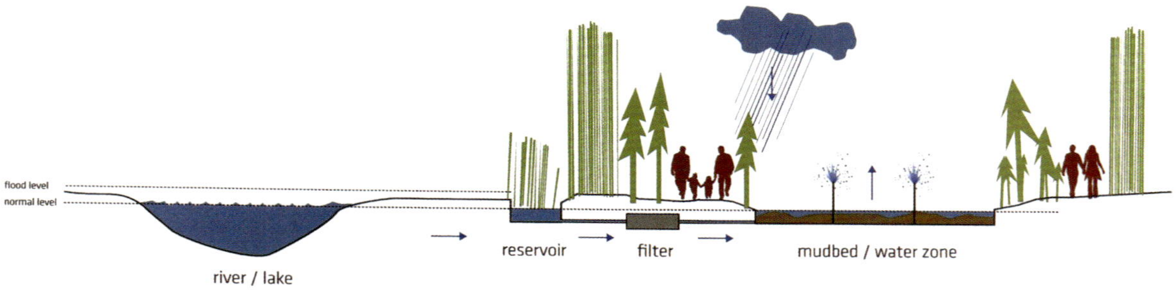

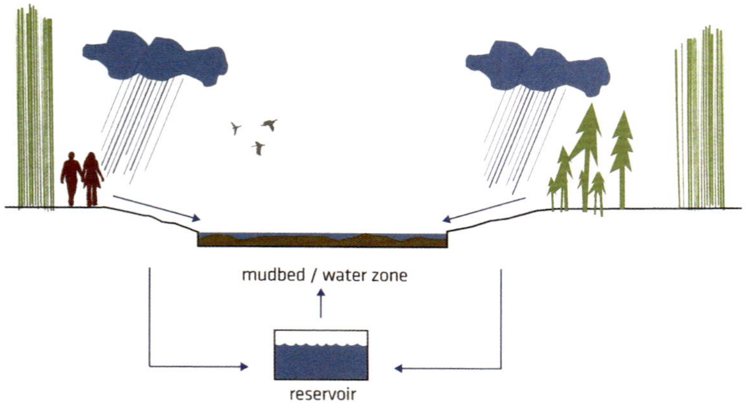

展园介绍：

在黄土园中，放眼望去所有的元素都与“土”有关。黄土园含有粘土元素以及一个不规则的平底水池。所有黄土都是从黄河收集而来。泥土的文明在于储藏的力量。项目还收集了来自西安北部 7 条河流的泥土以及中国其他省份的泥土，收集的泥土晒干之后，存放在当地仓库中。黄土园正是用这些收集的泥土对水池重新进行填埋。

最醒目的是一个不规则的平底水池。水池面积为 315 m^2，深 40 cm。当水池干涸时，黏土便开始开裂，形成月形表面。当喷泉湿润黏土时，黏土又重现，景象奇妙至极。而喷泉的用水均是从大自然收集的河水和雨水。

水池的底座堆砌着一些倾斜的石雕和垂直的镀金钢铁边缘，是艺术与现代的美妙结合。这些石材取自本地，钢铁边缘则是用垂直的材料互相堆砌形成的，保留最自然的原貌以形成蜿蜒曲折的形态，给水池增添了几许灵动与柔美。

花园里摆设的 9 个赤陶泥塑，均是用手工制作、经烘烤而成，延续了几百年以来的工艺传统。小径的沿途设有 3 处休息站，每一处休息站都设有砖块柱基，形状犹如雕刻在传统花园墙上的月亮洞，既可通行，也可透光。蜿蜒墙体上的几何窗洞能够把人们的视线引向独特的景色。

水池的四周是一条长 140m 的砖路，沿着这条路可以一直穿过花园。砖路时宽时窄，像一条蜿蜒的河流，绕过层层的障碍，流畅前进。花园是一个矩形围合，四周是浓密的竹林。园中有 7 棵保留下来的大树和 83 棵新植入的大树，设计根据观景点的不同，选取了 5 种外来开花植物，地面上则选择种植麦冬。

园中还设有一座小桥，这座桥是由闪闪发亮的黄色铁炉条建成的，分成 3 段，并在不同的方向连接着，在空间上形成了一个大尺度的“Z”字形，让人可体验到柔与刚的完美搭配。

黄土园中到了夜晚利用光纤技术进行照明。喷泉系统配有照明，让水流在白光中跳动。小径沿途设有低柱，黑色的细柱子安装有筒灯，用浅蓝色的光照，以突显小径的蜿蜒曲线，同时又照亮特定的树木以及整个地面的植被。带有过滤器的聚光灯还把青蛙的头像投影到泥层之上，金黄色的灯光更加重了黄土的气息，打造出自然的意境。

Big Trumpet Garden

大挖掘园

Chief Designer: Martin Cano

About the Designer: Martin Cano, one of chief designers of Topotek 1 which is the most progressive design corporation in Germany.

主要设计师：Martin Cano

设计师介绍：大挖掘园的设计师是德国 Topotek 1 事务所主创设计师 Martin Cano。Topoteck 1 是德国最先锋的设计公司之一。

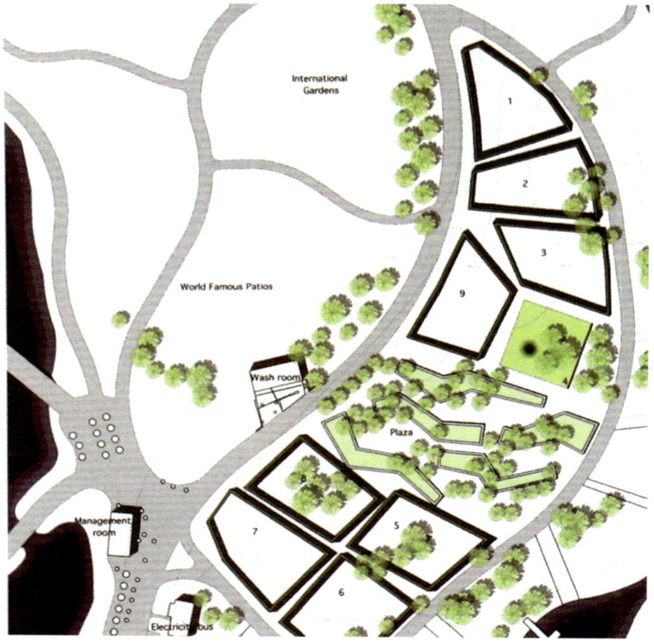

Introduction:

Big Trumpet Garden, also known as "the sound of the world", its theme is "listen to the voice on the other side of the world through the magical pit".

In Big Trumpet Garden, designer creates a structure of digging hole, to get to the opposite exit of the world. Look at this exit, it just like the mouth of a trumpet.

Designer takes the earth as a reference in this garden, with the help of submerged burrow, arousing your emotion to imagine to the other end of the world from here, providing you with a transnational atmosphere. It extends the traditional concept of garden, not only bring you the transition from interior to exterior, city to nature, but reflect the blend and transition from one culture to another.

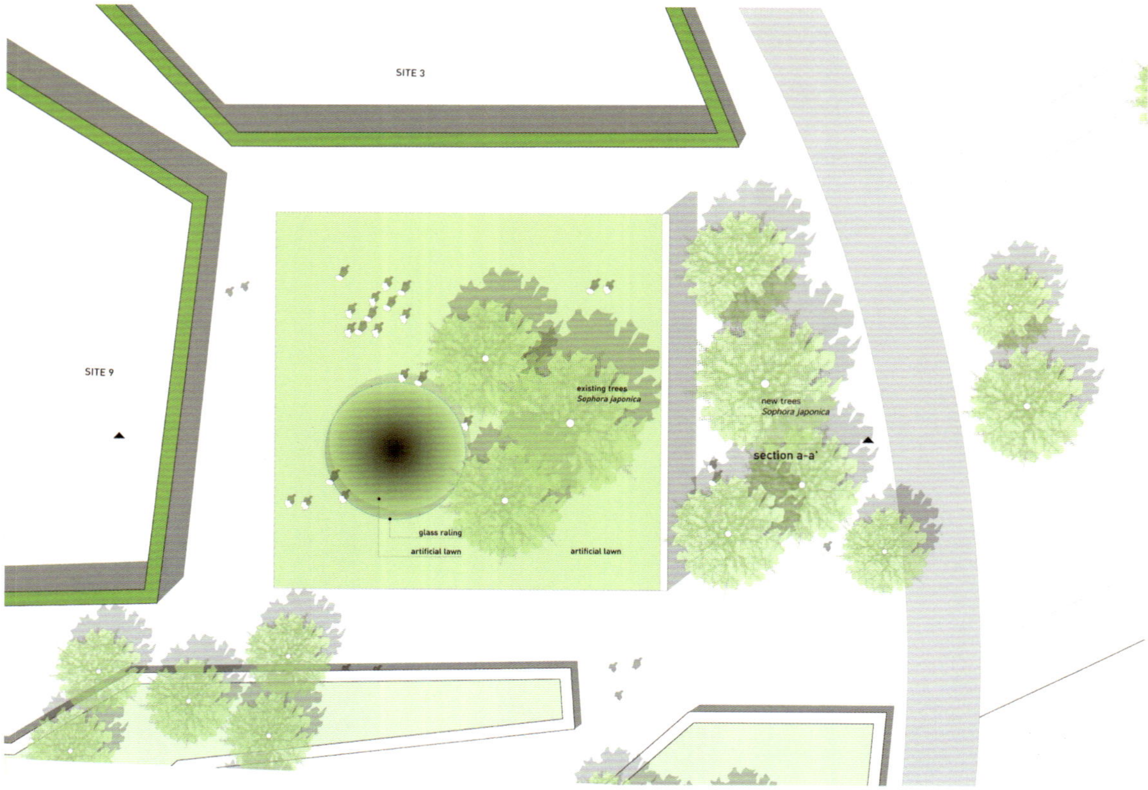

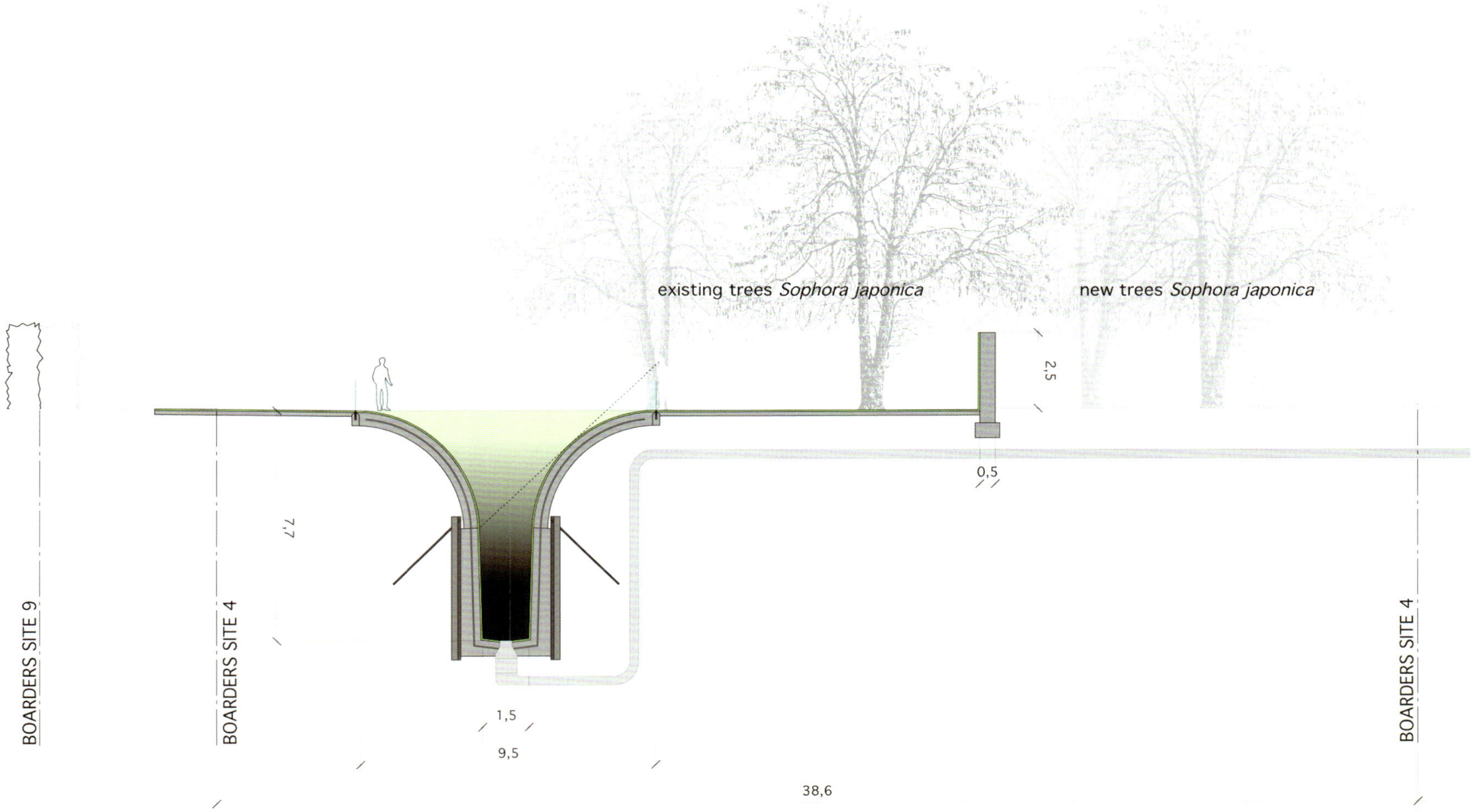

■摄影师：王艮

展园介绍：

大挖掘园，又名世界之声园，“神奇的大地洞，倾听世界那头的声音”是其主题。

在大挖掘园中，建筑大师创造了挖地洞的结局，即地洞到达地球另一端的那一个出口。在这个出口，地洞像一个喇叭口“喷涌”而出。

在这个展园中设计者以地球作为概念创意的参照物，借助下沉的地洞，激发游客想象，幻想由此到达世界的另一端，为游客营造跨国环境氛围。延伸了传统意义上花园的概念，不仅带给人们从室内到室外、从城市到自然的空间上的转换，同时体现了从一种文化到另一种文化的融合与过渡。

■摄影师：王良

Garden of Labyrinth

迷宫园

Chief Designer: Martha Schwartz

About the Designer: It is mainly designed by Martha Schwartz. She is the internationally renowned master hand and artist of landscape architecture, Lifetime Professor at Harvard University, honorary fellowship at Royal Institute of British Architects (RIBA), and the founder of Martha Schwartz Design Firm.

主要设计师：Martha Schwartz

设计师介绍：迷宫园由美国 MSP 事务所 Martha Schwartz 设计，她是世界著名景观建筑大师、艺术家，哈佛大学终身教授，英国皇家建筑师学会荣誉院士，Martha Schwartz 是设计事务所主创人。

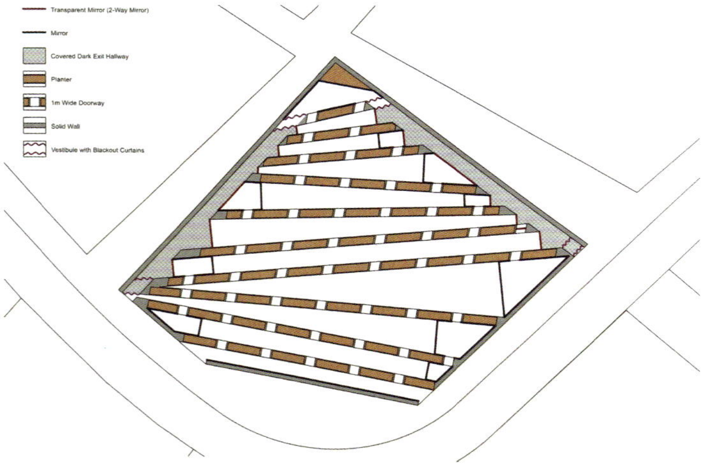

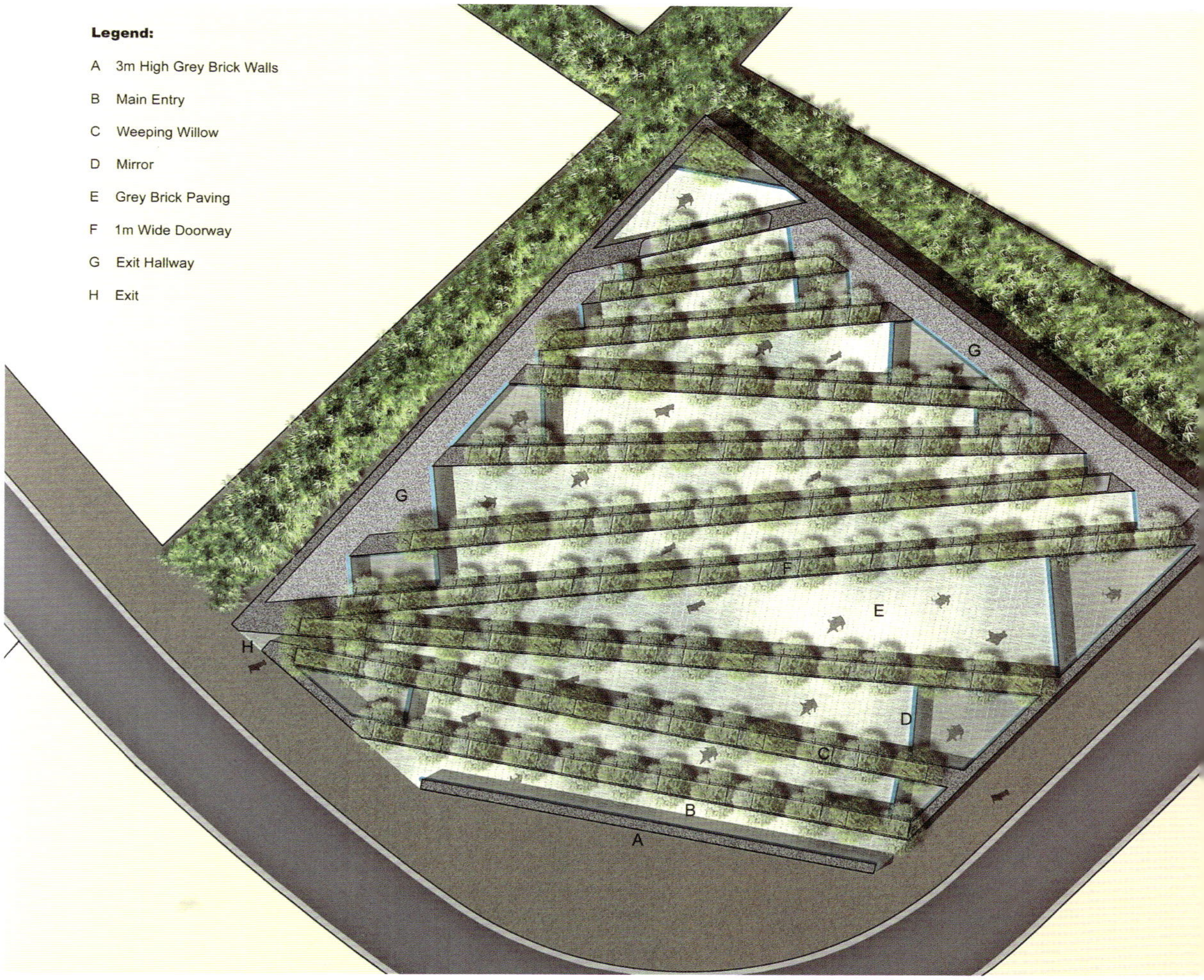

Introduction:

The Garden of Labyrinth is enclosed by 3m high traditional black-brick walls like a "big casket" seen from outside. In the middle of the garden, there are a series of narrow and long corridors without roofs for looking up to the blue sky and white cloud. With different widths, the interspaces between two walls of the corridors are planted with green weeping willows.

When entering into the Garden, people can pass through the archway easily and walk in different corridors. Some corridors are parallel and the distances between corridors are different.

At the end of the corridor, there is a 3m high mirror which extends the corridor limitlessly to the neighboring corridor. Sometimes, there will be a small room with mirrors all around or other unexpected scenes to enable you to experience the mysteries of the labyrinth.

■ 摄影师：王艮

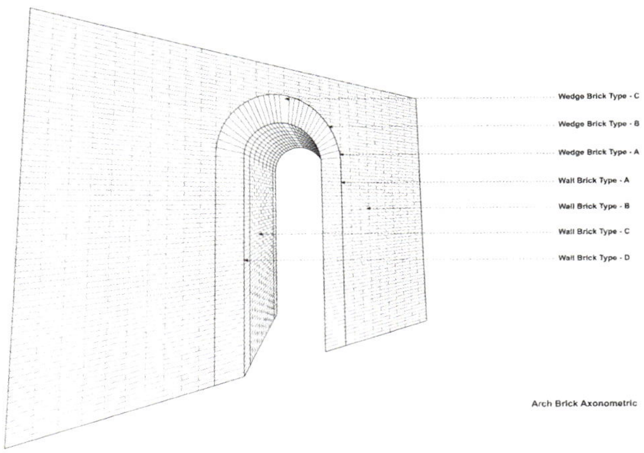

Arch Brick Axonometric

Arch Brick

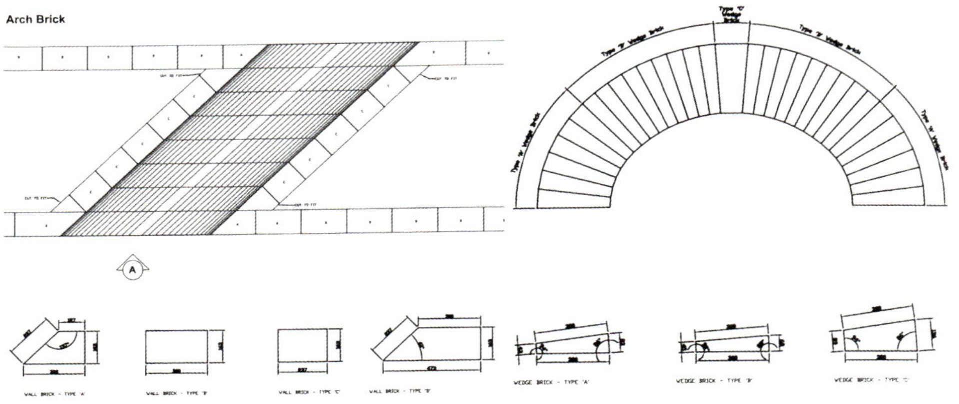

Arch Plan

Arch Elevation - A

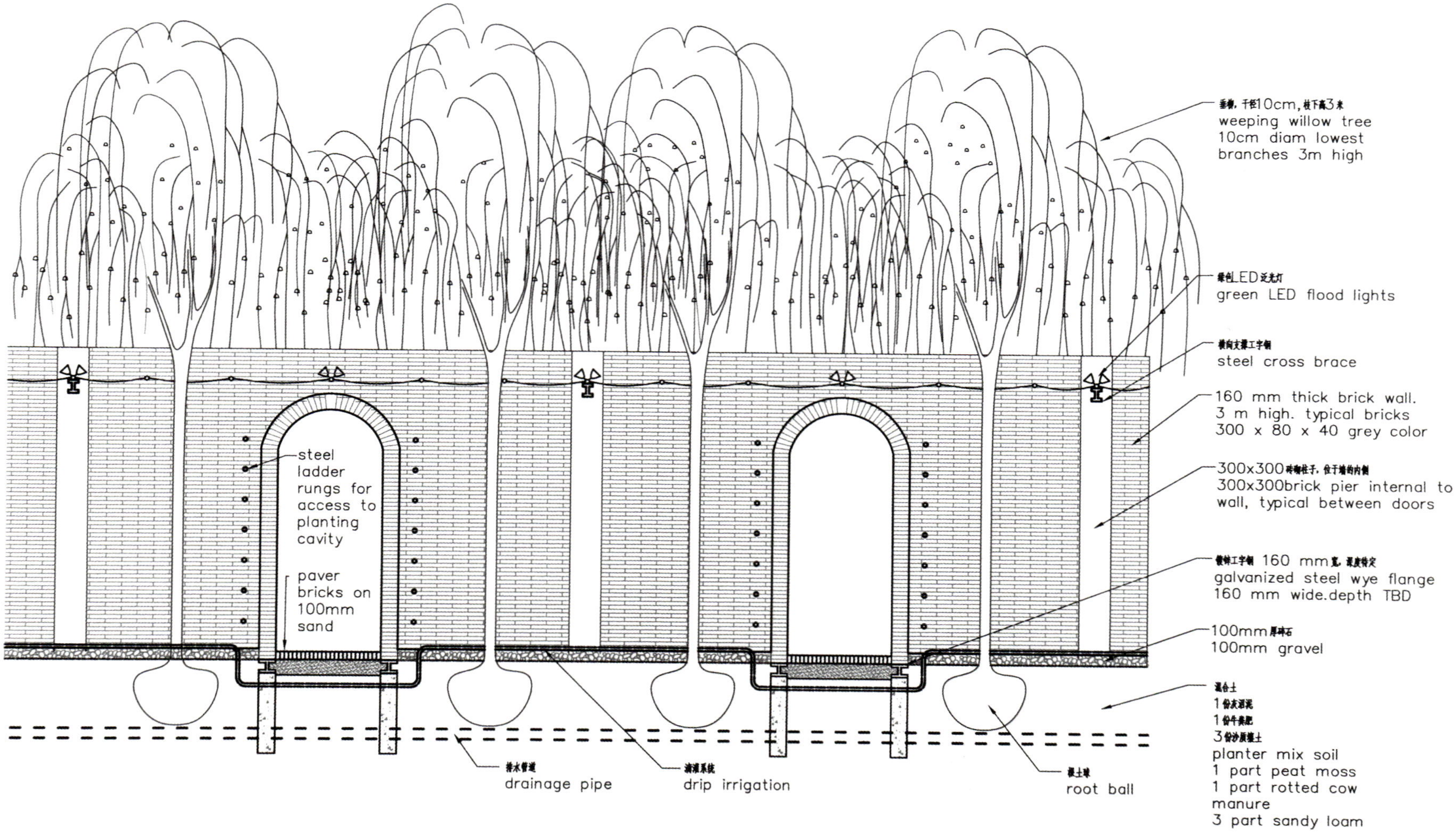

垂柳，干径10cm，枝下高3米
weeping willow tree
10cm diam lowest
branches 3m high
绿色LED泛光灯
green LED flood lights
横向支撑工字钢
steel cross brace
160 mm thick brick wall.
3 m high. typical bricks
300 x 80 x 40 grey color
300x300砖砌柱子，位于墙的内侧
300x300brick pier internal to
wall, typical between doors
镀锌工字钢 160 mm宽，深度待定
galvanized steel wye flange
160 mm wide.depth TBD
100mm厚碎石
100mm gravel
混合土
1份牛粪肥
3份沙质壤土
planter mix soil
1 part peat moss
1 part rotted cow
manure
3 part sandy loam
steel
ladder
rungs for
access to
planting
cavity
paver
bricks on
100mm
sand
排水管道
drainage pipe
滴灌系统
drip irrigation
根土球
root ball

展园介绍：

迷宫园由3m高的西安传统青砖墙构成，从园外只能看见一个3m高的青砖“大匣子”。花园中的空间被分成一系列狭长的走廊，这些走廊上空没有封顶，抬头可见蓝天与白云。走廊宽度不同，在走廊每两堵墙之间留有的空隙中，还种植着翠绿的垂柳。

走进园中，人们可以很方便地通过拱门，穿行于不同走廊。有的走廊是互相平行的，有的走廊间隔宽窄不一。

在走廊的末端，还设有一个3m高的大镜子；在镜子中，这些走廊的长度无限延伸下去——通常能够走进相邻的走廊，但有时又会闯入一个小的周围布满镜子的房间或是遇见其他意想不到的情景，迷宫带给你的神秘感会突然降临。

Landscape Garden of Map of China

山水·中国地图园

Chief Designer: Catherine Mosbach

About the Designer: Catherine Mosbach is the chief designer in French Mosbach Landscape Architecture Firm, founder and chief editor of 《Pages Paysages》, and the receiver of 2003 Rosa Barba European Landscape Design Award as well as the Grand Prize of American Society of Landscape Architects (ASLA).

主要设计师：Catherine Mosbach

设计师介绍：法国 Mosbach 风景园林事务所主创设计师 Catherine Mosbach 设计是《Pages Paysages》杂志创始人及主编，曾获 2003 Rosa Barba 欧洲景观设计奖，美国景观设计师协会 ASLA 大奖。

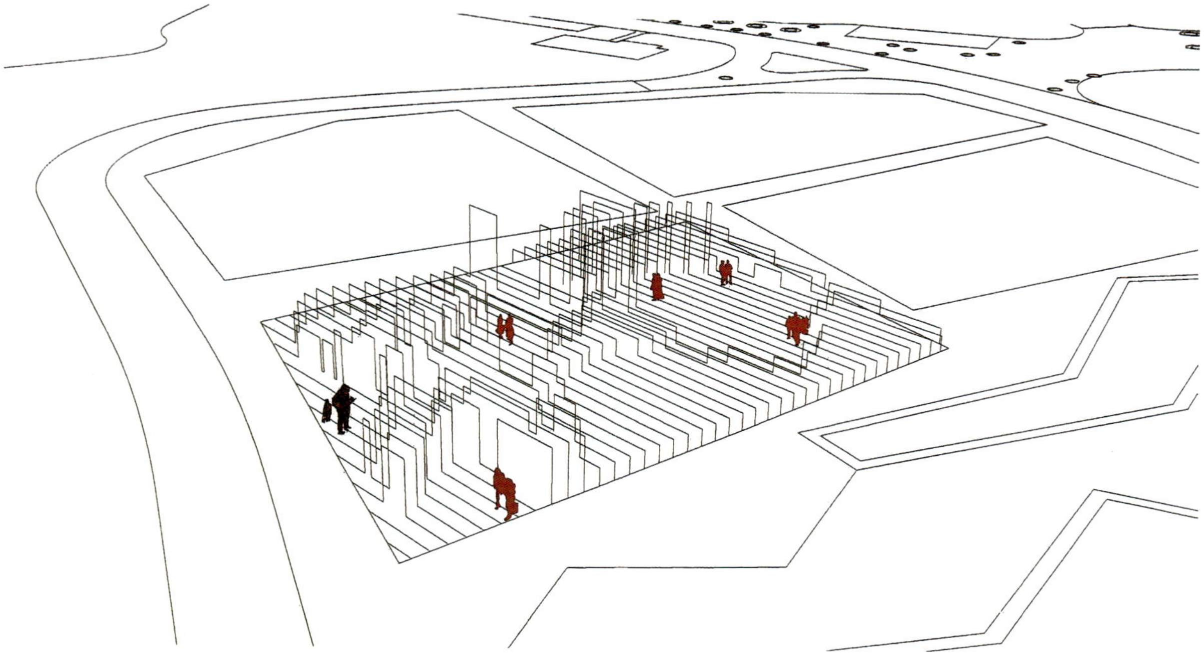

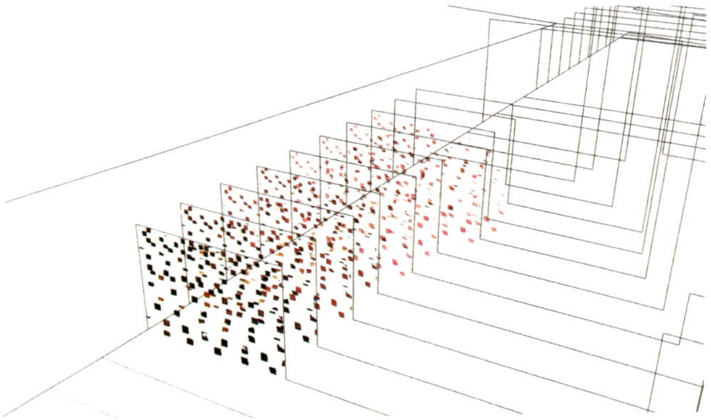

Introduction:

This garden expands the boundary to place the visitors in a microworld, interprets the traditional theme of garden art with modern style. It follows the principles of large royal garden to present a miniature map of China. It is not a miniature landscape in traditional sense, it transforms the rivers and lands in different height in the map into abstract lines, and sketches the outline with these lines. Later on, these outlines create a three-dimensional map, which endows the garden a form. Plant groups in different seasons give vitality to the map and artistic conception to Chinese landscape painting with their withering and flourishing states.

展园介绍：

山水·中国地图园通过拓展边界，让游客置身于一个微观世界，通过现代样式，对花园艺术的传统主题进行演绎。园子遵循大型皇家园林的法则，是微缩的中国地图。但项目并不是传统意义上的微缩景观，而是把地图中的河流、不同高度的大地，全部转换成抽象的线条，再借助线条勾勒出轮廓线。随后的这些轮廓线赋予花园形态，制造三维立体的山水·中国地图。加上精心布置着不同季节着色的植物组群，让植物的生长枯荣，给予地图生命力和中国山水画的意境。

Creativity Park

创意园

The Creativity Park covers about 10, 000m², inviting the design organizations of ten top universities from all over the world – the Architectural Association (London), Columbia University (USA), University of California Berkeley, University of Southern California, Universidad Torcuato Di Tella (Argentina), University of Toronto (Canada), University of Saint Joseph (Macau), Feng Chia University (Taiwan), Peking University, Hong Kong University. Therefore, it is also called "University Park". The Creativity Park is designed in accordance with the concept of Chinese classical garden and the analysis of the field conditions and the vegetation resouces.

The Sky City designed by University of Southern California has reflected people's expectation to the nature through the setting of sunshine courtyard, reflection garden and mist park; the radioactive greening spool thread designed by University of Saint Joseph (Macau) has interweaved the visiting lines and the green plants to form the interfering layout and present the close relation between city landscape and human in Macau City; Universidad Torcuato Di Tella (Argentina) is trying to build a "Pampas Impression" in Xi'an where is different from Argentina on physics, geography and culture…The innovative and unconventional designs have raised people's deep thought about human and nature.

创意园占地面积约1万m²，邀请来自全球包括伦敦建筑联盟学院、美国哥伦比亚大学、伯克利大学、南加州大学、阿根廷托尔夸托迪特加大学、加拿大多伦多大学、澳门圣约瑟大学、台湾逢甲大学、北京大学、香港大学等10所顶尖大学的设计机构完成设计规划，因此又给人称为"大学园"。创意园的设计师秉承中国古典园林的设计理念，对现场条件和植被资源进行了分析，在此基础上进行设计。美国南加州大学设计的天空之城通过设置日光中庭、反射花园、云雾园反映了人们对自然的美好憧憬；澳门圣约瑟大学设计着重以放射性的绿轴将参观流线与绿植交织，形成互相干涉的格局来表现澳门中城市景观与人类的密切关系；阿根廷托尔夸托迪特加大学正试图在物理、地理、文化风韵与阿根廷不尽相同的西安创造出局部"潘帕斯印象"……新鲜而别致的设计引发了人们对人与自然关系的思考。

Weaving Nature

编织自然

Designed by: Peking University

Design feature: The designer has placed mankind and nature on two layers by using a knitted flimsy web.

设计学院：北京大学

设计特色：利用一层编织的薄网创造了一种共生的和谐。

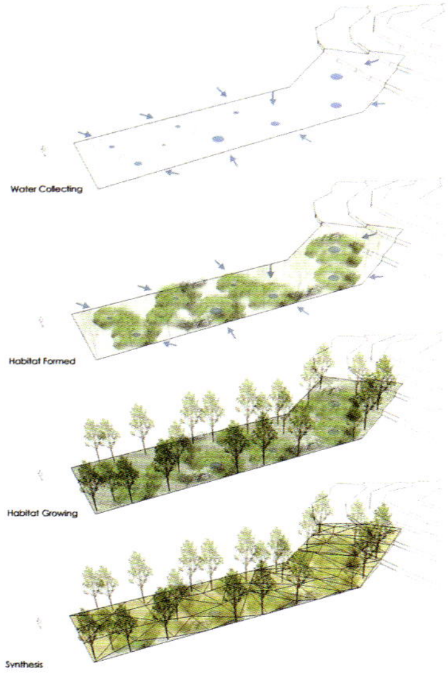

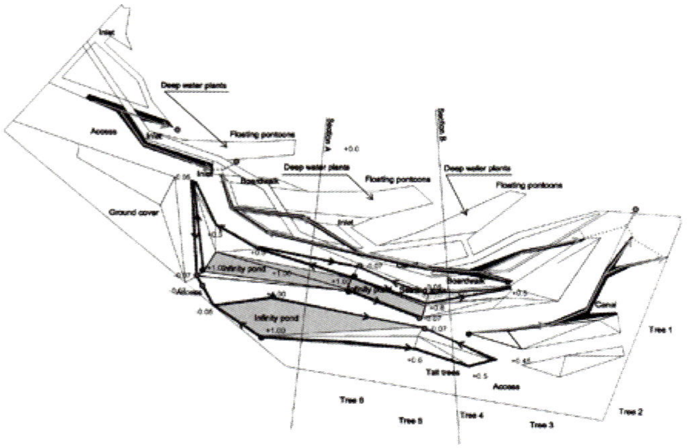

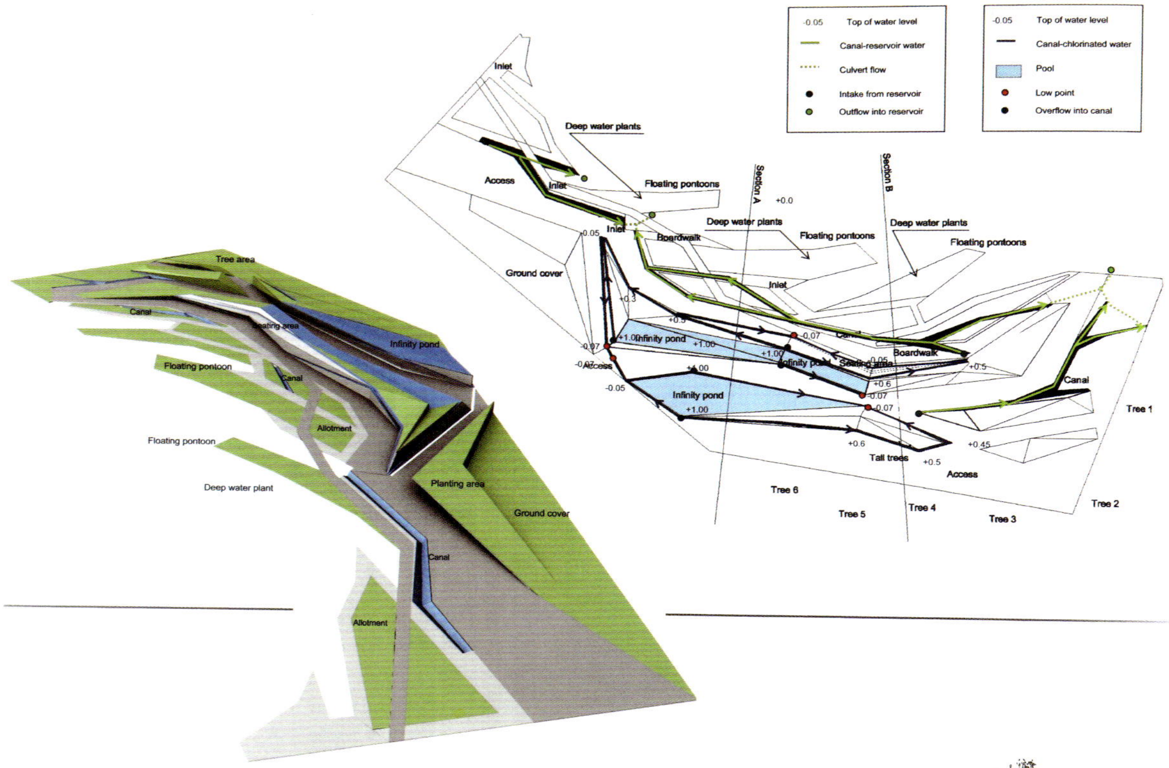

Introduction:

A special way of interaction and communication between mankind and nature is provided. The designer has placed mankind and nature on two layers by using a knitted flimsy web. The upper layer is where people are and the lower layer is where illumination system is installed with plants around it to form an interaction. The design of the flimsy web greatly reduces the destruction to nature by human beings and creates a harmony of co-existence. The purpose of the design is to achieve the barrier-free communication between visitors and nature, and reduce the negative effect caused by human activities to make the landscape develop naturally.

展园介绍：

编制自然为人们提供一种特殊的人与自然互动交流的方式。设计者利用一层编织的薄网将人与自然划分成两个层面，薄网上层属于人的活动空间，下层安置照明系统，植物穿梭其间从而形成一种互动。薄网的设计大大减少人对自然的破坏，创造一种共生的和谐。该园的设计主旨正是让参观者与自然能够彼此无障碍地互相交流，同时减少人类活动对景观的负面影响，使景观自然地成长。

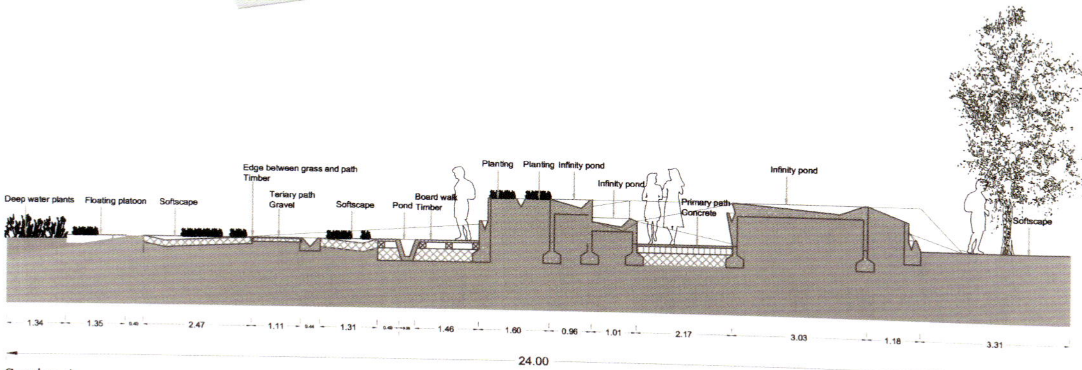

Section A

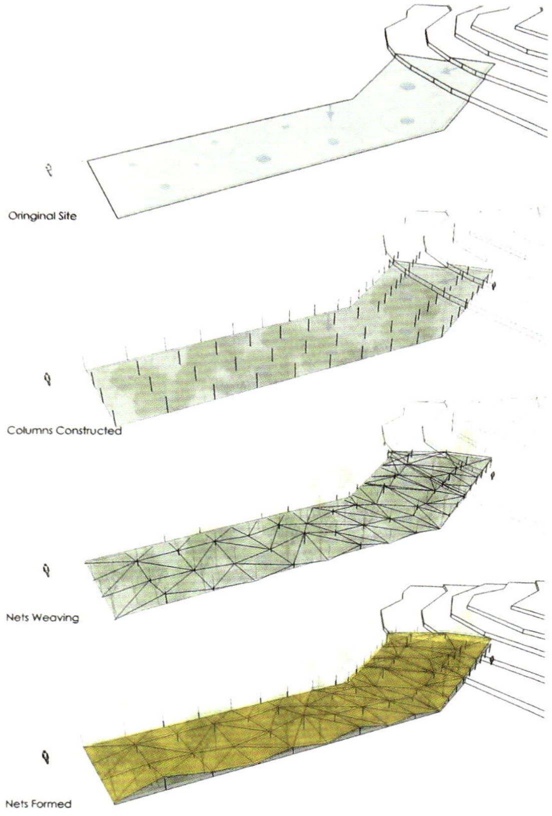

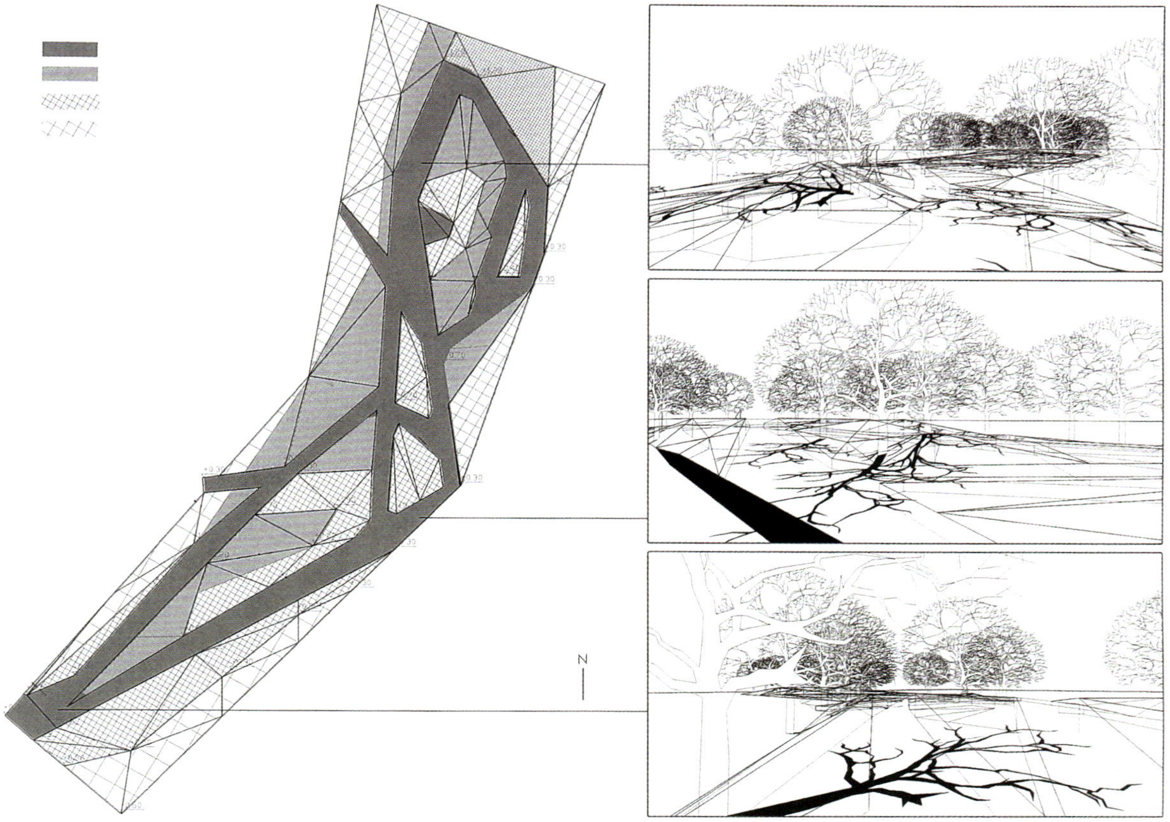

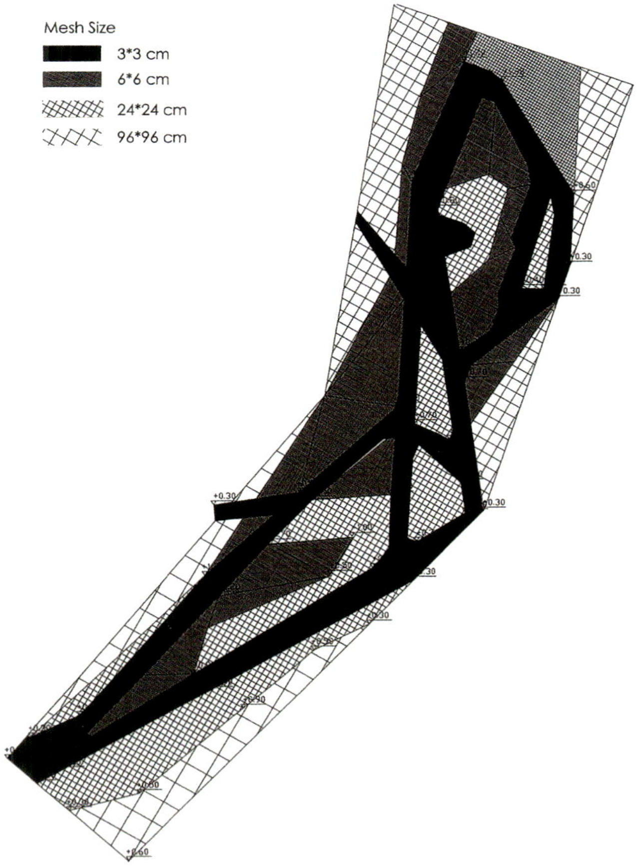

Different Types of Weaving Density and programs

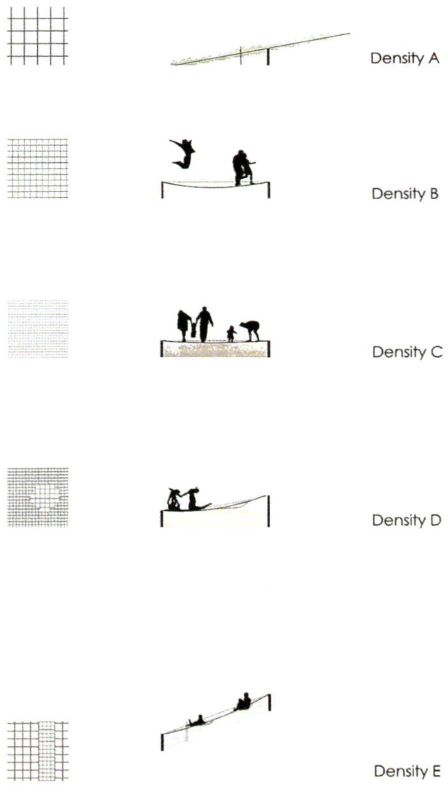

Aromatic Garden

芬香花园

Designed by: University of Toronto, Canada

Design feature: The aroma of plants is used to create a garden with sense of smell, sense of sight and sense of touching.

设计学院：加拿大多伦多大学

设计特色：利用植物的香味营造出具有嗅觉、视觉、触觉的庭院。

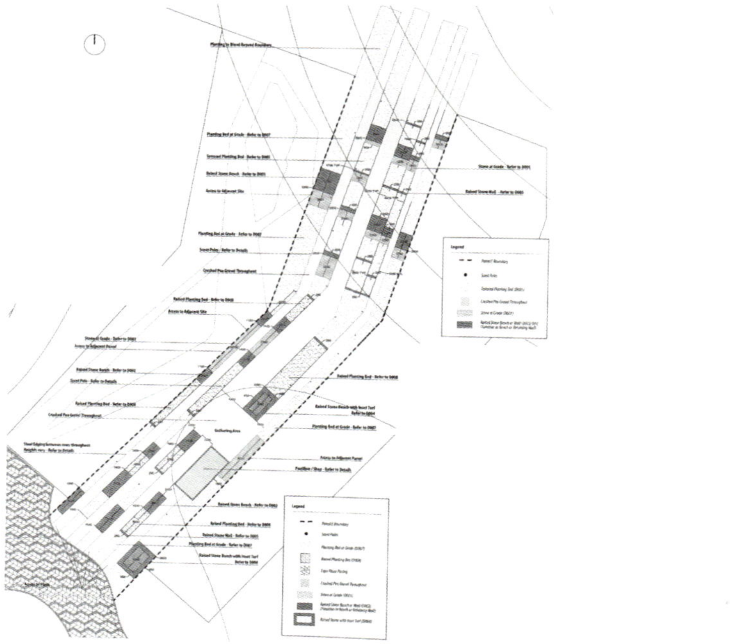

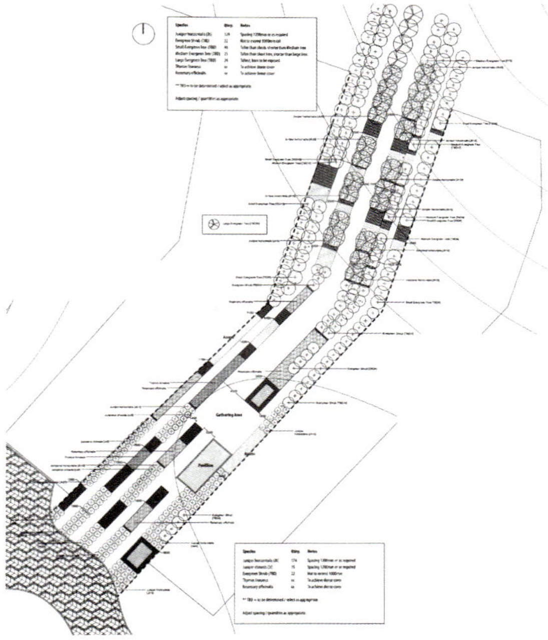

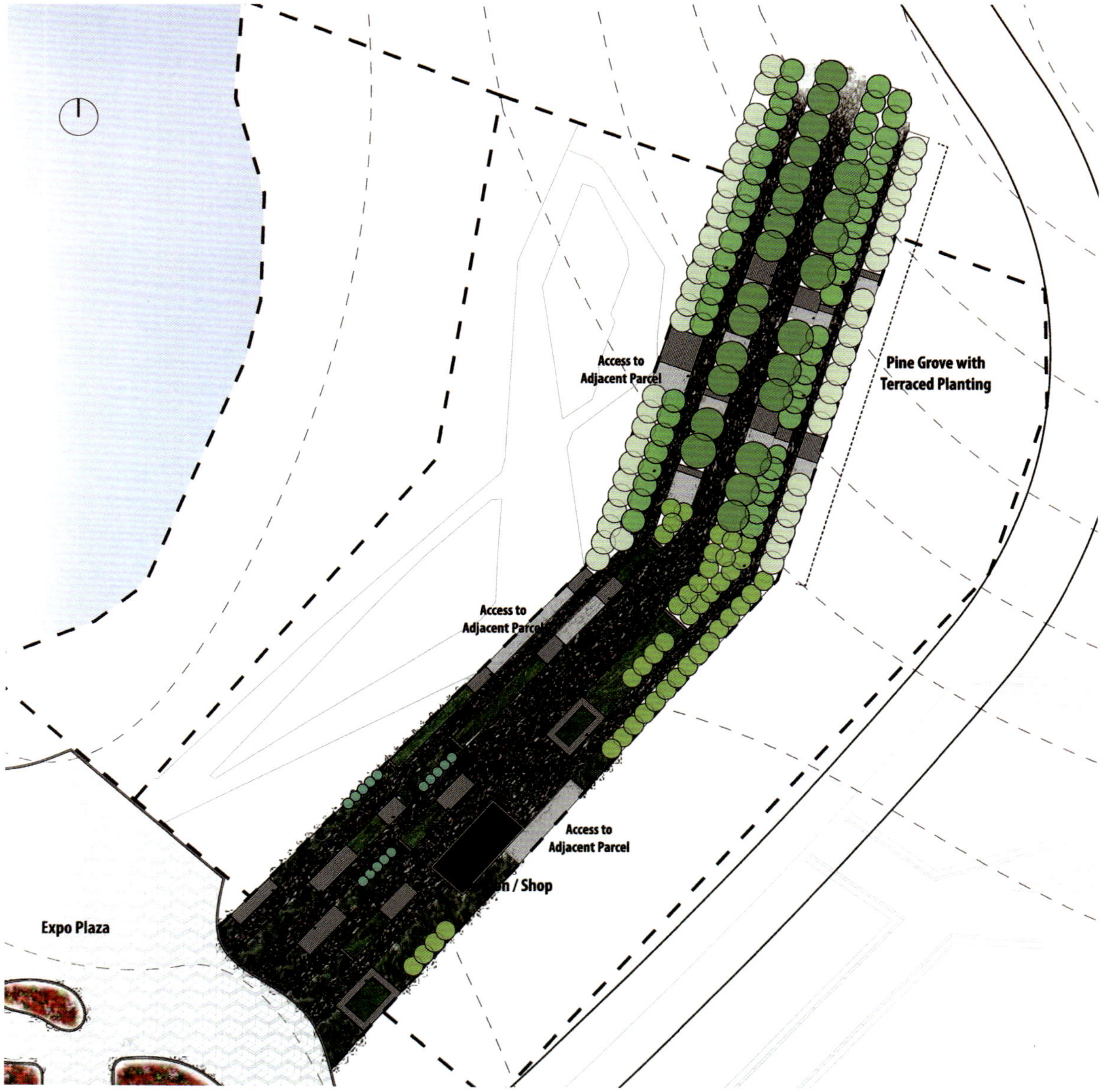

Introduction:

The aroma of plants is used to create a garden with sense of smell, sense of sight and sense of touching. The designer attracts the visitors by the aroma wafted by the unique aroma column. Except planting the local coniferous trees for the sense of smell, there is also "aroma column" which consists of varied pollens and perfumes from other areas of China to remind people of other scenes and expand the garden to some extent.

展园介绍：

芳香花园是利用植物的香味营造出具有嗅觉、视觉、触觉的庭院，设计者通过独特的香味柱所散出来的香气来吸引游客。除了栽种西安当地针叶林树种来营造嗅觉感观外，园区更通过多个来自中国不同地区的花粉香料组成的"芳香柱"来诱发人们对对景象的联想，并在一定程度上扩充了外延。

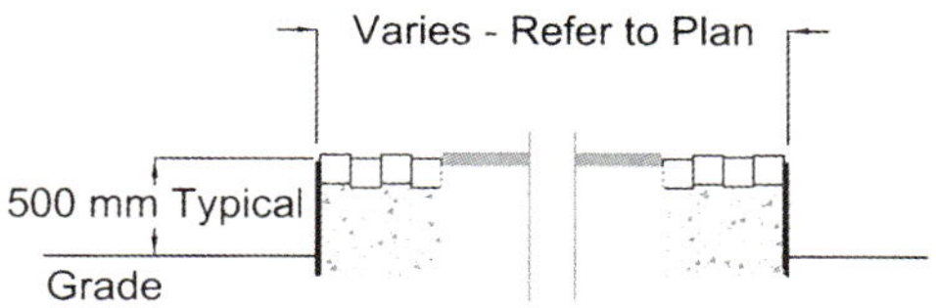

Section View

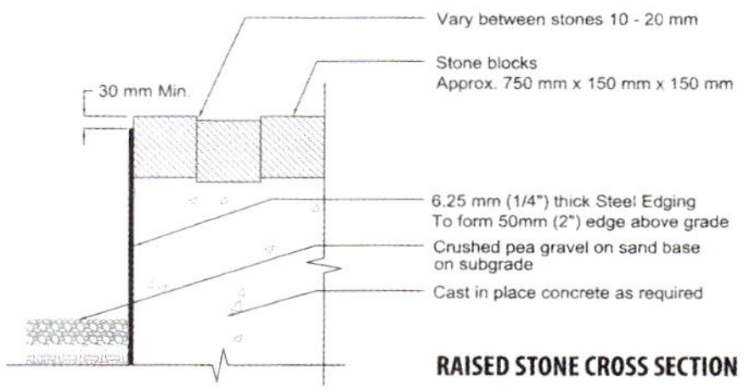

RAISED STONE CROSS SECTION

Section View

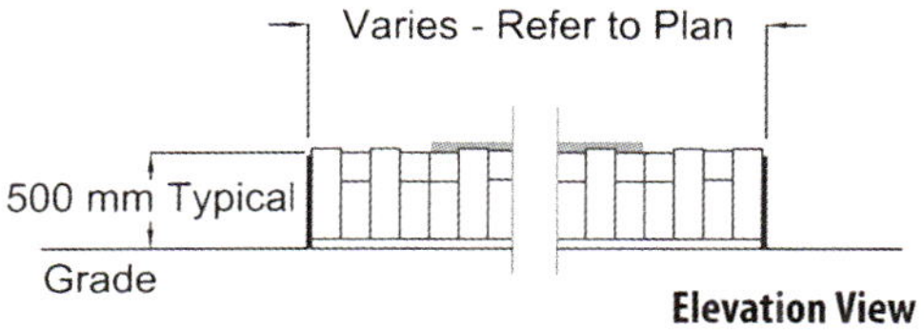

Elevation View

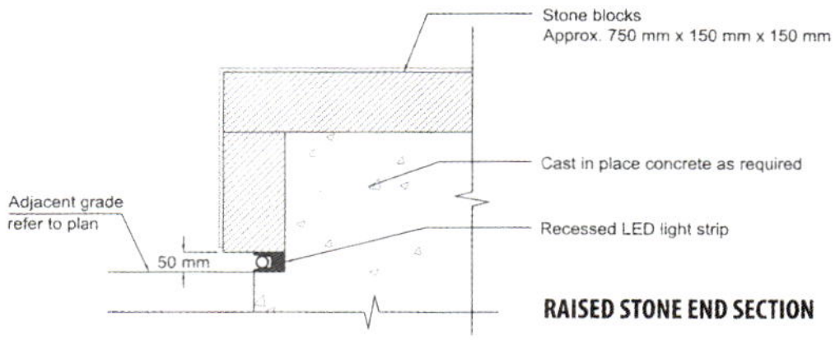

RAISED STONE END SECTION

Elevation View

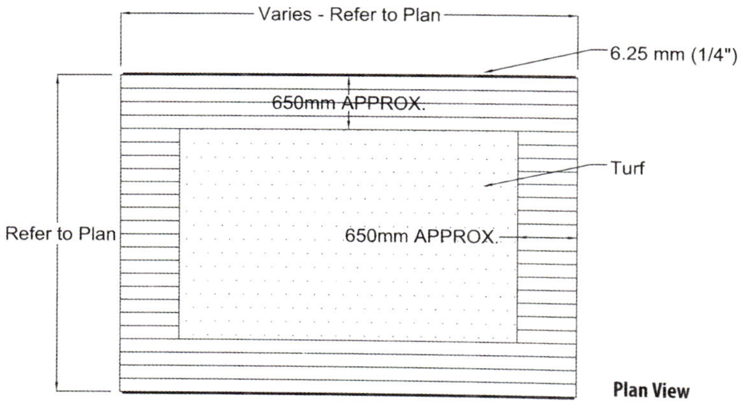

Plan View

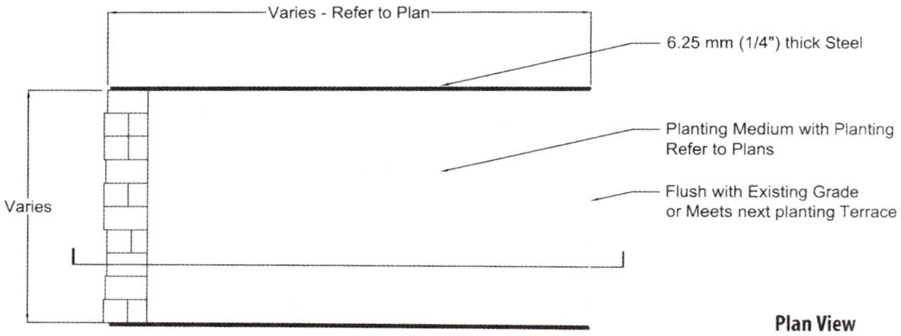

Plan View

Section View

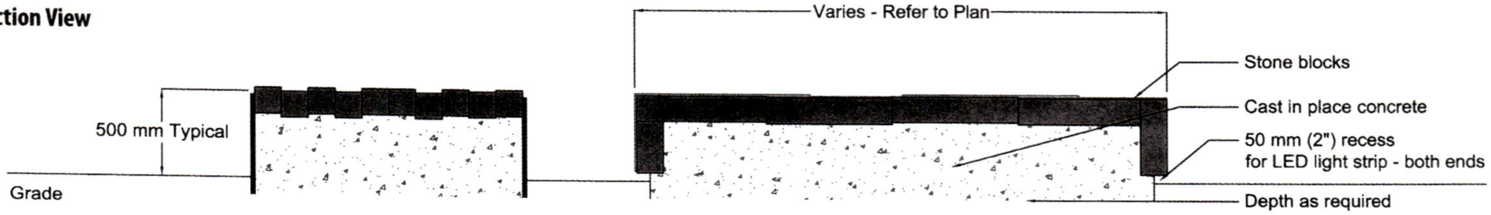

Elevation View

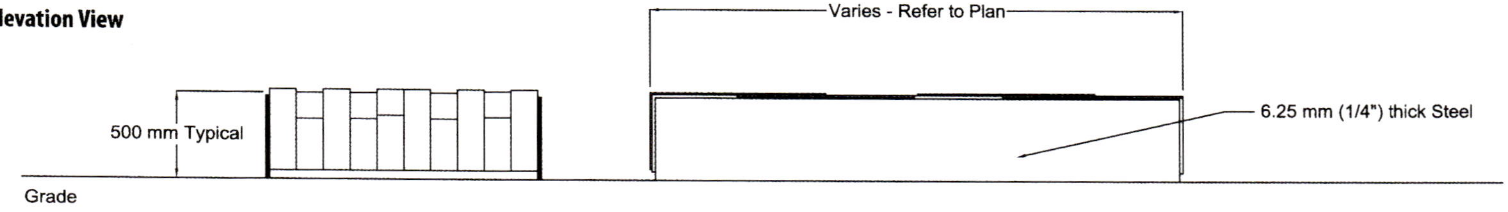

Section View

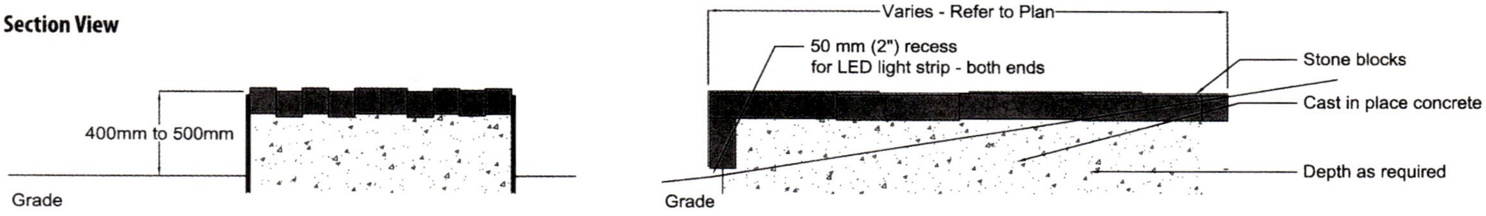

Elevation View

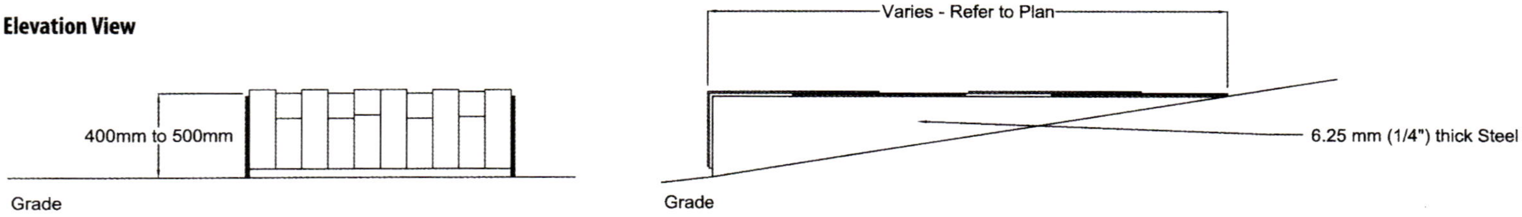

Plan View

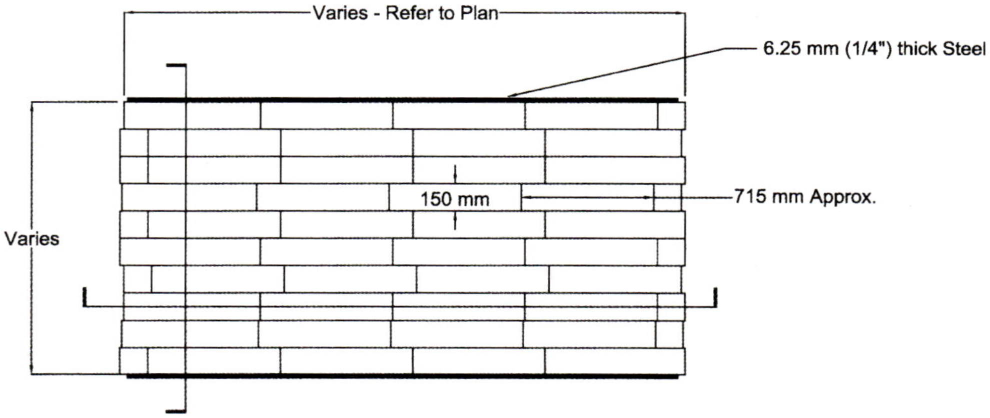

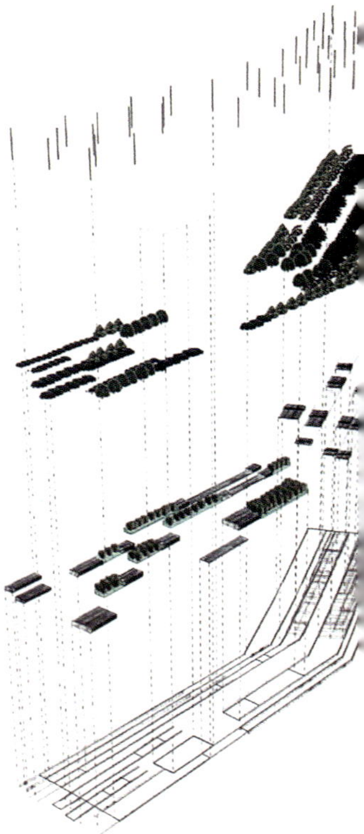

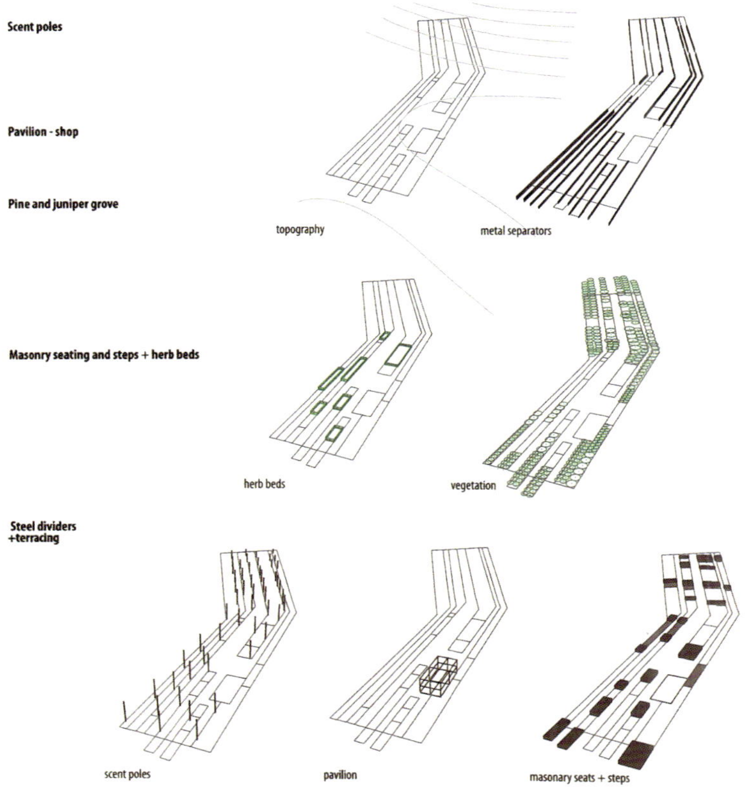
Scent poles
Pavilion - shop
Pine and juniper grove
topography
metal separators
Masonry seating and steps + herb beds
herb beds
vegetation
Steel dividers +terracing
scent poles
pavilion
masonary seats + steps

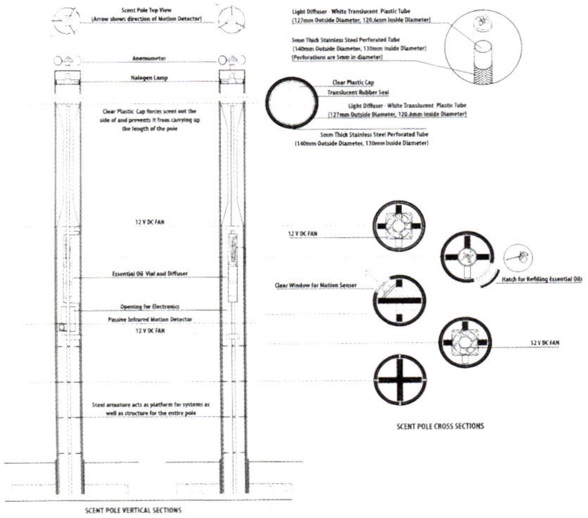
12 V DC FAN
SCENT POLE VERTICAL SECTIONS
SCENT POLE CROSS SECTIONS

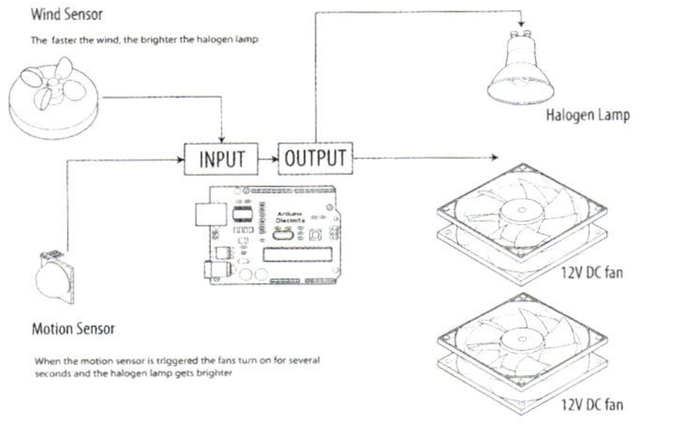
Wind Sensor
The faster the wind, the brighter the halogen lamp
INPUT
OUTPUT
Halogen Lamp
12V DC fan
Motion Sensor
When the motion sensor is triggered the fans turn on for several seconds and the halogen lamp gets brighter
12V DC fan

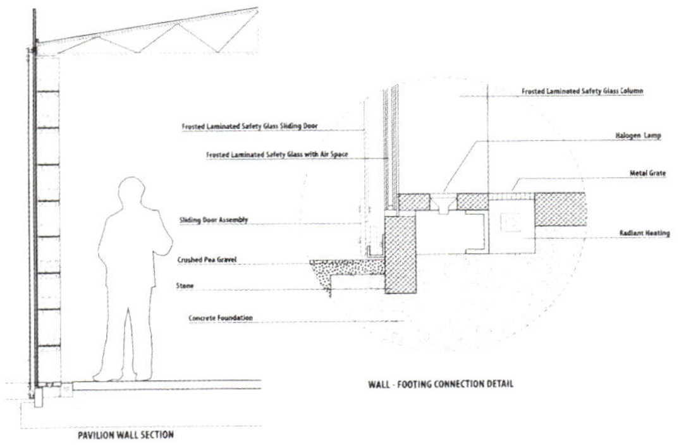
WALL - FOOTING CONNECTION DETAIL
PAVILION WALL SECTION

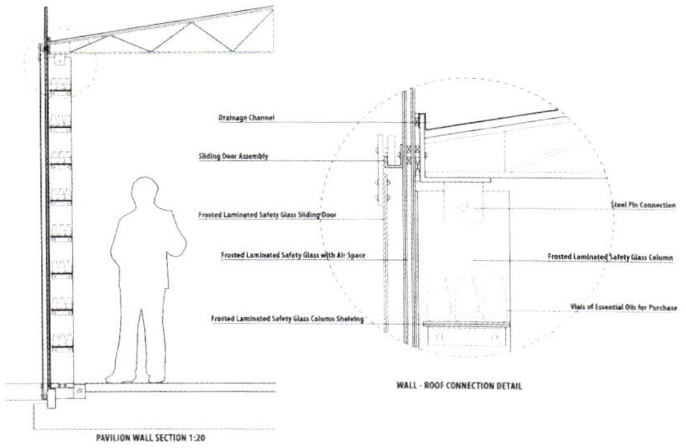
WALL - ROOF CONNECTION DETAIL
PAVILION WALL SECTION 1:20

Floating Garden

流园

Designed by: University of California Berkeley

Design feature: With the theme of "flowing water", it has reflected the uncertainty of the branches of flowing water.

设计学院：加州大学伯克利分校

设计特色：以"水流"为主题，将自然的水流分支产生的不确定性发挥到极致。

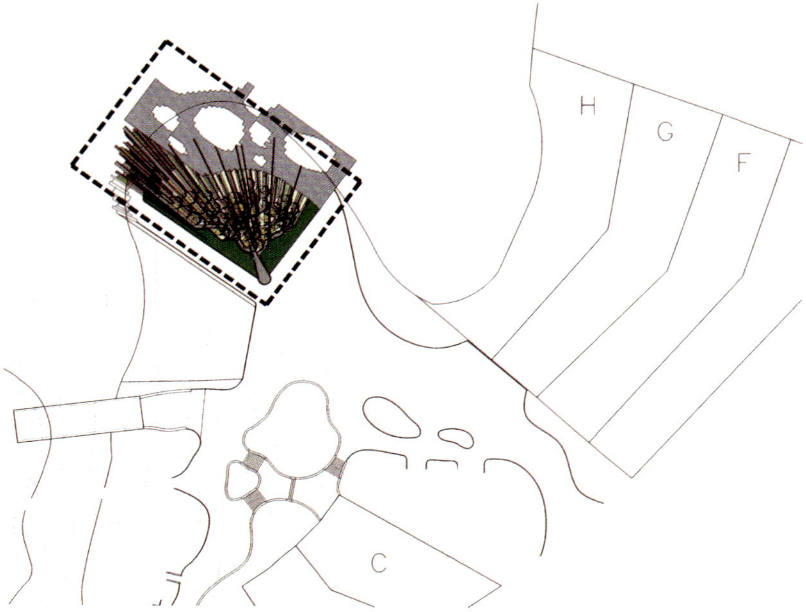

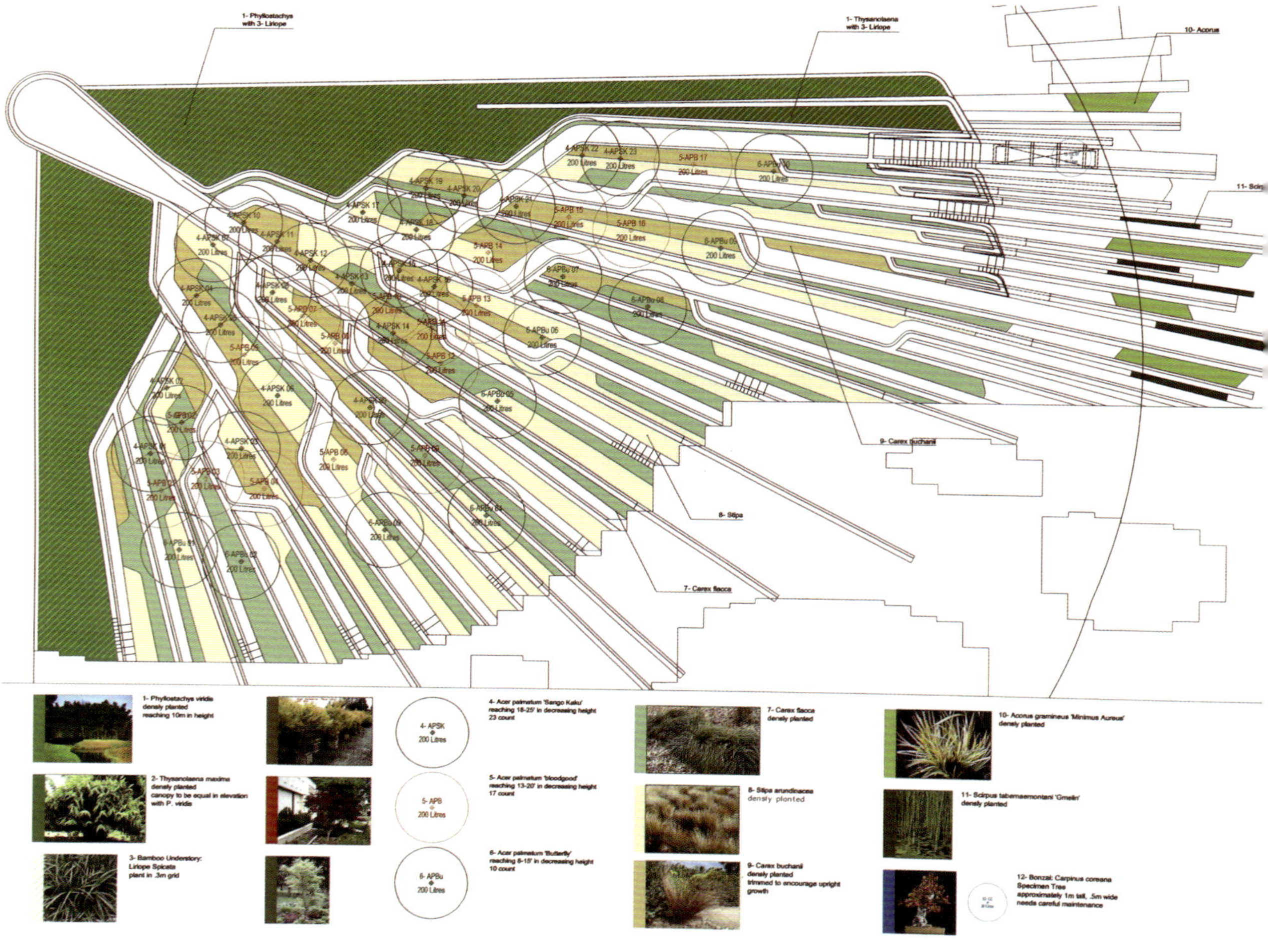

Introduction:

The designer provides an itinerary for visitors to visit freely. Each itinerary will bring different visiting experience. With the theme of "flowing water", it has reflected the capricious social status of current society where paths are selected, abandoned and re-selected and the complex mentality of the people living under such circumstance by the uncertainty of the branches of flowing water. The logical relation between the branches of the path is led by flowing water. People can also choose different sequences and paths freely.

展园介绍：

设计为参观者提供了一个自由参观的线路，每条线路都会带来不同的参观感受。这个作品设计以"水流"为主题，将自然的水流分支产生的不确定性发挥到极致，表达了当今社会，选择路线，逆转、反复无常的社会形态，以及在这种状态下生存的人们所展现的复杂心里。当然路径分叉之间的逻辑关系通过水流来引导，人们还可以根据自己的需要选择不同的序列、不同的游径。

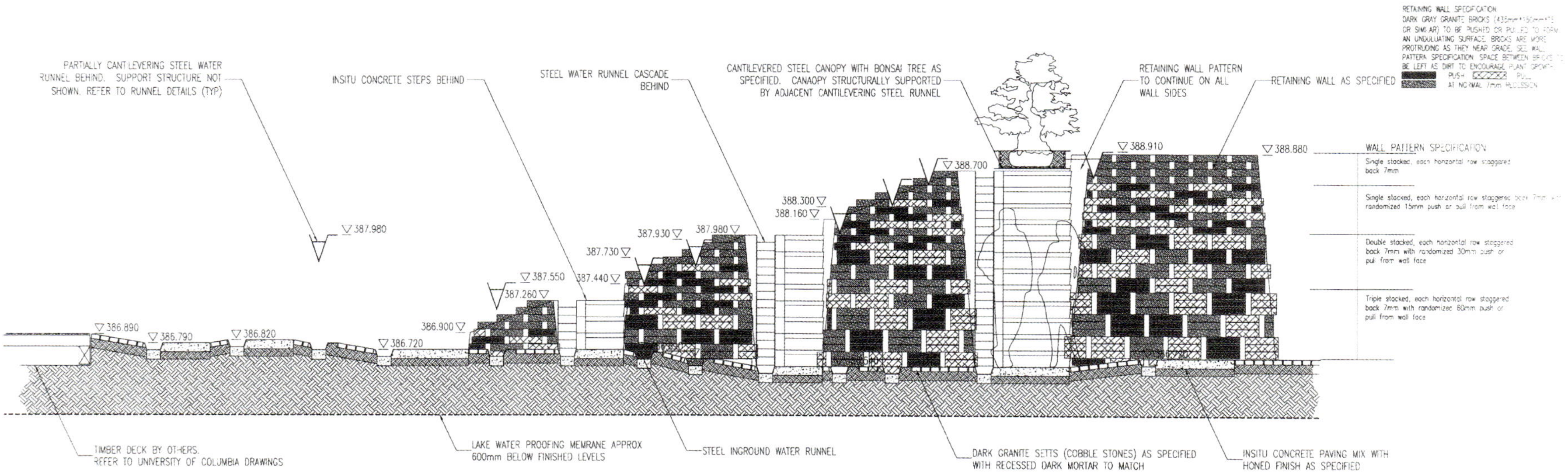

D01 / B.02 sectional elevation RETAINING WALL WATERFRONT

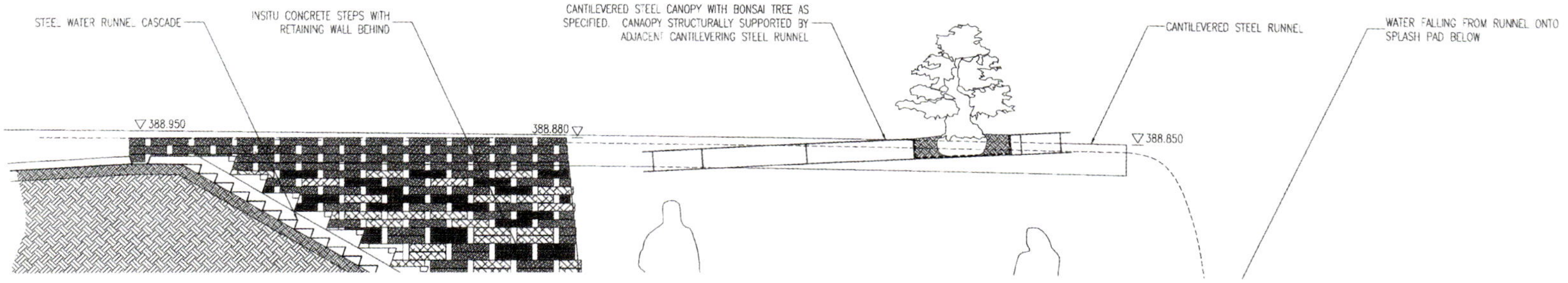

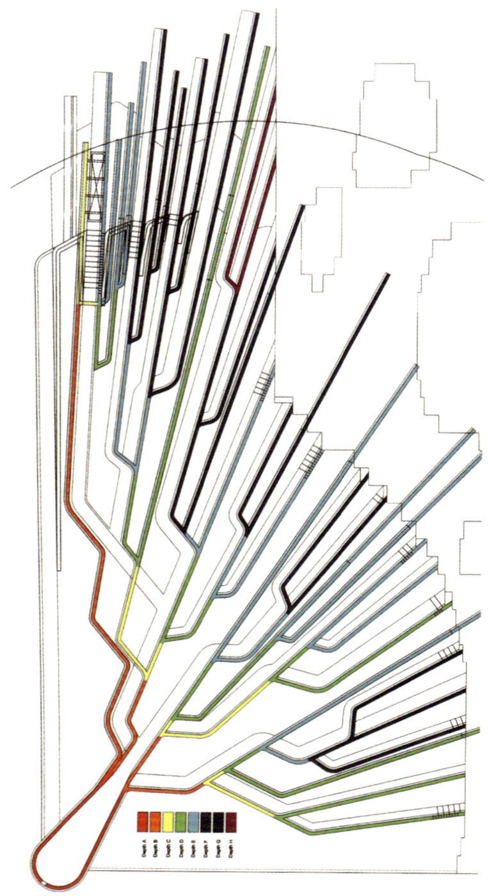

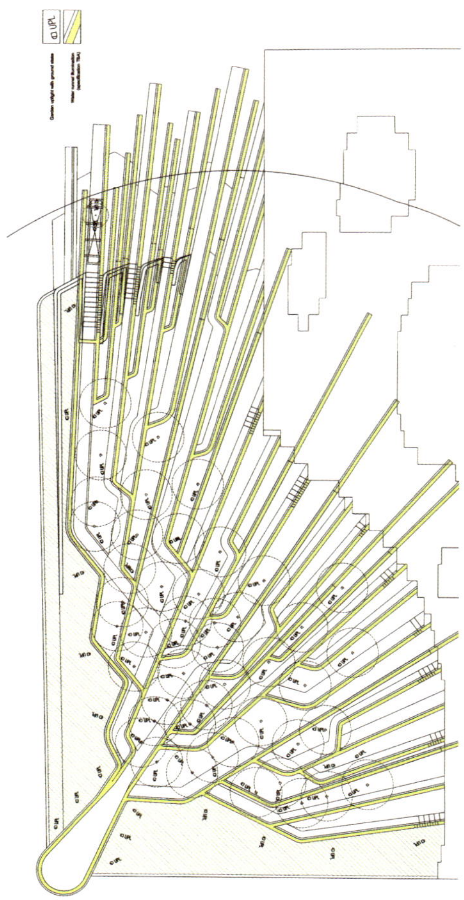

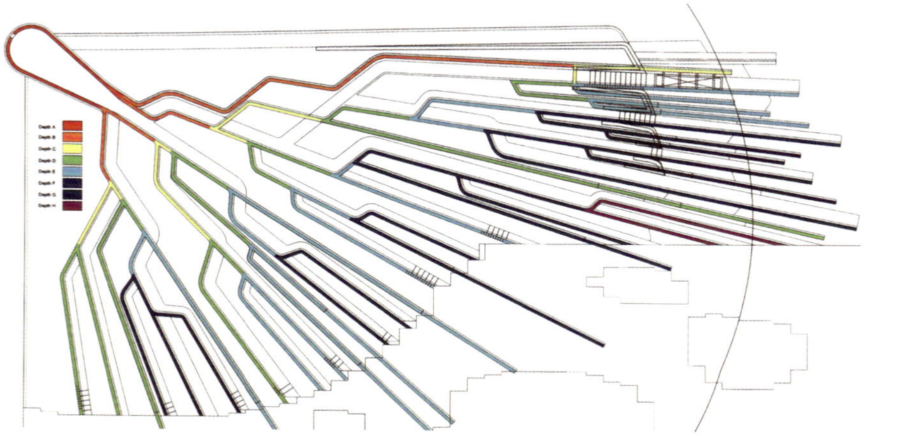

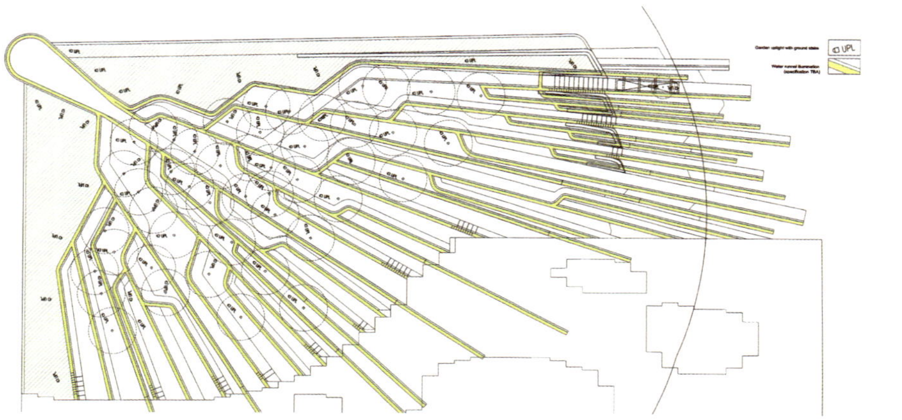

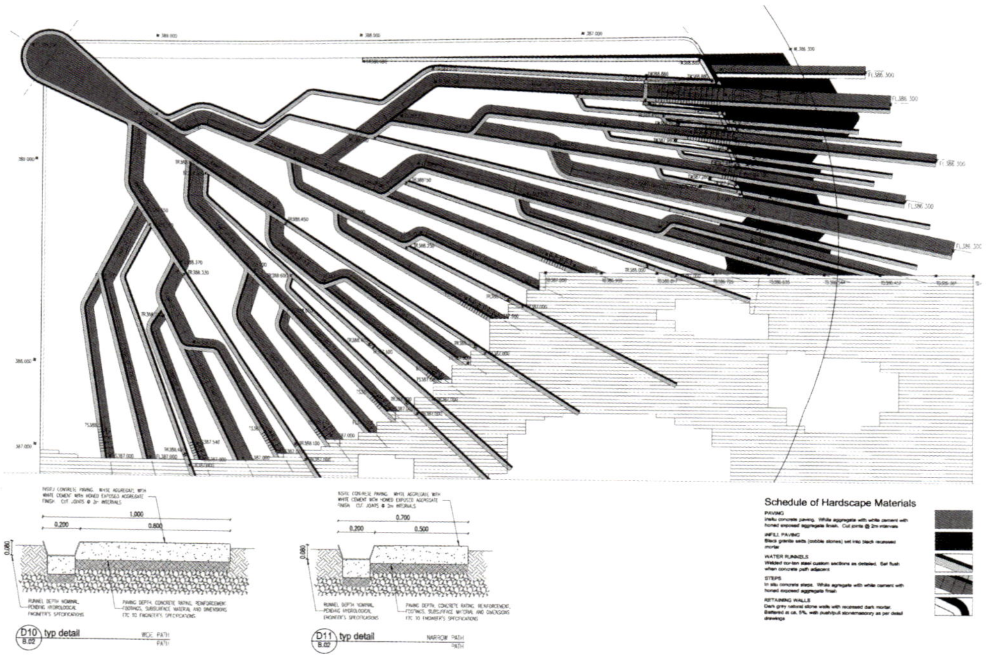
Schedule of Hardscape Materials
PAVING
Insitu concrete paving. White aggregate with white cement with honed exposed aggregate finish. Cut joints @ 2m intervals
INFILL PAVING
Black granite setts (cobble stones) set into black recessed mortar
WATER RUNNELS
Welded cor-ten steel custom sections as detailed. Set flush when concrete path adjacent
STEPS
In situ concrete steps. White aggregate with white cement with honed exposed aggregate finish
RETAINING WALLS
Dark grey natural stone walls with recessed dark mortar. Battered at ca. 5%, with push/pull stonemasonry as per detail drawings
D10 typ detail
D11 typ detail

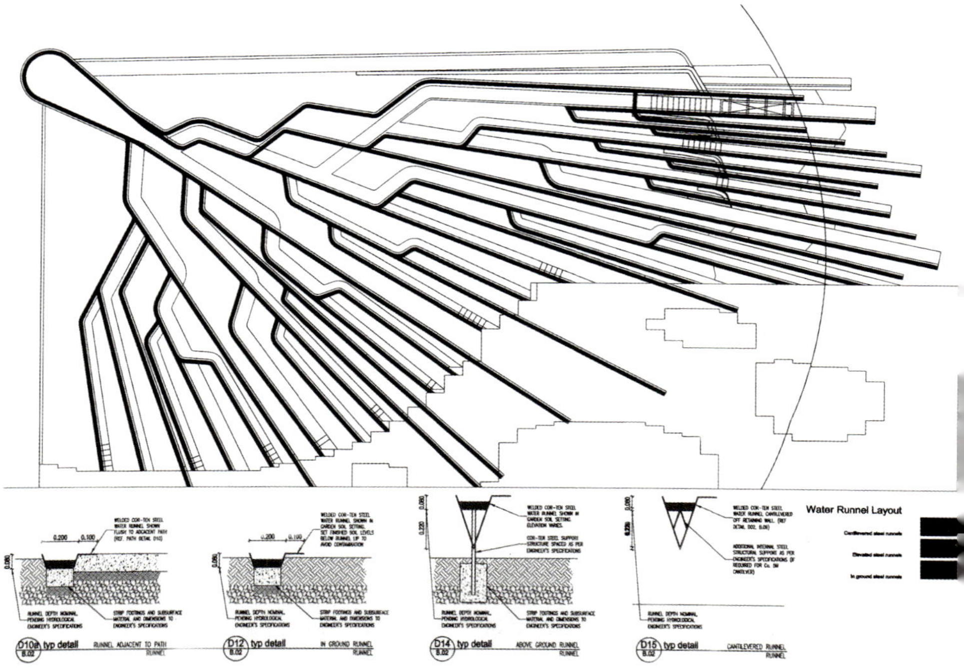
Water Runnel Layout
D10a typ detail RUNNEL ADJACENT TO PATH
D12 typ detail IN GROUND RUNNEL
D14 typ detail ABOVE GROUND RUNNEL
D15 typ detail CANTILEVERED RUNNEL

Reinforced Riverside

强化的河畔

Designed by: the Architectural Association (London)

Design Feature: Themed as "reinforced riverside", it is reflected by the transition belt made by the alternation of solid and liquid.

设计学院：伦敦建筑联盟学院

设计特色：以"强化的河畔"作为设计主题，以固体液体轮流交替转化的过渡地带进行表现。

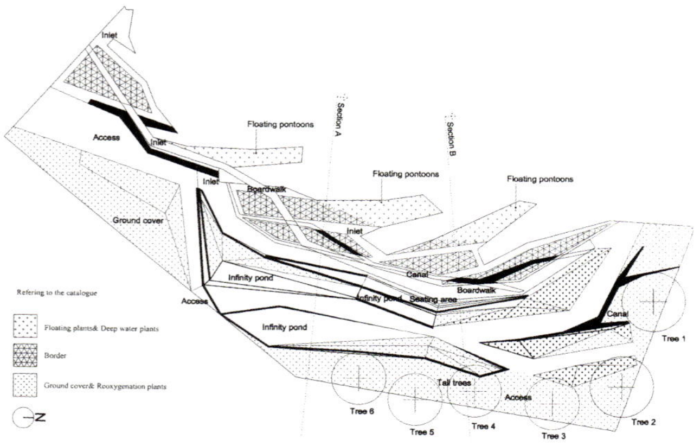

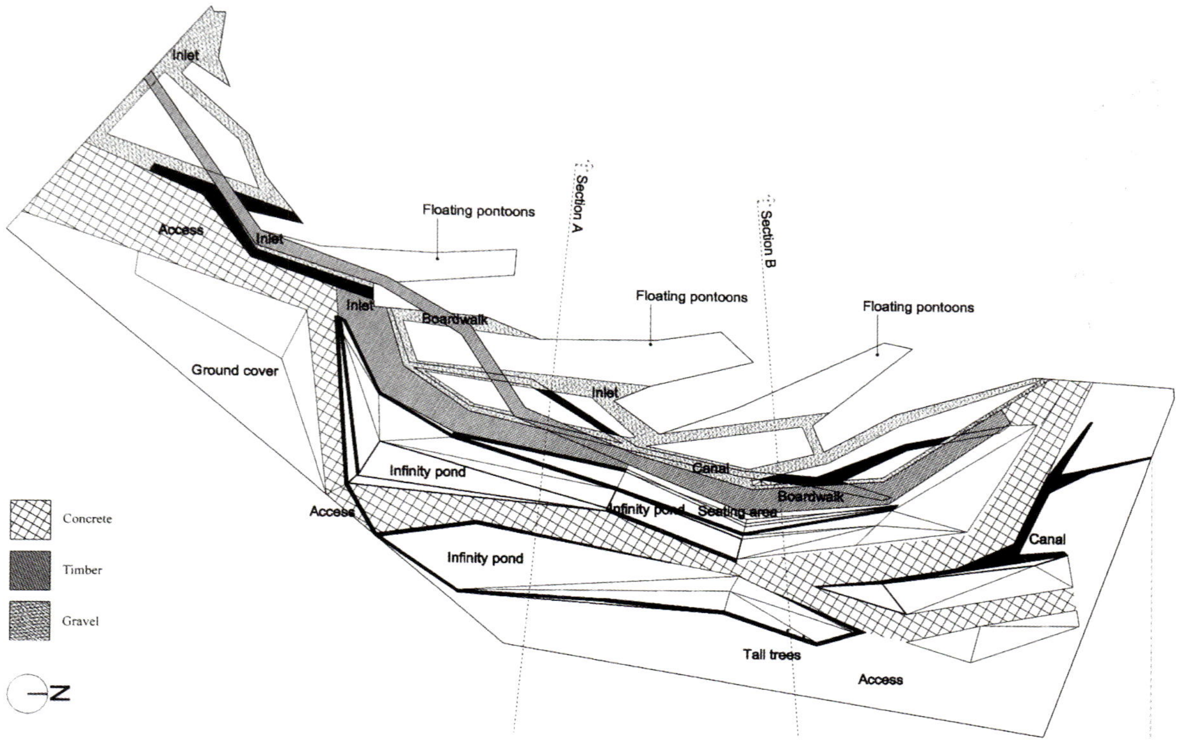

Introduction:

Themed as "reinforced riverside", this garden takes riverside advantages to probe into the way of handling the common boundary between water and land, which is reflected by the transition belt made by the alternation of solid and liquid. The infrastructure made of concrete and timber, and the unique landscape created by pontoon bridge, land and aquatic plants blur the boundary between lake area and garden. The acclivitous space form breaks the horizontal and vertical space dimensions that people familiar with, furthermore, it highlights the uncertainty, fuzziness and transparency of the whole.

展园介绍：

该园以"强化的河畔"作为设计主题，利用地处水边的优势和特点，探讨水陆交界的处理手法，以固体液体轮流交替转化的过渡地带进行表现。混凝土和木材的基础设施，浮桥、陆地和水生植物创造出别致水景从而模糊了湖区与庭院的边界。以倾斜的空间形式打破了人们所熟悉的水平和纵向的空间维度，并突出了整体的不明确性、模糊性和透明性。

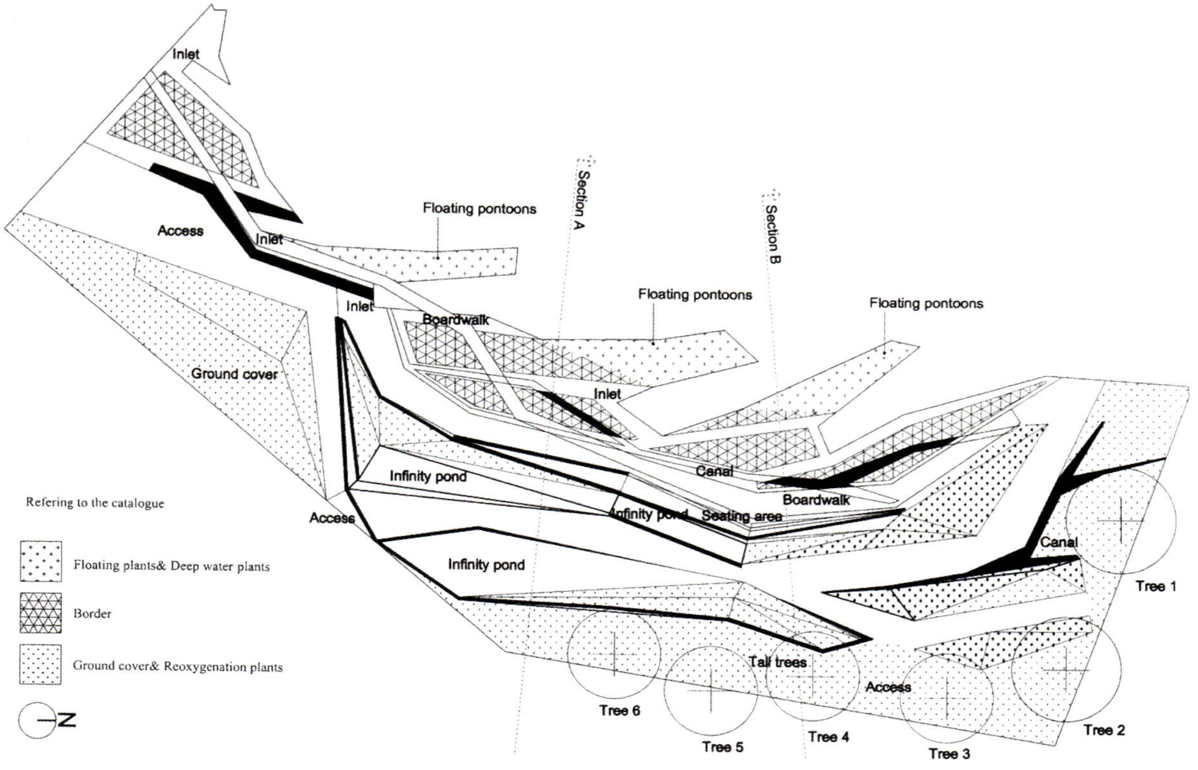

CHANEL	P1	P2	h1	h2	Dh	Length	Slope	Cooment
			m	m	m	m	%	
1	2	1	-0.05	-0.07	0.02	11	0.2%	Stepped finishing
2	3	2	1	-0.05	1.05	10.5	10.0%	Stepped finishing
3	4	2	0.3	-0.05	0.35	5.7	6.1%	Stepped finishing
4	5	4	0.5	0.3	0.2	5.4	3.7%	Stepped finishing
5	8	5	1	0.5	0.5	11.2	4.5%	Stepped finishing
6	5	9	0.5	-0.07	0.57	11.7	4.9%	Stepped finishing
7	6	7	1	-0.07	1.07	8.6	12.4%	Stepped finishing
8	8	10	1	0.6	0.4	8.9	4.5%	Stepped finishing
9	11	9	0.3	-0.07	0.37	8.1	4.6%	Stepped finishing
10	10	12	0.6	0.5	0.1	9	1.1%	Rough bottom canal
11	12	11	0.5	0.3	0.2	9.1	2.2%	Rough bottom canal
12	14	13	-0.05	-0.07	0.02	4.1	0.5%	Rough bottom canal
13	15	14	1	-0.05	1.05	4.9	21.4%	Stepped finishing
14	16	14	1	-0.05	1.05	6.4	16.4%	Stepped finishing
15	16	17	1	-0.07	1.07	17.4	6.1%	Stepped finishing
16	15	18	1	0.6	0.4	17.4	2.3%	Rough bottom canal
17	20	17	0.45	-0.07	0.52	8.7	6.0%	Stepped finishing
18	18	19	0.6	0.5	0.1	6	1.7%	Rough bottom canal
19	19	20	0.5	0.45	0.05	2.8	1.8%	Rough bottom canal

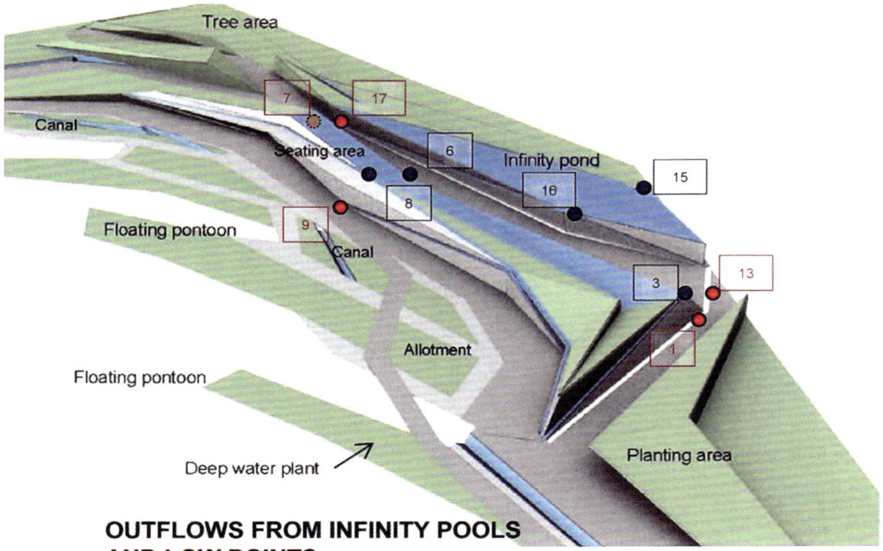

OUTFLOWS FROM INFINITY POOLS AND LOW POINTS

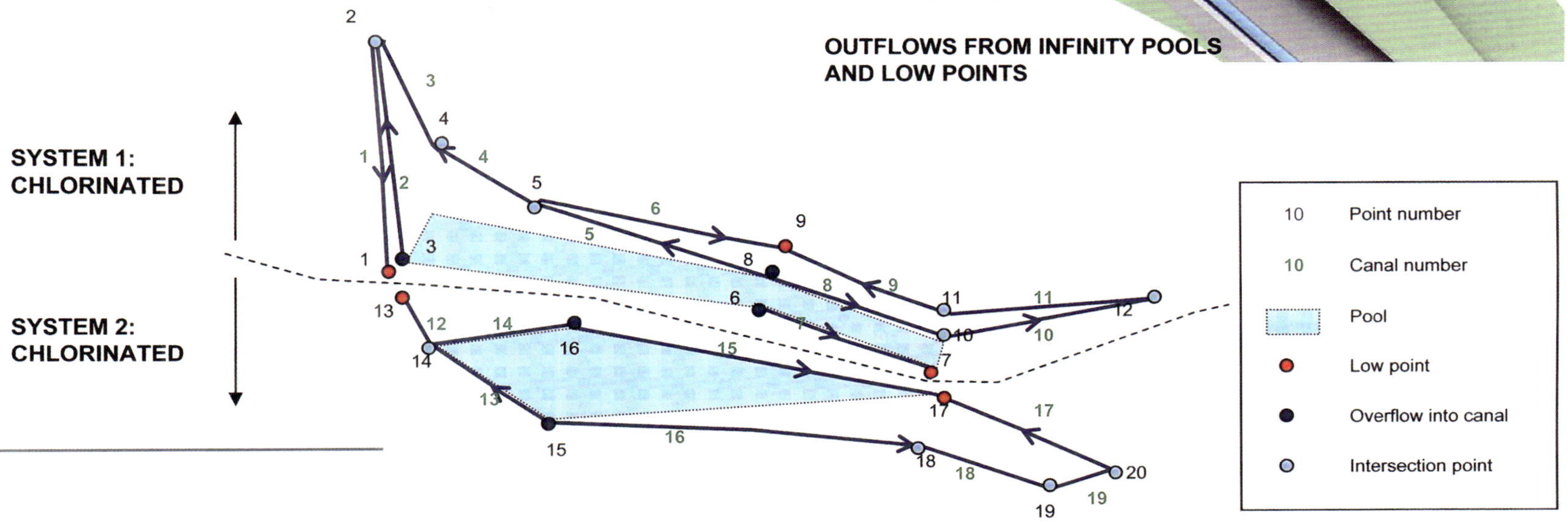

POINT AND CHANNEL NUMBERING

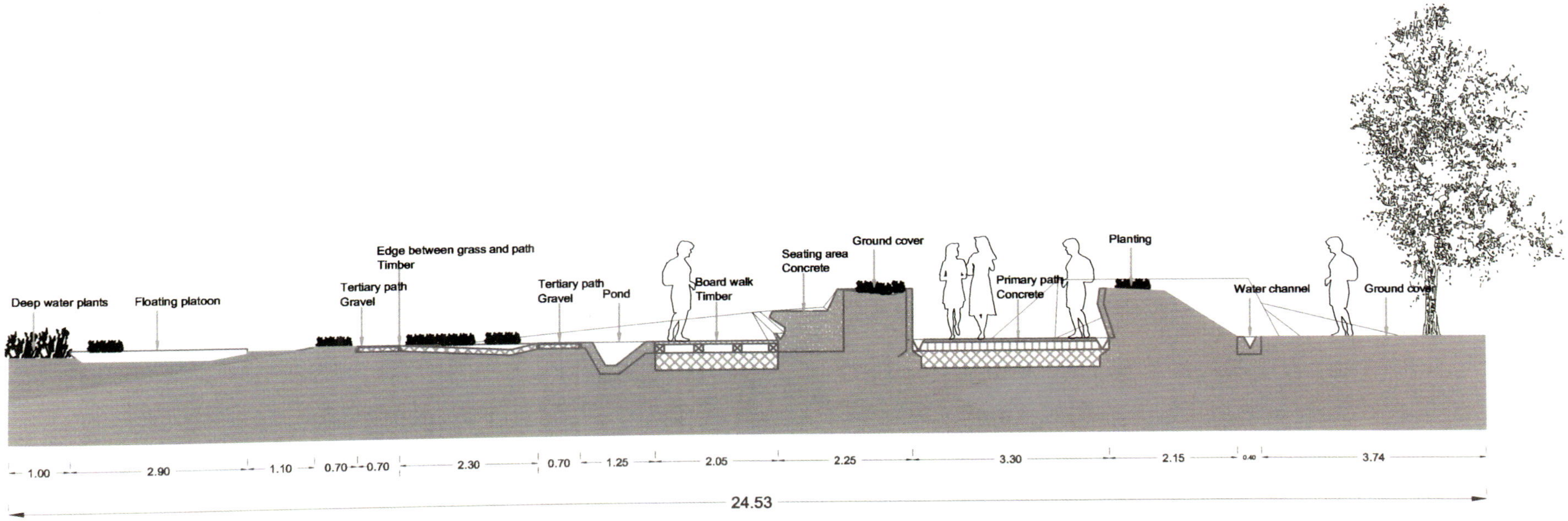

Section B

Ecological Platform

生态平台

Designed by: Columbia University (USA)

Design Feature: It is designed by using many uncertain factors within designated design scope.

设计学院：美国哥伦比亚大学

设计特色：利用在制定的设计范围内诸多不确定的因素所带来的变化而进行设计。

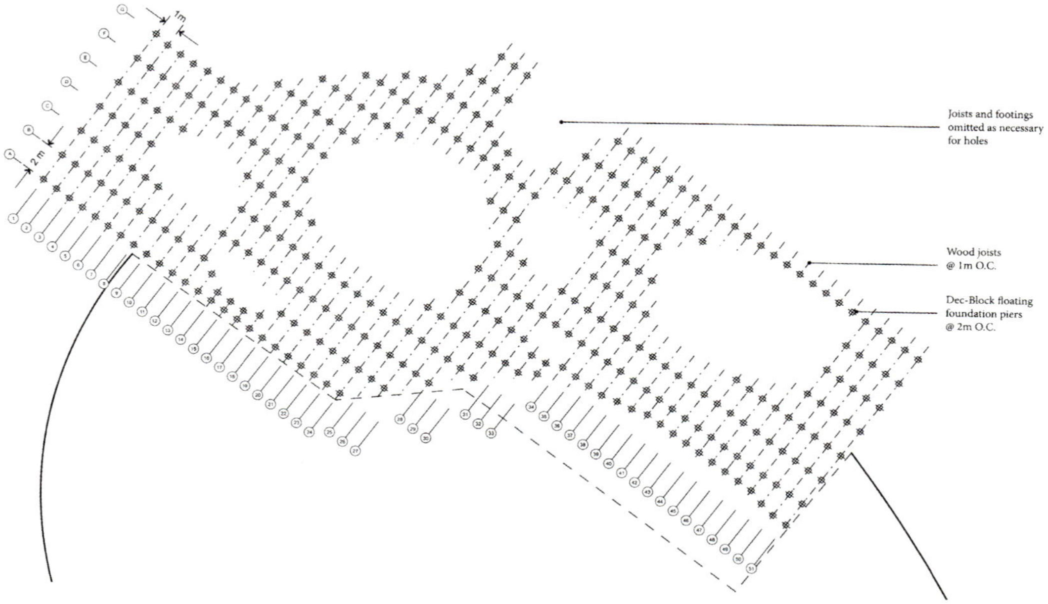

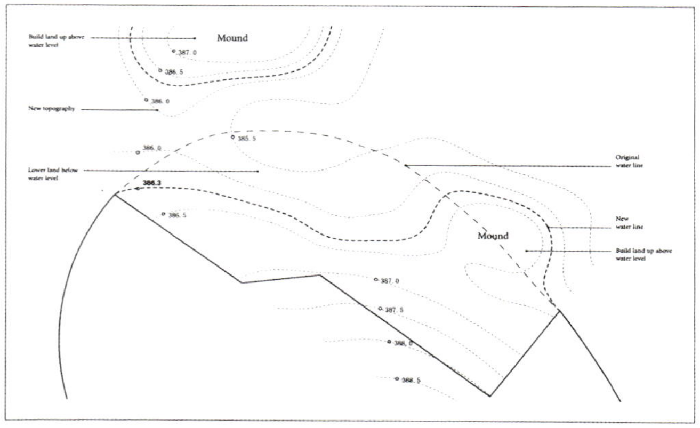

Introduction:

Landscape design, as a dynamic design, is affected by weather change, season change and other factors. "Ecological platform" is designed by using many uncertain factors within designated design scope. The designer creates a multidimensional ecological swamp by building a riverside platform and displays different landscape effects by seasonal plants change as well as the rise and fall of water level. This kind of design is different to the used one that tames nature, by doing so, it gets easier for the biocenosis to adapt the unceasing reformation, thus to achieve the real harmony of man and nature.

展园介绍：

景观设计是一种动态的设计，它的形式是受到天气的变化、季节的交替等因素影响。"生态平台"是利用在制定的设计范围内诸多不确定的因素所带来的变化而进行设计的。设计者通过构建一个淡水平台，创造一个多维的生态系统湿地，利用植物季相的变化、水位的上升和下降展现出不同的景观效果。这种"顺其自然"的设计态度与过去试图"驯服"自然不同，这样更易于生物群落在不断改造的过程中适应，从而达到人与自然真正的和谐。

City of Sky

天空之城

Designed by: University of Southern California

Design Feature: With sunshine atrium, reflection garden and cloudy and misty garden, it conveys a concept that people yearn for beautiful nature.

设计学院：南加州大学

设计特色：运用日光中庭、反射花园和云雾园诠释出一种人对自然美好憧憬的理念。

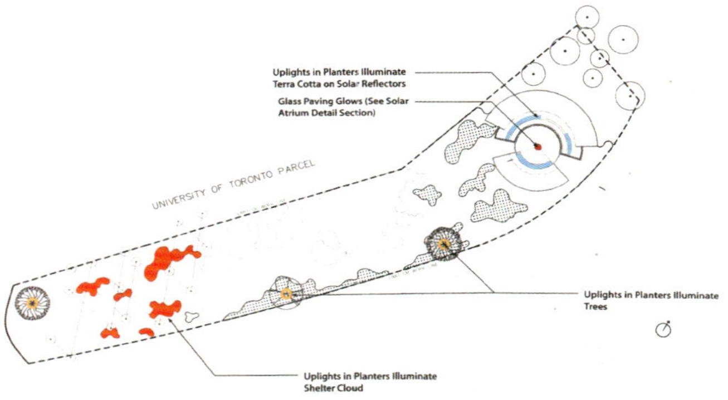

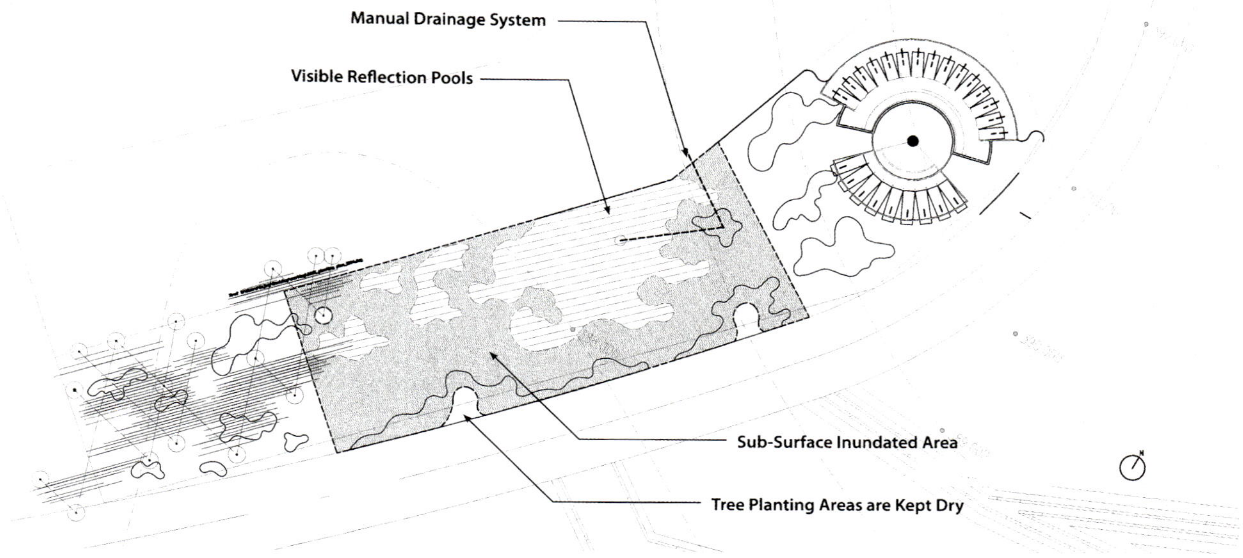

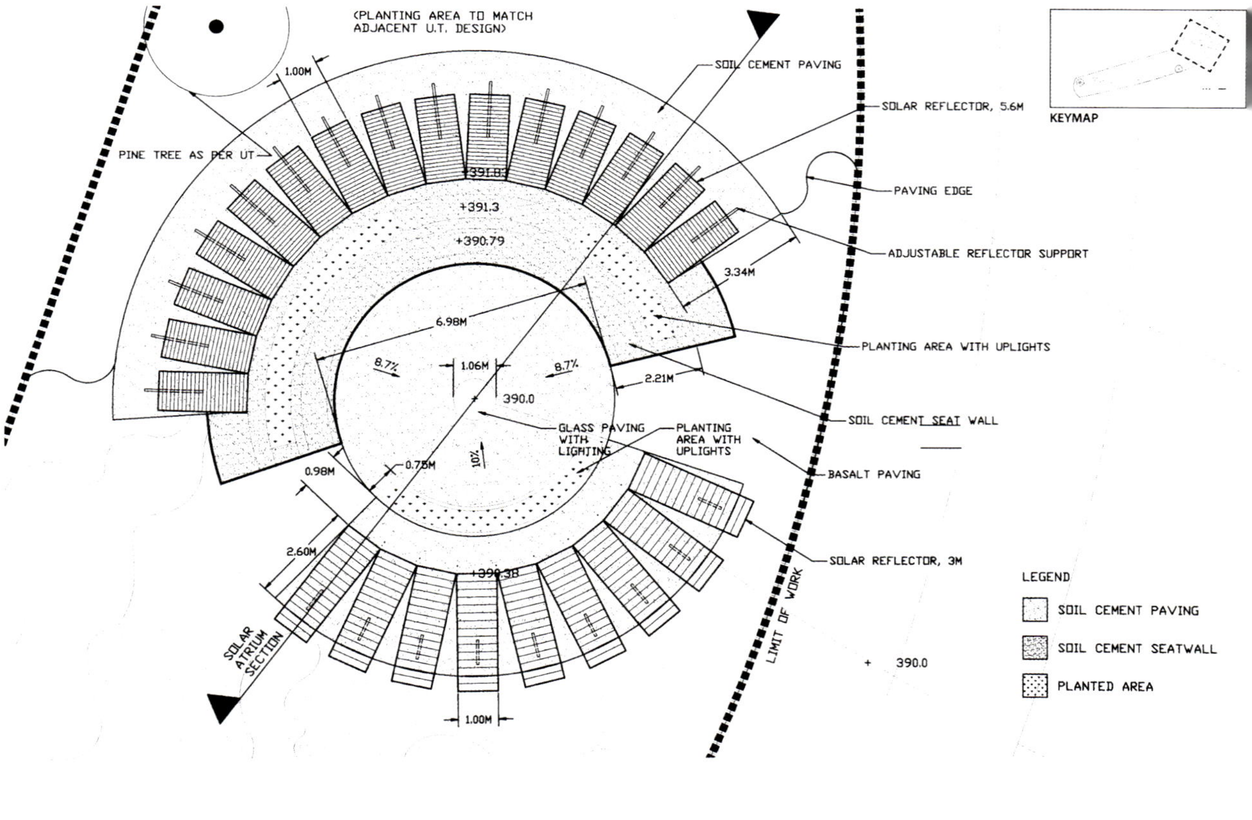

Introduction:

It conveys a concept that people yearn for beautiful nature. The three main characteristics are sunshine atrium, reflection garden and cloudy and misty garden, which respectively reflect the basic relations between mankind and sky: direct perception, indirect reflection and roaming. The City of Sky makes you lost yourself and brings you a visiting experience—increases the humidity of the air through spraying system and creates a mist-shrouded impression, which enables you to perceive the connotation in the bewildering sky and consider the relationship between mankind and nature.

展园介绍：

天空之城诠释出一种人对自然美好憧憬的理念。其三个主要特点分别是日光中庭、反射花园和云雾园，当中均分别反映出人与天空的基本关系是直接感知、间接反射、畅想遨游。天空之城让游客迷失在天空之中，同时带来参观体验——通过喷雾系统加大了空气的湿度并且营造出云雾的观感，让游客在迷雾中参与到整体景观之中，感受其中的蕴藏，思考人与自然的关系。

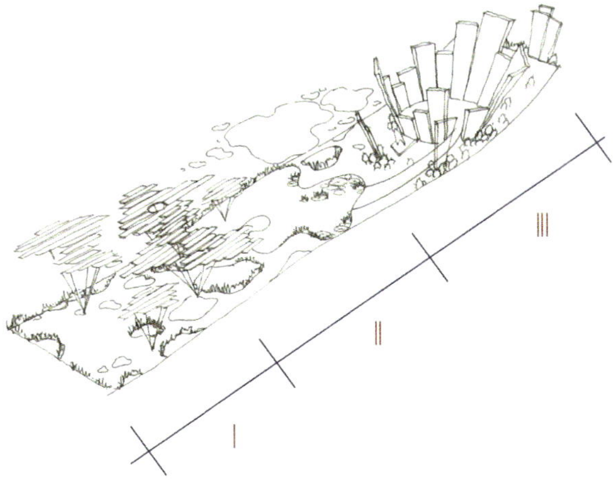

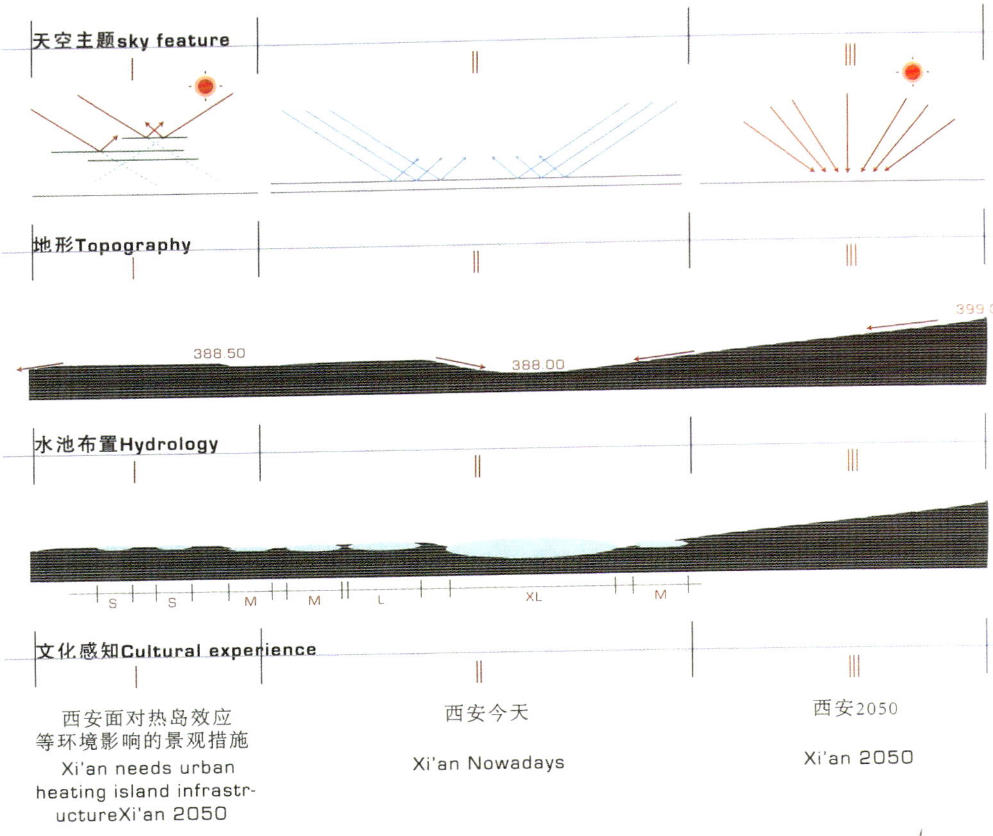

景观剖面：元素构成

Components

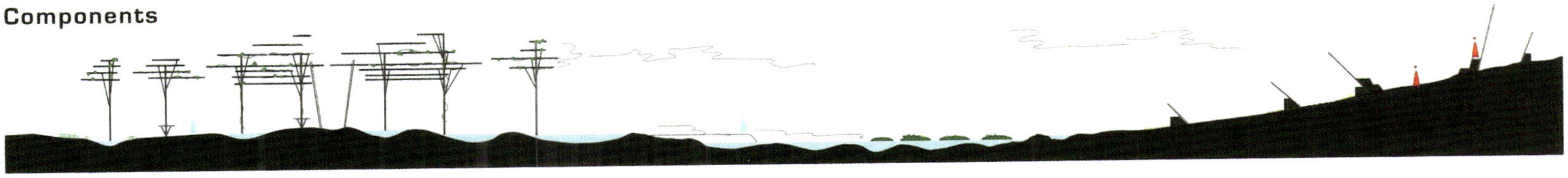

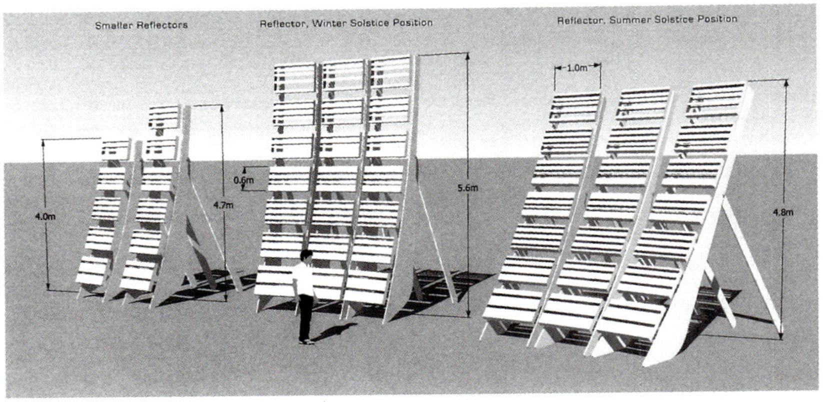

Conceptual Visualization of Reflector

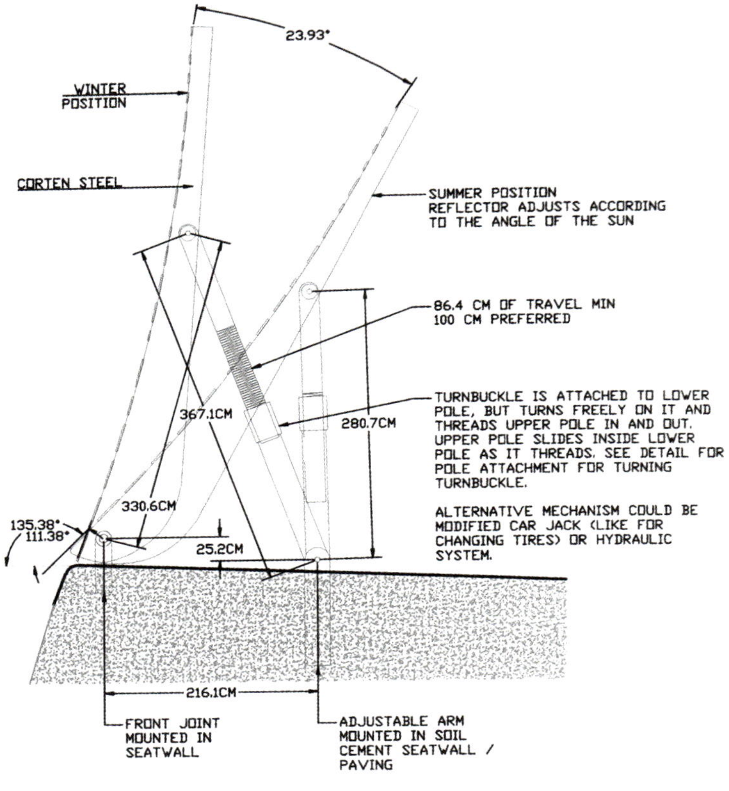

Large Solar Reflector Detail Side Section

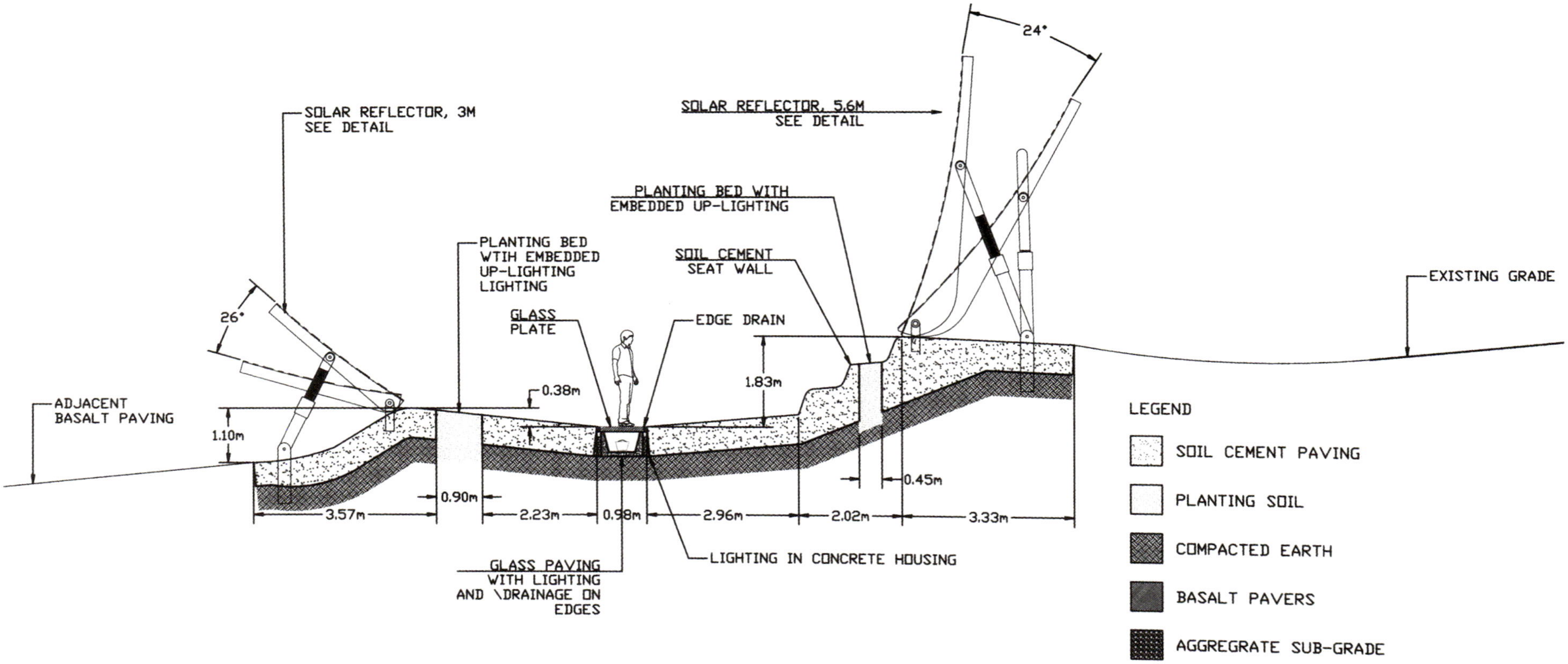
SOLAR REFLECTOR, 3M
SEE DETAIL
26°
ADJACENT
BASALT PAVING
1.10m
PLANTING BED
WTIH EMBEDDED
UP-LIGHTING
LIGHTING
GLASS
PLATE
0.38m
EDGE DRAIN
PLANTING BED WITH
EMBEDDED UP-LIGHTING
SOIL CEMENT
SEAT WALL
1.83m
SOLAR REFLECTOR, 5.6M
SEE DETAIL
24°
EXISTING GRADE
0.45m
3.57m
0.90m
2.23m
0.98m
2.96m
2.02m
3.33m
GLASS PAVING
WITH LIGHTING
AND \DRAINAGE ON
EDGES
LIGHTING IN CONCRETE HOUSING
LEGEND
SOIL CEMENT PAVING
PLANTING SOIL
COMPACTED EARTH
BASALT PAVERS
AGGREGRATE SUB-GRADE

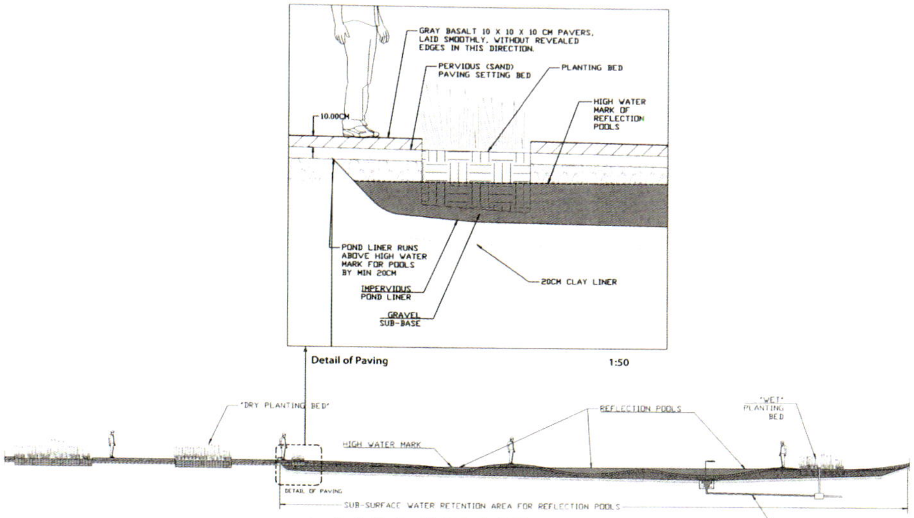

Detail of Paving 1:50

Longtitudinal Section of Reflection Pools & Paving

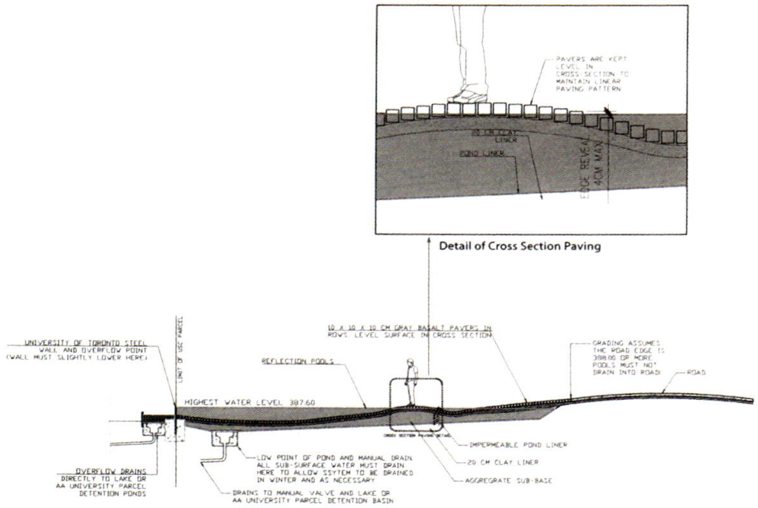

Detail of Cross Section Paving

Cross Section

Layer Structure:
pole + rod + mister +planter + steel rope

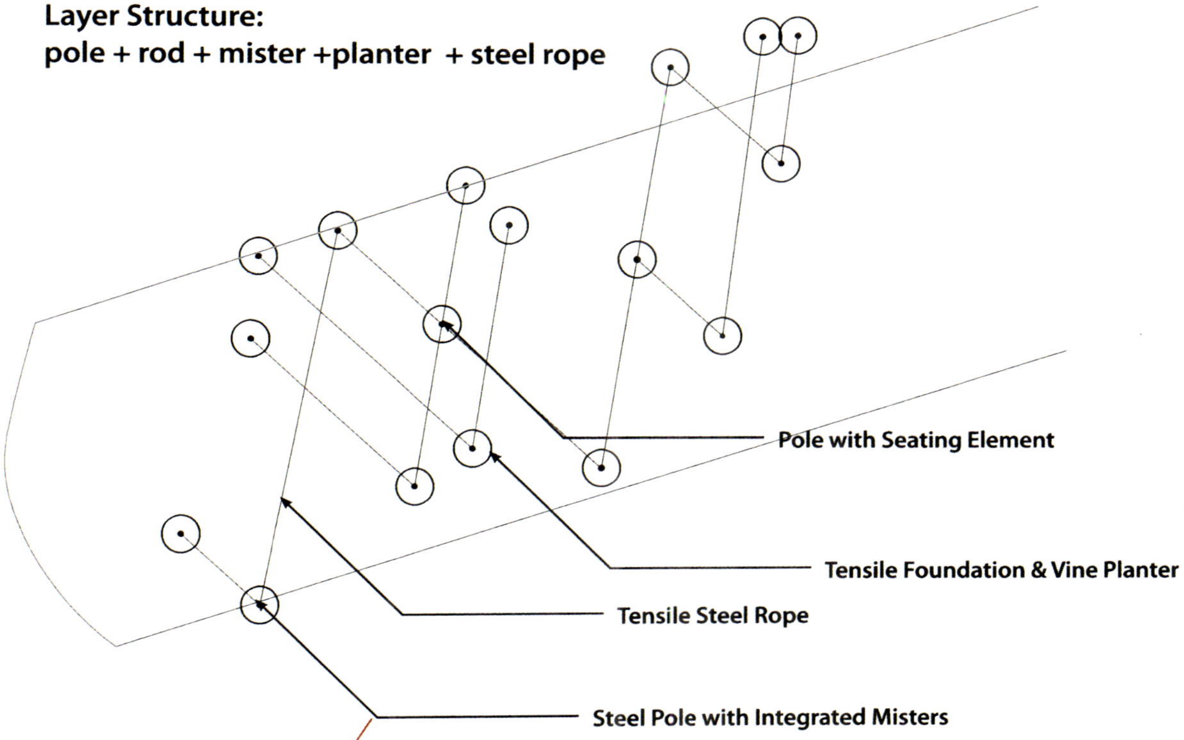

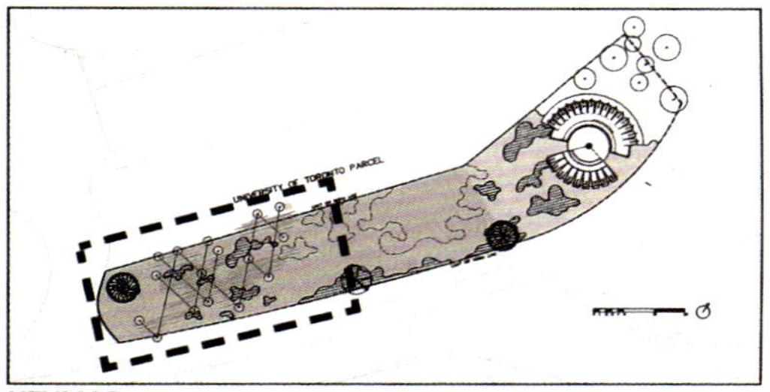

KEYMAP

Conceptual Visualization

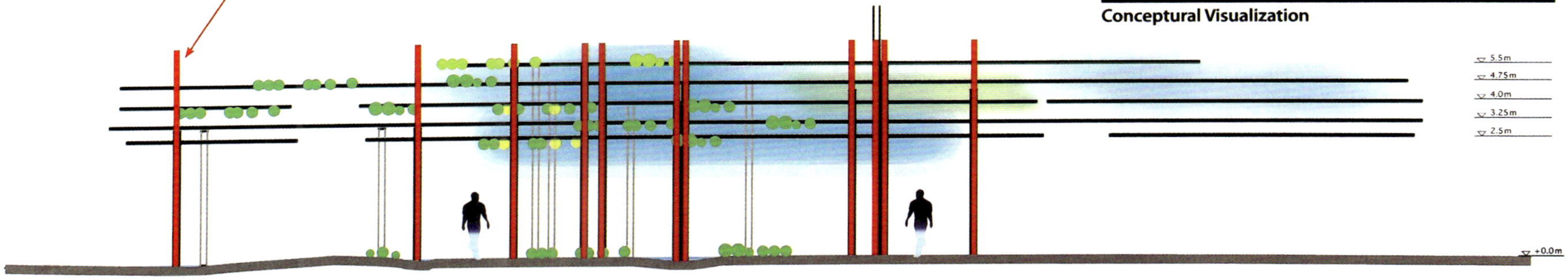

Ecology-Solar Term

生态 - 节气

Designed by: Feng Chia University, Taiwan

Design feature: Starting from solar term, it has created a deteriorating micro climate.

设计学院：台湾逢甲大学

设计特色：从“二十四”节气出发，创造出一个持续变化的微气候环境。

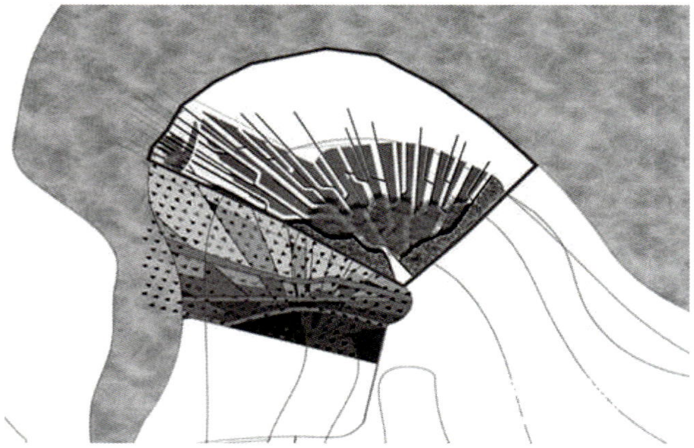

Introduction:

Ecology–Solar Term uses the concept of sundial to define the sequential period pattern of Xi’an. The angles and lengths of the shades generated by the intensively distributed equidistant green columns create a deteriorating micro climate. Based on the intensively distributed equidistant form, a changeful micro climate is created to enrich visitors’ experience with the concept of solar term, the limitation of the environment, the topography, the paths, the diversity of vegetation, the climate condition and the behaviors of the visitors.

展园介绍：

设计从“二十四”节气出发，利用日晷的概念来定义西安的时序周期形态，利用高密度等间距的绿柱所产生阴影的角度和长度来创造出一个持续变化的微气候环境。该园以高密度等间距的形态作为布局的基础，根据节气原理、基地环境限制、地形、路径、植被多样性、气候条件和参观者行为的不可预测性，创造出一个持续变化的微气候环境，来丰富参观者的体验。

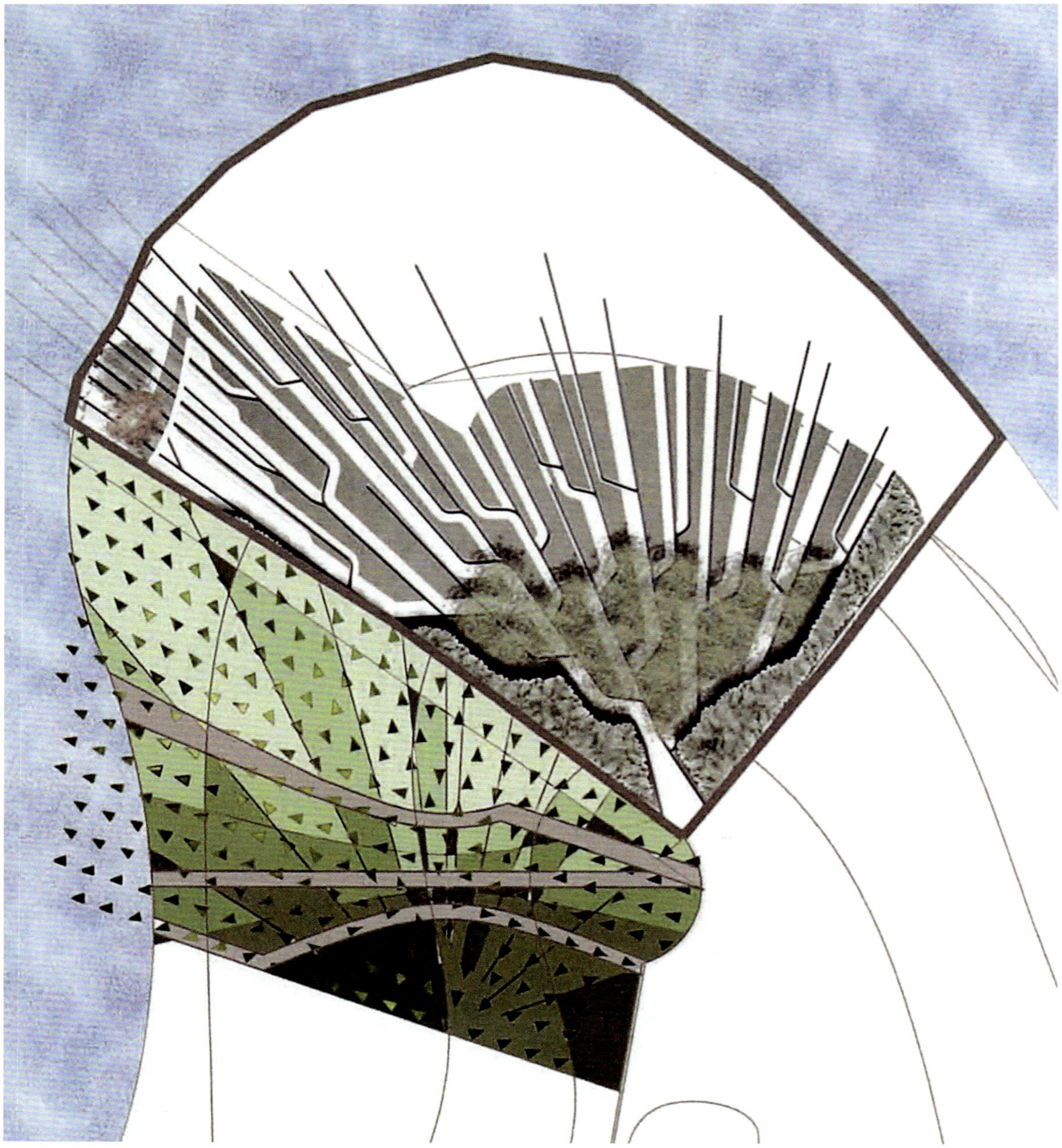

面　积：600平方米
设计主题：生态 节气
概况介绍：
生态节气是利用日晷的
间距的绿柱所产生阴影的角
参展单位：北京市花木公司
AREA: 600 m²
DESIGN THEME: Ecology-
INTRODUCTION:
Ecology-solar term uses the
period pattern of Xi'an. The
the equidistant green colum
PARTICIPANT: Beijing Flowe

City Landscape Intervention

城市景观干预

Designed by: University of Saint Joseph, Macau

Design feature: The designer uses the architectural form of hanging bridge which is peculiar to Macau to reflect the cultural background of Macau.

设计学院：澳门圣约瑟大学

设计特色：利用澳门特有的吊桥建筑形式反映出澳门的文化背景。

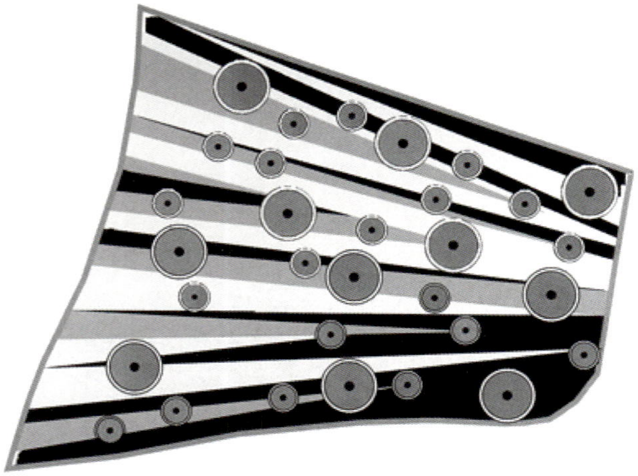

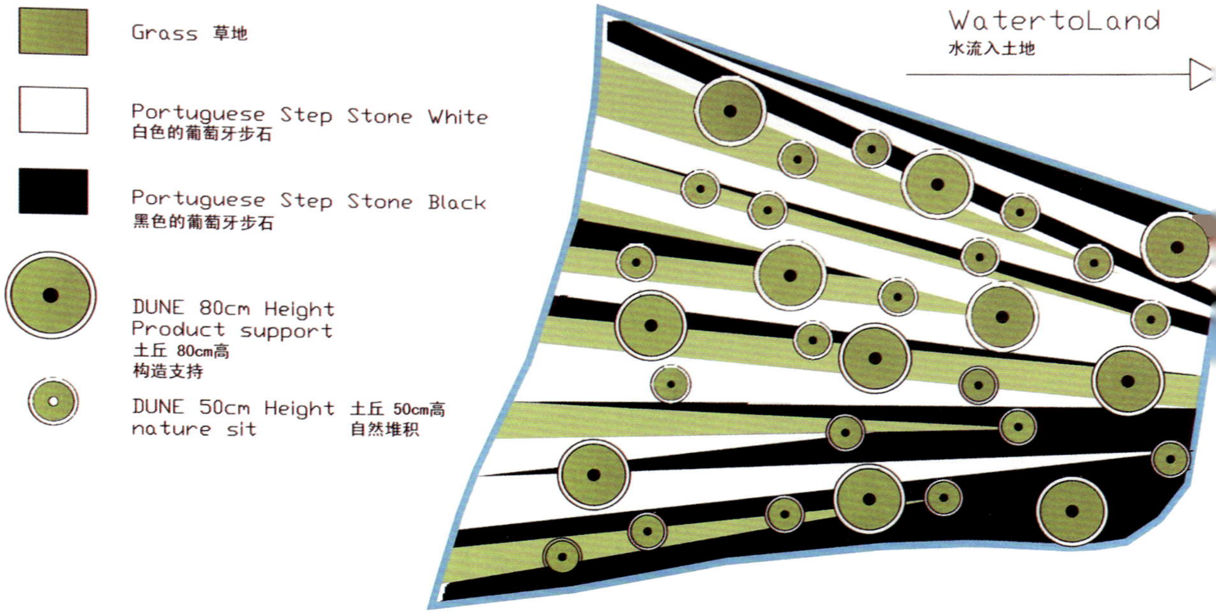

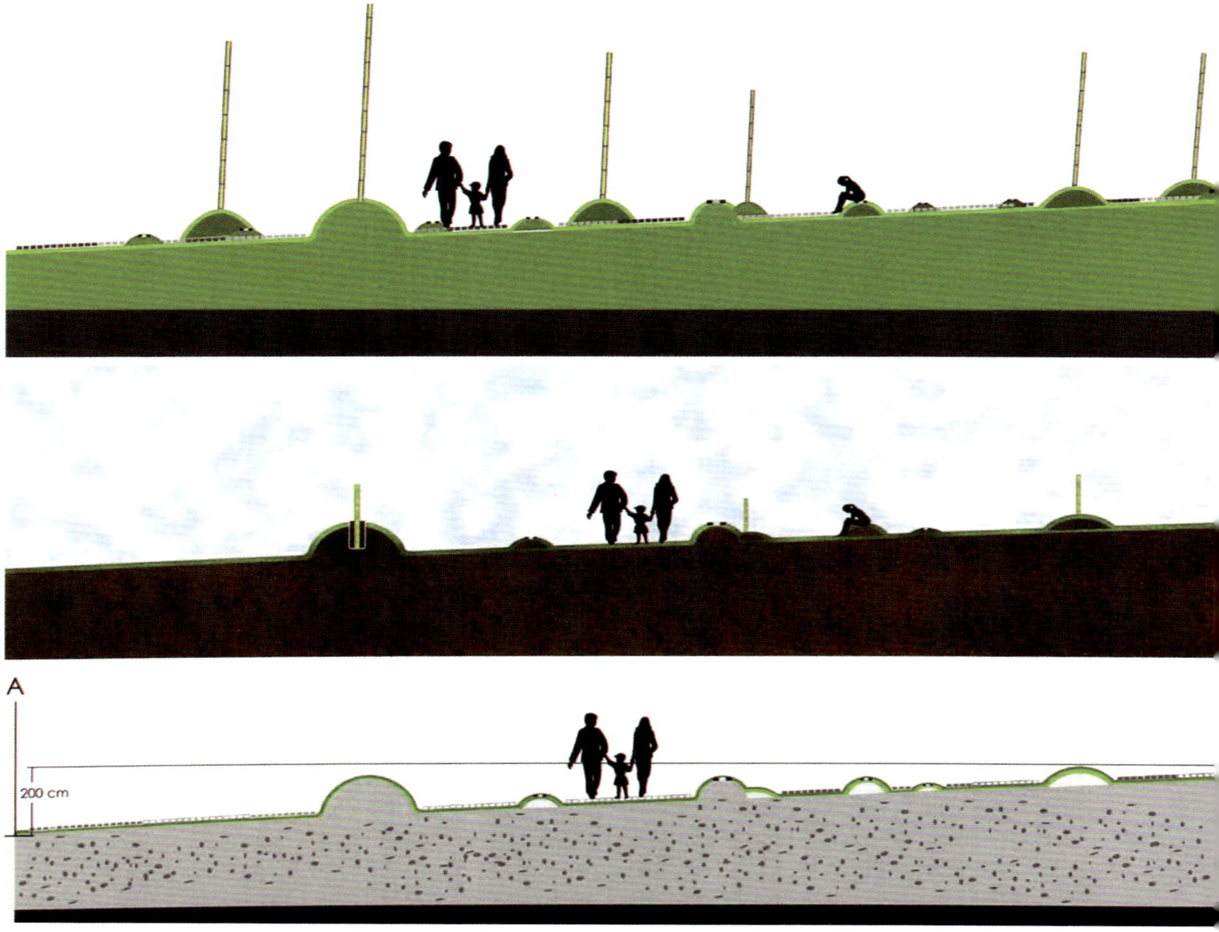

Introduction:

City landscape intervention. The designer uses the architectural form of hanging bridge which is peculiar to Macau to reflect the cultural background of Macau, connecting the past and present of Macau as well as the past and present of Macau and mainland China. The typical fishing village, wigwams and hanging bridge in the past of Macau are replaced by today's high-rise complexes. The garden has used emissive green axis to connect the visit route with green plants and form a mutual interferential layout which showcases the close relationship between urban landscape and human as well as echoes the theme of "creation and nature" for 2011 Xi'an International Horticultural Exposition.

展园介绍：

城市景观干预，设计者利用澳门特有的吊桥建筑形式反映出澳门的文化背景，同时连接了澳门的过去和现在、连接了澳门与祖国大陆的过去和今朝。当年澳门地域特色的渔村、棚屋、吊桥已经被如今的高层建筑综合体所取代了。该园是着重以放射性的绿轴将参观流线与绿植交织形成互相干涉的格局，来表现澳门中城市景观与人类的密切关系的同时，呼应了世园会创意自然的主题。

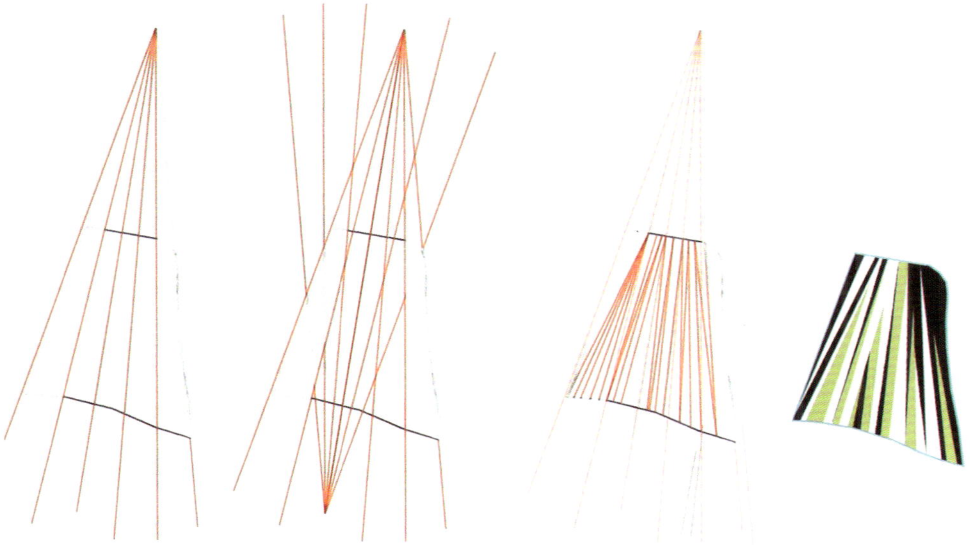

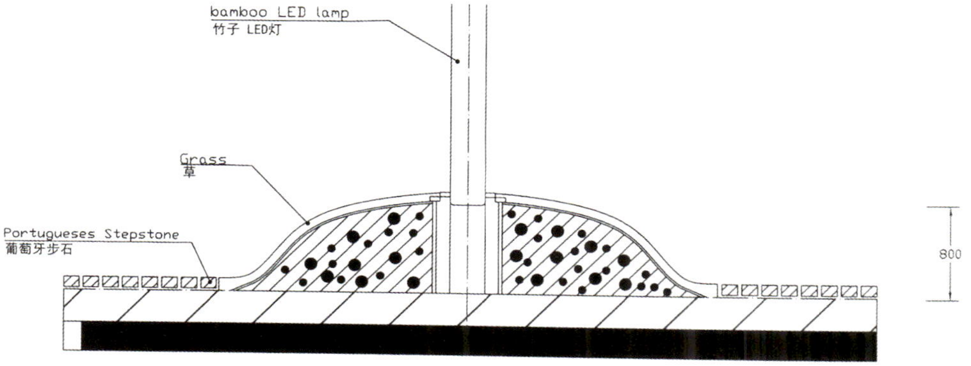
bamboo LED lamp
竹子 LED灯
Grass
草
Portugueses Stepstone
葡萄牙步石
800

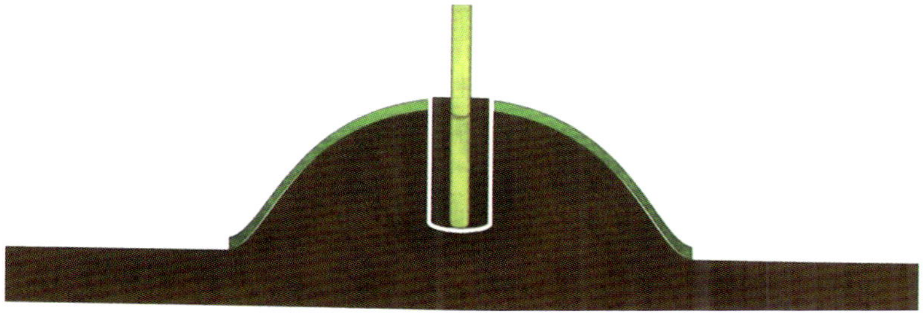

Poem of Song

风的诗歌

Designed by: Hong Kong University

Design feature: With "wind" as the main element, the Garden is used to reflect the vitality of Hong Kong.

设计学院：香港大学

设计特色：以"风"作为设计使用的主要元素，以庭院来表现香港的城市活力。

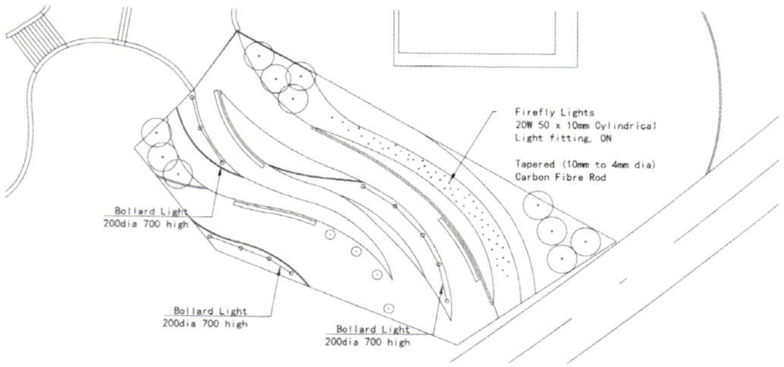

Introduction:

Garden is used to reflect the vitality of Hong Kong. "Poem of Song" tries to express the feeling of living in the city by a garden. The constant energy flow is shown by the flowing of wind. The wind wheel which is running unceasingly day and night is used to symbolize the vitality of the city, and display the characteristics of Hong Kong by creating two kinds of winds: unperceivable wind and perceivable wind.

With "wind" as the main element, the garden has responded to the theme of environmental protection advocated in New Times. Windtower is built to create a vertical landscape system.

展园介绍：

展园以庭院来表现香港的城市活力。"风的诗歌"力求在一座庭院园中表达出身处城市环境的感受，城市中恒定的能量流，显示为风的不断流动，而设计中的风车装置日夜转动，是充满活力的展现，同时营造隐蔽风和感受风两种不同的场所展现香港的城市特点。

以"风"作为设计使用的主要元素是展园对于新时代倡导的环抱主题作出的回应。搭建风塔等对于塑造展园立体的景观层次也起到良好的作用。

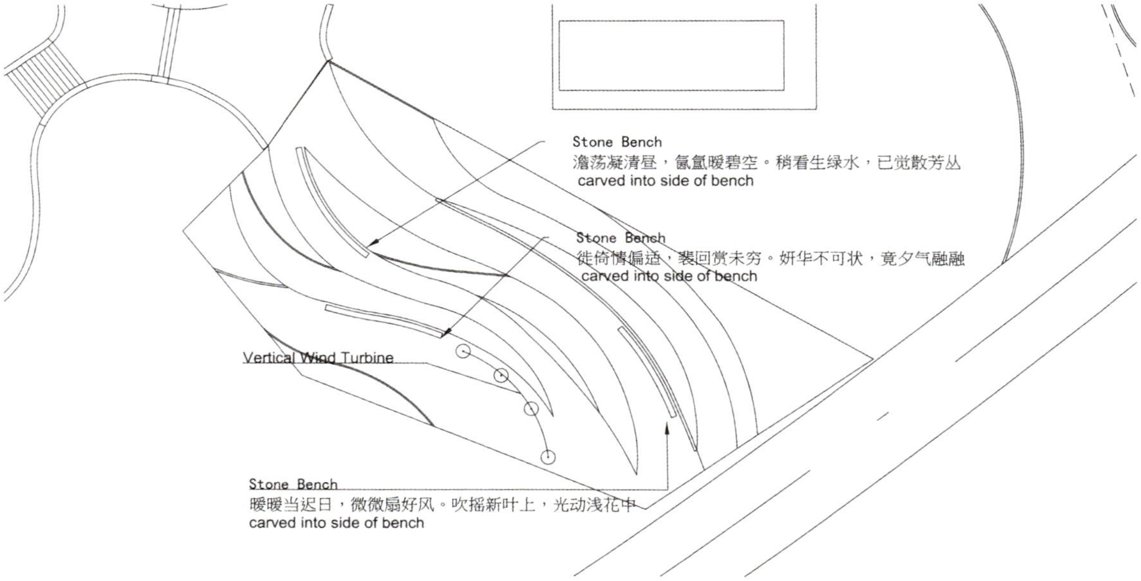

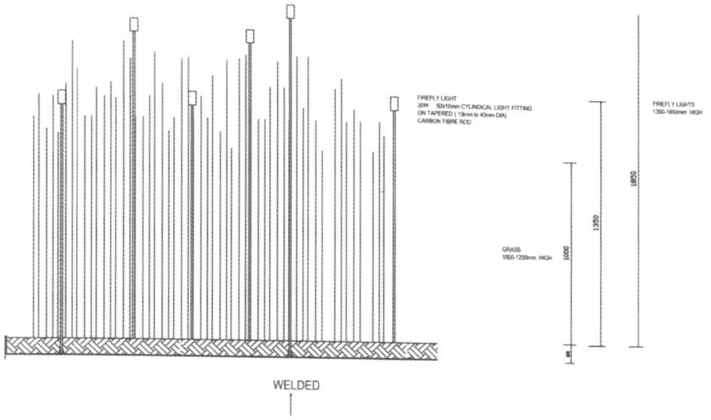

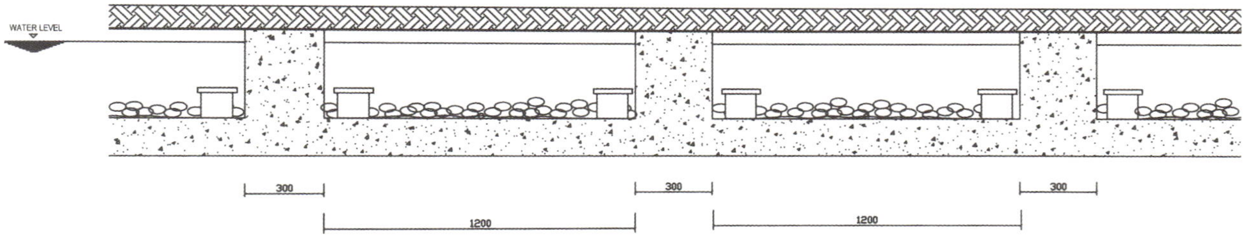

SPAN BETWEEN SUPPORT 1200mm

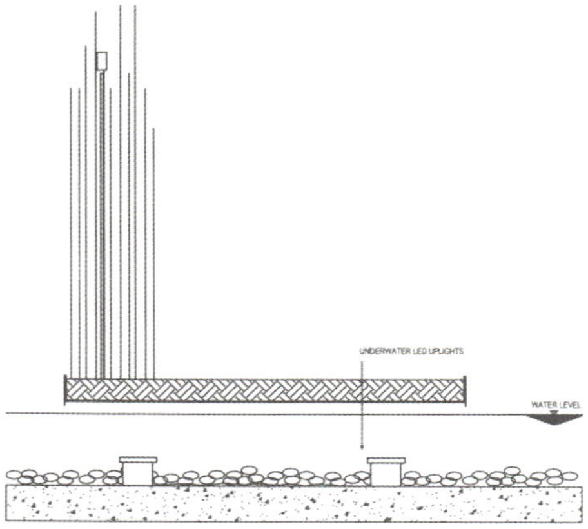

SECTION-MID SPAN

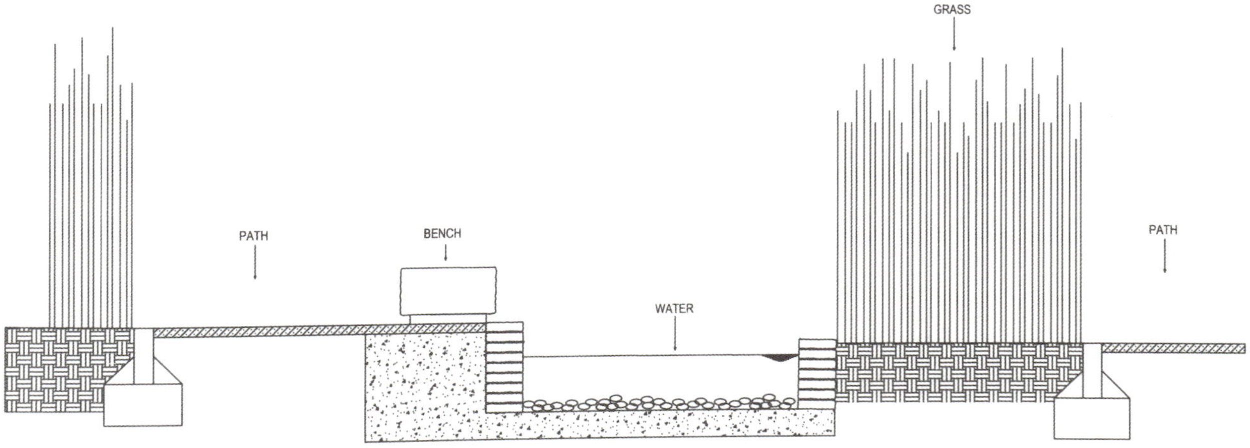

SECTION - GENERAL ARRANGEMENT

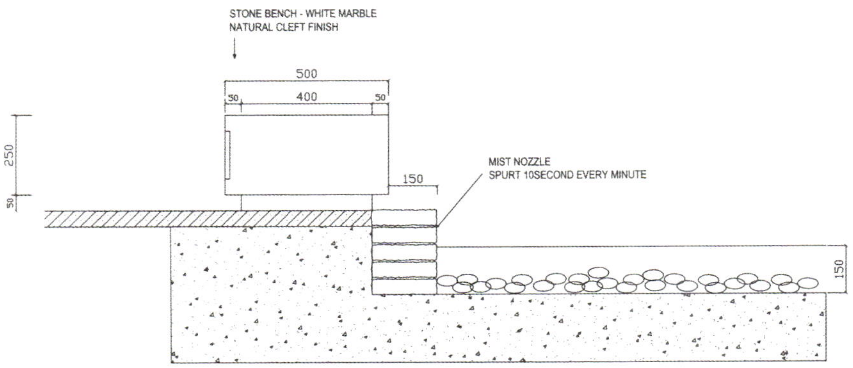
STONE BENCH - WHITE MARBLE
NATURAL CLEFT FINISH
500
400
50
50
250
50
150
MIST NOZZLE
SPURT 10SECOND EVERY MINUTE
150

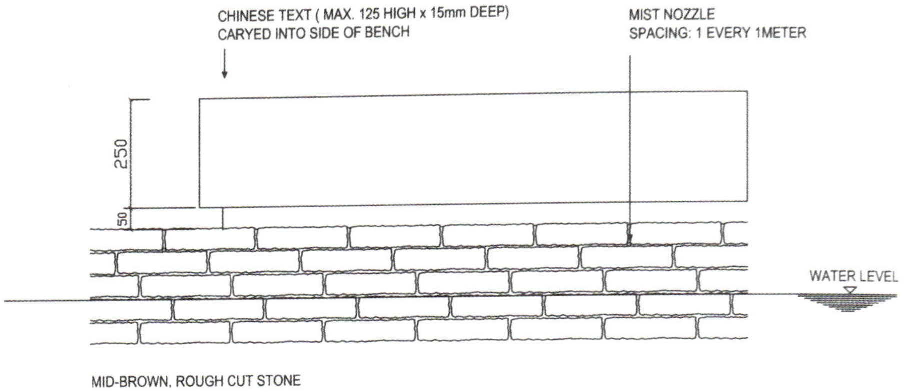
CHINESE TEXT (MAX. 125 HIGH x 15mm DEEP)
CARYED INTO SIDE OF BENCH
MIST NOZZLE
SPACING: 1 EVERY 1METER
250
50
WATER LEVEL
MID-BROWN, ROUGH CUT STONE

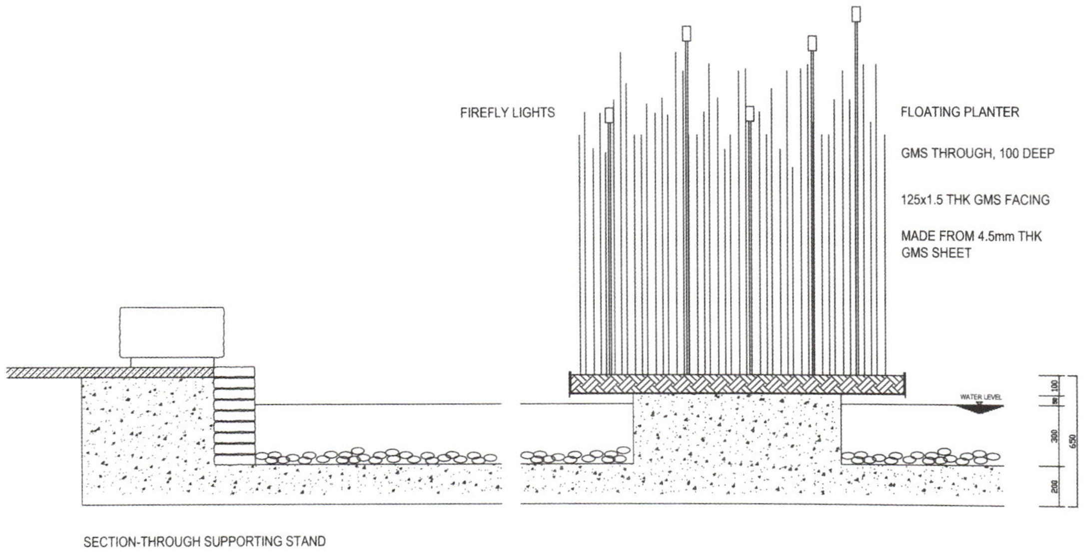
FIREFLY LIGHTS
FLOATING PLANTER
GMS THROUGH, 100 DEEP
125x1.5 THK GMS FACING
MADE FROM 4.5mm THK
GMS SHEET
WATER LEVEL
SECTION-THROUGH SUPPORTING STAND

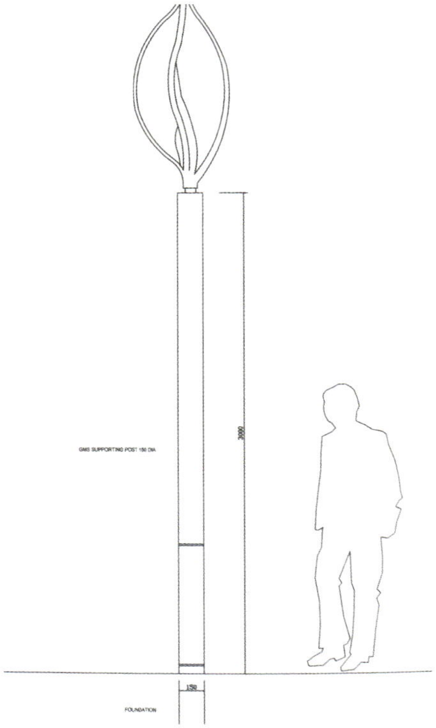

092–093

Pampas Impression

潘帕斯印象

Designed by: Universidad Torcuato Di Tella, Argentina

Design feature: The designer attempts to use the physical, geographic and cultural epitome of Pampas to create a sense of Pampas in International Horticultural Exposition 2011 Xi' an.

设计学院：阿根廷托尔夸托迪特加大学

设计特色：试图在物理、地理、文化风韵与阿根廷不尽相同的西安创造出局部“潘帕斯印象”。

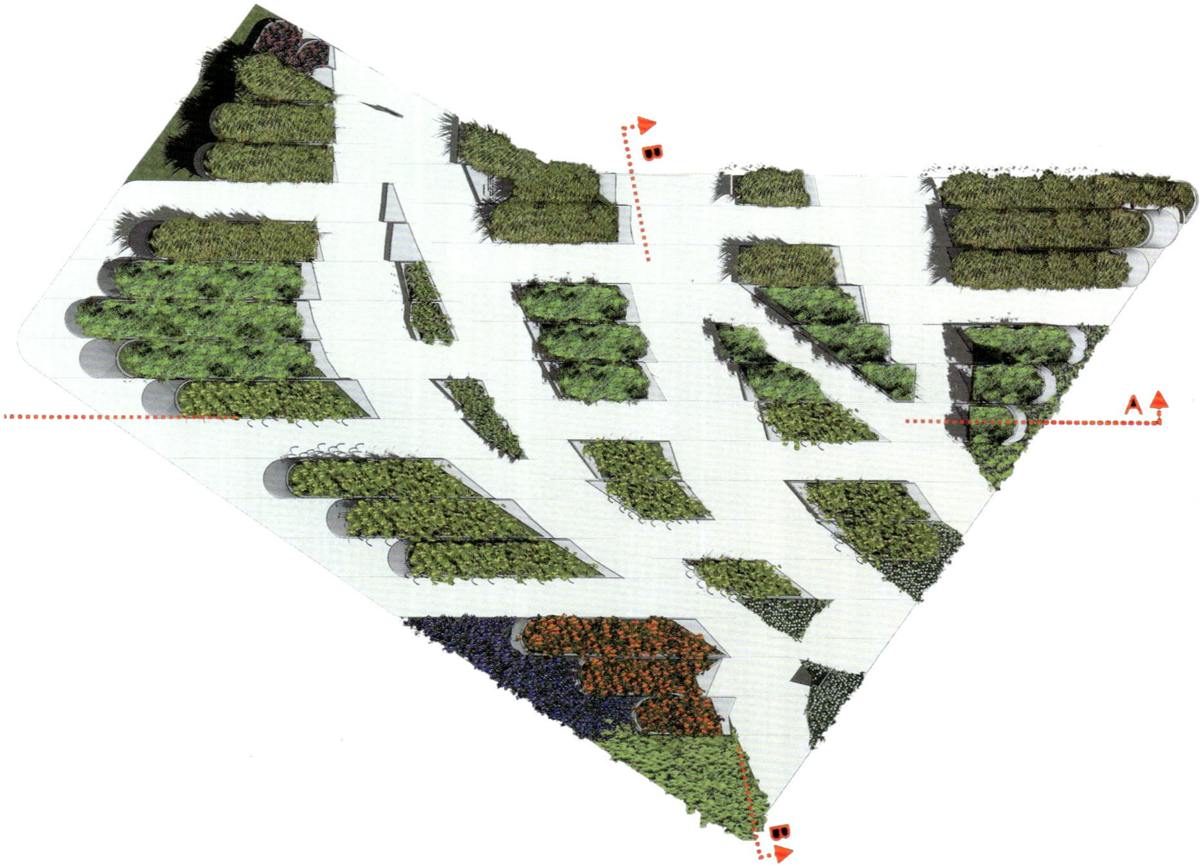

Introduction:

Pampas is a vast and fertile plain of Argentina which is also called South American Grassland or Argentina Grassland, with an area of 76,000,000 m^2. It seems to have nothing in common with Chinese garden. However, the one who knows theme will get a common image from them – which comes from the root of the memory but not only from the real scene. The designer attempts to use the physical, geographic and cultural epitome of Pampas to create a sense of Pampas in International Horticultural Exposition 2011 Xi' an.

展园介绍：

潘帕斯草原是阿根廷一片非常广阔肥沃的低地，又称南美草原或阿根廷草原。面积约 76 000 000 m^2。其看起来似乎与中国的庭院完全没有任何共同之处，但是了解它们的人却能从中获得一种共有的影像——一种基于记忆本源而非仅仅从对现实存在的景象的理解。设计者正试图在物理、地理、文化风韵与阿根廷不尽相同的西安创造出局部“潘帕斯印象”。

World Garden

世界庭院

World Garden is located on the island on the south of the exposition area, surrounded by water. it is an independent exhibition garden displaying the development systems, gardening arts and garden patterns of western gardening.

On this 18, 000m^2 island, the development systems, gardening arts and garden patterns of western gardening are displayed by assembling Ancient Greek garden, Italian Terrace Garden, French Classical Garden, British Natural Landscape Garden and Spanish Islamic Garden. The visitors would feel like staying in Penisola Appenninica, British Isles, Scandinavia or Seine River. In accordance with the space pattern and gardening elements of different styles in every stage, the whole garden is arranged exquisitely with classical and typical works, full of exotic sentiment.

世界庭院位于园区南部，坐落在四面环水的小岛上，是一个以展示西方园林的发展体系、造园艺术、园林形式的独立展园。

正是在这个约 18 000 m^2 的四面环水的小岛上，展示了西方园林的发展体系、造园艺术、园林形式，集合了古希腊园林、意大利台地园、法国勒诺特尔园林、英国自然风景园、西班牙伊斯兰园林，让人仿佛亲临亚平宁半岛、不列颠群岛、斯堪的纳维亚半岛，或是柔美的塞纳河畔……整个园林根据各时期不同风格的空间形式，造园要素等特征，以某个经典且能代表当时园林特点的作品片段为骨架，融合所在时代和国家的经典造园元素，悉心地整理组合——于是，浓郁的异国风情扑面而至。

French Le Notre Garden
法国勒诺特尔园林

French Le Notre Garden is a symbol of absolute monarchical autocracy in the 17th century and a master in terms of French classical gardens.

该园是法国古典主义园林的集大成者，同时亦是十七世纪绝对君权专制政体的象征。

Introduction:

French Le Notre Garden is a symbol of absolute monarchical autocracy in the 17th century. Le Notre was a master in terms of French classical gardens. The garden pattern shows the authority of king. Le Notre Garden adopts the geometric grids with clear subordination and rigorous order. The mansion is located at the highest point of the garden, standing as the command. The central axis through the garden heavily decorated by flowerbeds, statues, fountains and pools is the visual center of the entire garden. In terms of waterscape, Le Notre Garden mainly displays the static waterscape and focuses on creating the mirror-like waterscape.

Classical Versailles Palace' s embroidered flowerbed is shown in this garden. Le Notre Garden widely uses plenty of broad-leaved trees and embroidered flowerbeds. Embroidered flowerbeds that consist of 100-meter stone, grass and flowers are set on both sides of the central axis. Such an arrangement can be found in the Versailles Palace. Outside the embroidered flowerbeds are plants used as the background. The tree-shaded paths cross through the plants in the middle. There are four miniature pools symbolizing "Spring", "Summer", "Autumn" and "Winter" at the ends of the shady avenues. In each pool center is an exquisite goldern statue. The western end of the axis extends into the water and pulls people' s attention to the distance, while there are two fine white marble statues on each side of the eastern end of the axis. The central waterscape is three copper-head-sculpture wall fountains with the ripple of the water.

展园介绍：

法国勒诺特尔是法国古典主义园林的集大成者，同时亦是十七世纪绝对君权专制政体的象征。勒诺特尔式园林的格局表现为君王权威。园林采用主从分明、秩序严谨的几何网格。府邸位于最高处，起着统率的作用。同时，在贯穿全园的中轴线上，用花坛、雕像、泉池等加以重点装饰，形成全园的视觉中心。在水景方面，勒诺特尔式园林主要展示静态水景，以形成镜面似的水景效果为主。

在植物种植方面，是典型凡尔赛宫中刺绣花坛展示。勒诺特尔式广泛采用丰富的阔叶乔木和刺绣花坛。设计将凡尔赛宫里那种由百米石、草地、花卉组成的刺绣花坛分放于中轴线两侧，再往外用植物做背景，中间林荫小道。林荫道两端缩小版的象征“春”“夏”“秋”“冬”的四个水池。水池的中央各放有一座金色精致的雕像。轴线的西端伸入水中将实现延伸到远处，轴线的东端两侧各有两精细的汉白玉雕塑，中心水景是三个铜头像壁泉，水声潺潺可听。

Ancient Greek Garden

古希腊园

The Garden has not only followed the principle of the regular garden design of the Western World but also integrated the artistic features of Greek mythology.

该园既遵循西方规则式园林的设计原则，又融入了希腊神话的艺术特征。

Introduction:

The reason for choosing Ancient Greek Garden at the entrance of the World Garden is as follows: The form of this kind of garden is at a relatively simple initial stage to be the embryo of the later European garden. With the development and influence of mathematics, geometry and corresponding philosophy of that time, the ancient Greek garden emphasized the formal style, which laid the foundation of Western formal garden.

Ancient Greek Garden is located to the north of the World Garden with an area of about 2,000 m^2. The garden consists of three parts: Soros Square, Sacred Wood, and Lion Gate from west to east. It takes the square of Greek Delphi site (i.e., Soros Square) as its main body to show the application of perfect proportion and Greek column style with a far-reaching influence on Western architecture. Considering Greek myths has a huge impact on Western garden, painting, literature and other arts, twelve statues of Greek main gods are scattered among the shady trees. Moreover, there is a holy forest surrounded by palm trees.

The world' s famous Lion Gate, the entrance to ancient Mycenae' s acropolis, is placed at the east entrance. Two solid stone columns carry the heavy and slightly arched stone beam, on which is a huge triangle-shaped stone. The soul of the Lion Gate is on this stone. There is a group of reliefs in the front of the huge stone: two symmetric lions, one on the left side and the other on the right side, step on the altar with their front feet. A pillar stands between two lions, which is a symbol of the palace. The corbel arch on the Lion Gate is one of the world' s earliest arch-style structure sites and has a profound influence on the later architecture.

展园介绍：

因为古希腊园林在形式上处于初始阶段，是后世欧洲园林的雏形；且由于当时数学、几何学以及相应哲学的发展与影响，古希腊园林强调规则的形式，奠定了西方规则式园林的基础。所以将古希腊园设置为西安世园会世界庭院区的首个展园。

古希腊园位于世界庭院的北侧，大约 2 000 m^2。希腊园分 3 个部分，从西向东依次为索罗斯广场、圣林、狮子门。希腊园以希腊德尔斐遗址广场（即索罗斯广场）为主体，展示了对西方建筑影响深远的希腊柱式以及完美比例的运用。在葱郁的树丛里散置着的 12 个希腊神话主神雕塑，均出于对西方园林、绘画、文学等艺术方面有着巨大的影响的希腊神话。此外还特别修建了一个被棕榈树环绕的圣林。

在东入口的地方设置文明世界的狮子门，它是古迈锡尼卫城的入口。两根稳健的石柱承载着笨重而略成弓形的石制横梁，当中恰好嵌入一块三角形巨石，狮子门的精华和灵魂就在这块三角形巨石上。巨石的正面是一组浮雕：两只对称的狮子一左一右以前足踏立在祭坛上，中间立着一根象征着宫殿的柱子。狮子门上的叠涩券，是世界上最早的券式结构遗迹之一，对以后的建筑有着深远的影响。

Spanish Islamic Garden

西班牙伊斯兰园

The Moors of Spain have created a new culture and a new environment on the separated southern Mediterranean coast, which have reflected in the garden design.

该园将西班牙摩尔人在南部地中海岸这片分离的土地上创造的一种新文化、以及与之相应的新环境反映于园林设计理念之中。

Introduction:

The Moors of Spain have created a new culture and a new environment on the separated southern Mediterranean coast, which have reflected in the garden design: the internal courtyard is enclosed by solid castle-type architectures; waters and plants are used to adjust the temperature of the courtyard and the architecture. Now this kind of design has appeared in 2011 Xi' an International Horticultural Exposition.

Spanish Islamic Garden is also known as Moorish Garden. Affected deeply by ancient Roma Manors, the Moorish garden is specially built on the hillside. The slope will be transformed into a series of terraces. When high walls enclose the slope, a closed space will be formed. In the closed space, the waterscapes such as cross or parallel canals, channels, and fountains arranged along the axis of the garden have divided the garden space. Moorish Garden generally applies a large area of colorful ceramic mosaic for its pavement and creates ornamental perforated windows on the walls for view borrowing.

This garden' s prototype is famous Court of Lions in Alhambra Palace, Spain. The garden is a rectangular palace, about 35m long and 20m wide, surrounding by a marble-column covered corridor. There is pavilion-type architecture in the garden, which is built according to Cistercian clean hand room. The round roof is decorated with the beautiful filigree patterns while the ground is ornamented with Arab image-shaped or words-shaped seasonal

flowers. The cross-like channels in the middle of the garden bring vitality to the entire garden.

In the center of the garden is a fountain surrounded and supported by 12 lions. In the past, 12 lions could count time and there would be a lion to spray water every hour. The crossroads in the courtyard symbolize four roads in heaven, which stands for harmony, balance and year rolling in Moorish religion.

展园介绍：

西班牙摩尔人在南部地中海岸这片分离的土地上创造了一种新文化、以及与之相应的新环境，并将其反映于当时的园林设计理念之中：由厚实坚固的城堡式建筑围合而成的内庭院；利用水体和大量的植被来调节园庭和建筑的温度。如今这种设计理念正呈现在西安世园会上。

西安世园会上的西班牙伊斯兰园林又被称为摩尔园林。摩尔园林深受古罗马庄园的影响，主要建造在山坡上，将斜坡改造成一系列的台地，围以高墙，形成封闭的空间。在墙内，沿着轴线布置水景，如交叉或者平行的运河、水渠、喷泉等，以水体来分割园林空间。摩尔园林中通常采用大面积的色彩鲜艳的陶瓷马赛克铺装。同时，在墙面上开有装饰性的漏窗，起到借墙外景色的作用。

本展园的设计原型来自西班牙著名的阿尔罕布拉宫狮子院，园子为一长方形宫院，长约35m，宽20m，周围环绕一排大理石圆柱的俏巧游廊，中间有模仿西妥教团净手间形式的亭式建筑，轻灵的圆形屋顶饰有金银丝镶嵌细工的精美图案。地面用时花按阿拉伯图文摆放，中间十字水渠给整个园子带来了活力。

在园子正中，立着狮子环绕的庭院的喷水池，由12头狮子支撑。过去12头狮子可以计时，每过一钟头就有一头狮子喷水。庭院中十字道路象征天堂里的四条路，在摩尔人宗教里有象征祥和、平衡、岁月流转的意义。

Italian Terrace Garden

意大利台地园

The Italian Terrace garden became popular during the Renaissance, to meet the desire of the rich and powerful for their outdoor life in the scenic environment.

该园是文艺复兴时期盛行开来，为满足人们对风景秀丽环境中的户外生活的渴望所设计的。

Introduction:

Throughout the island's east-west axis are the Italian Terrace Garden and the French Le Notre-style Garden. The Italian Terrace garden became popular during the Renaissance, generally built on the hills in the outskirts to meet the desire of the rich and powerful for their outdoor life in the scenic environment.

To fit the mountainous terrain of the Apennines Peninsula, the terrace garden adopts the layout of continuous layers of terraces, which also brings attractive effect for the garden. Landform determines the arrangement of the garden's important axes. In early mansion, each layer of terrace had its single axis but there was no axis to link all terraces. The central axis throughout the entire garden appeared in middle stage. Meanwhile, waterscape, statue and so on were arranged on the axis to enrich the garden's axis.

In terms of designing the Italian Terrance Garden, we selected the part of Villa Palazzina Farnese, one of three villas in the peak of Roman villa building, as the main skeleton. The garden adopts the central axis symmetric layout. A wide gentle slope from the platform to the viewing corridor constitutes the garden's central axis. Dolphin cascade fountains lines on both sides of the centipede-shaped stone sinks, which constitutes a series of cascade water landscape. The second layer is an elliptical square. The curved steps on both sides of the square surround a transparent water-spraying ball that jets water to the central water tray. There is a huge water bowl in the center, from which the curtain-style waterfall flows down to the water tray. Two Achelous

statues respectively lean on the left side and the right side of the water bowl, with horns in their hands, guarding waterscape and the viewing corridor.

On both sides of the gentle slope is "Garden of Love" of the Villandry Castle that illustrates people's emotional lives by flowers and shows four love themes. The pattern of "Sad Love" is the cutting edge of a dagger or sword when men used to fight a duel. The red flowers in the middle of the pattern symbolize the blood splattered during the duel. The pattern of "Tender Love" consists of many heart-shaped designs—the symbol of tender love. The pattern of "Flaming Love" is heart-shaped but many broken "hearts". As to the pattern of "Frivolous Love", fans in the four corners symbolize frivolity. The ox horns among the fans stand for false love. Buxus-Arten hedge constitutes plant flowerbed. Two horse sculptures make the garden lively. The low walls on both sides of the garden limit the space and function as benches too. There are sixteen statue columns on the walls. Fine work, flowing water, and geometric plants show all garden's classic elements.

展园介绍：

在整个小岛的东西轴线上分别是意大利台地园和法国勒诺特园林。意大利台地园是文艺复兴时期盛行开来的。多建造于城郊的丘陵地形上，用来满足富豪权贵们对风景秀丽环境中的户外生活的渴望。

为了适应亚平宁半岛多山的地形，台地园采取了连续几层台地的布局方式。地形决定了园区重要轴线的安排。早期庄园中，各台层有独立的轴线而无联系各层之间的轴线；中期开始，则开始有贯穿全园的中轴线。同时，在轴线上安排水景、雕塑等，以丰富轴线。

在意大利台地园的设计上，截取了罗马庄园建造盛期的三大庄园之一的法尔奈斯庄园的部分为主要骨架，采用了中轴对称的构图形式。中间轴线是一条宽大的缓坡，从平台到观景廊。海豚叠水分列两侧，中间是蜈蚣形的石砌水槽，构成系列跌水景观。第二层是椭圆形广场，两侧弧形台阶环绕着透光的喷水球，向中间水盘喷水。

缓坡的两侧是在维兰迪城堡的“爱园”，以花语表达了四个爱情主题。“悲情之爱”以男人决斗时所使用的匕首或剑的锋刃为图案，中间种植着红色的花，象征着决斗中四溅的血液；“温柔之爱”像星星点点的火焰一样分割出许多象征温柔之爱的心形图案；“炽热之爱”虽然是心形图案，但却是一颗颗破碎的心；“轻浮的爱”位于四角的扇子象征着轻薄，扇子之间摆放着代表虚假爱情的牛角。

另外，黄杨篱组成的植物花坛、两座马匹塑像用来活跃气氛恰到好处。花园的两面均有矮墙，既限定了空间，又用作坐凳。墙上有 16 根神像柱。精细的做工，流动的水系以及，几何的植物展示了园林的经典要素。

British Natural Landscape Garden

英国自然式风景园

This landscape garden is characterized by the wide grass land, natural planted bushes and winding paths. The humid and cloudy climate condition of Britain islands makes people chase the bright landscape.

该园以开阔的草地、自然式种植的树丛、蜿蜒的小径为特色、潮湿多云的气候条件促使人们追求开朗、明快的自然风景。

Introduction:

On the south of the island, there is the British Natural Landscape Garden covering 4, 000m^2 land area. It is originated from the natural landscape gardens rising in the 18th century. This landscape garden is characterized by the wide grass land, natural planted bushes and winding paths. The humid and cloudy climate condition of Britain islands makes people chase the bright landscape. The undulating terrain with hills and the large area of pasture landscape provide clear example for the gardening pattern.

These conditions bring about the unique British garden. The gardening ideas emerged from the group of the British politicians, thinkers and scholars in the 17th century while the gardening practice started in the 18th century.

The British gardener Bridgeman adopted the unsymmetrical trees planting models for the first time. William Kent developed Bridgeman' s gardening skills and emphasized completely imitating the nature and reproducing the natural gardening principles. Lancelot Brown further developed the gardening skills using the landscaping elements like slope grassland, natural lake bank, natural waterfall and snake-shaped garden road. Humphry Repton, a famous gardener after Brown, held that the regular elements should be kept near the constructions as the transition from the construction to the natural garden. After Repton, William Chambers raised that there should be entertaining and resting places in the garden. The elements like artificial hills, grass land, pavilions, flower zones and rocks integrate together and form a natural and rural scenery with the waters around.

展园介绍：

在岛的南侧是英国自然式风景园，其占地约4 000 m^2，原意是指英国在18世纪发展起来的自然风景园。这种风景园以开阔的草地，自然式种植的树丛，蜿蜒的小径为特色。不列颠群岛潮湿多云的气候条件，促使人们追求开朗、明快的自然风景。英国本土丘陵起伏的地形和大面积的牧场风光为园林形式提供了直接的范例。

这些条件促成了独具一格的英国式园林的出现。造园思想首先是在十七世纪的英国政治家、思想家和文人圈子中产生的，而造园实践开始于十八世纪。

英国造园师布里奇曼首次采用了非行列式的不对称的树木种植方式。威廉·肯特发展了布里奇曼的造园手法，强调完全模仿自然、再现自然的造园准则。而郎斯洛特·布朗进一步发展了造园手法，缓坡草地、自然式湖岸、自然式瀑布、蛇形园路等都是常用的造景要素。胡弗莱·雷普顿是继布朗之后著名的造园师，主张在建筑附近保留规则式的元素，作为建筑与自然式园林之间的过渡。在雷普顿之后，威廉·钱伯斯提出了造园还应该使其成为供人娱乐休息的地方。展园中将假山、草地、亭子、花境、岩石等元素融合在一起，与周边的水面形成一个自然意境，一幅田园风光显露于眼前。

Four Landmarks

四大标志建筑

The permanent constructions "One Axis and Four Musuems" of Shanghai World Expo has become the new landmark of Shanghai. In Xi' an Horticultural Expo, there are also four landmarks: Chang' an Tower, Guangyun Entrance, Greenhouse and Theme Pavilion. After the Horticultural Expo, they would be kept as the new landmark of Xi' an.

Guangyun Entrance, Theme Pavilion and Greenhouse are designed by Eva, chief designer of Plasma and world leader of Landscape Urbanism, manifesting the advanced, drastic and abstract design style. A comprehensive conversation between Xi' an and the world is in process here, through which the collide between tradition and modern brings innovation to reality.

The three drastic works of Eva have created the beauty of harmony with Chang' an Tower instead of abrupt intenseness. The exhibit recruiting personnel speak highly of the standing Chang' an Tower and the three lying construction: " 'Standing' stands for subjectivity which presents the figure of the nation."

上海世博会永久性建筑"一轴四馆"已成为上海的新地标。西安世园会也有长安塔、广运门、自然馆、创意馆四大标志性建筑，世园会后这四大建筑也将保留成为西安的新地标。

广运门、创意馆与自然馆均由全球"景观都市主义"领军人物英国普拉斯马公司首席设计师伊娃设计，彰显前卫、激进、抽象的设计风格。"这次西安世园会没有选用兵马俑、华清池这些经典符号，而是突出'景观都市主义'的风格，说明西安与世界充分展开交流对话，在这里，传统与现代产生碰撞从而实现了创新。"

而外表传统的长安塔却与伊娃激进的三大设计之间，没有造成一种突兀的紧张感，反而营造了和谐之美。关于"竖立"的长安塔与"躺着"的三大建筑，西安世园会招展处人员是这样阐述的："竖立代表了主体性，是民族形象和国家形象的彰显。"

Chang'an Tower

长安塔

Design Theme:

Eternal peace & harmony between nature & mankind, nurturing the future earth.

设计主题:

天人长安，创意自然。

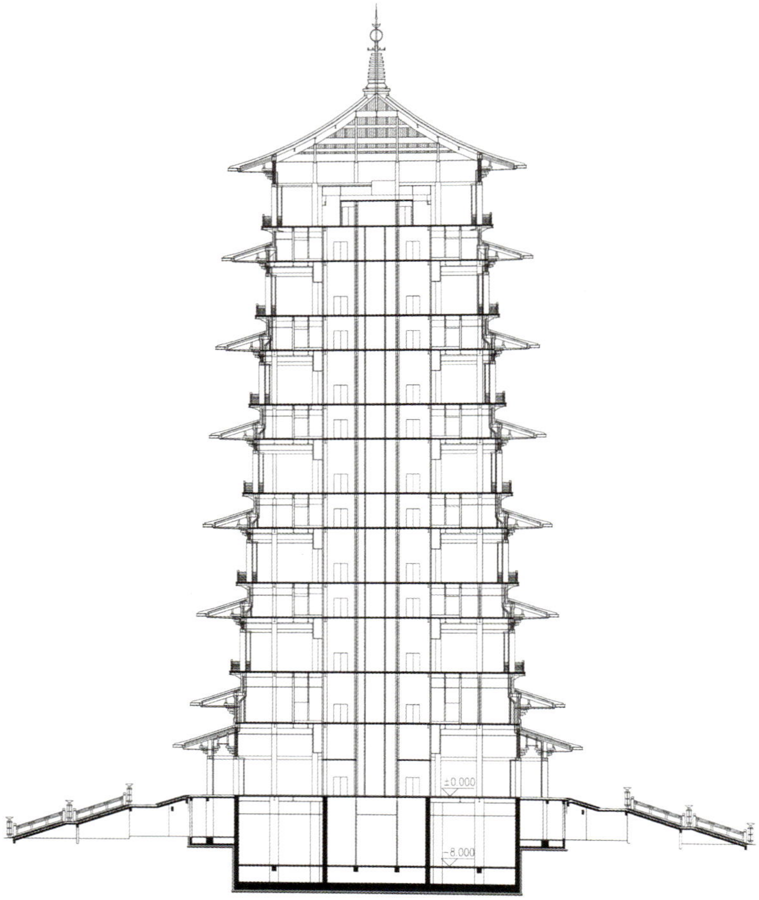

Introduction:

The Chang' an Tower, an observation tower and the symbol of the exposition, is of commanding height at the exposition site—Small Zhongnan Mountain. Climbing up to the top of the tower, tourists may have a panoramic view of the whole site.

The design of the tower incorporates modern elements while keeping the main form of an antique square tower of the Sui and Tang Dynasties (581–907). It is a reflection of both the essence of Chinese architectural culture and modern, trendy urban features. It also represents a perfect combination of green architectural technology and art, exemplifying ecological practice.

This tower has a symbolic cultural meaning of "Nature and People in One in Chang' an", and its image has the characteristics of "Chang' an". The quadrate design of the tower is based on the characteristics of pagodas in Chang' an in the Tang Dynasty. The exterior styling of the tower has the characteristics of the timber-structure pagodas in the Tang Dynasty. The tower embodies the coexistence between nature and people from Chang' an in history. Hence, the tower is a symbol of culture.

The entire tower adopted advanced inside-steel-frame support structure with light self-weight and fast construction.

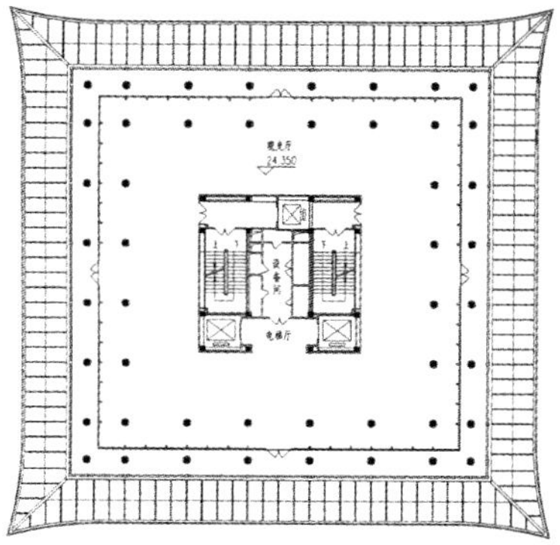

三层平面

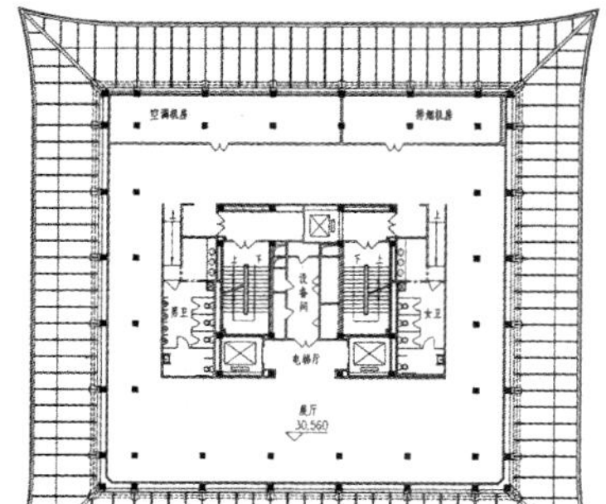

三层暗层平面

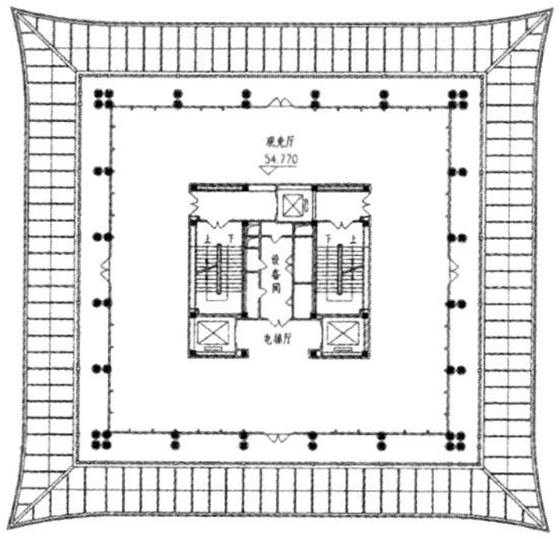

六层平面

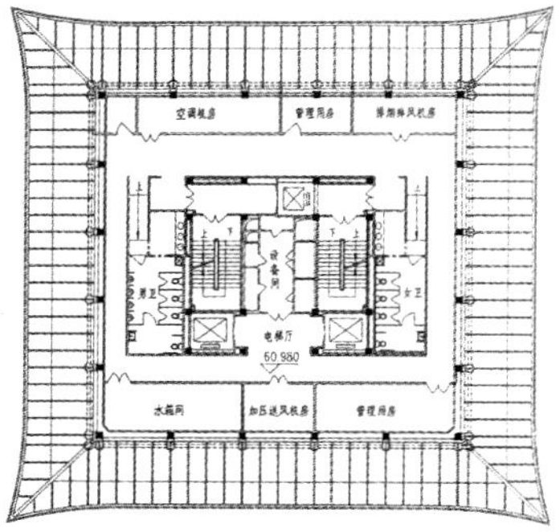

六层暗层平面

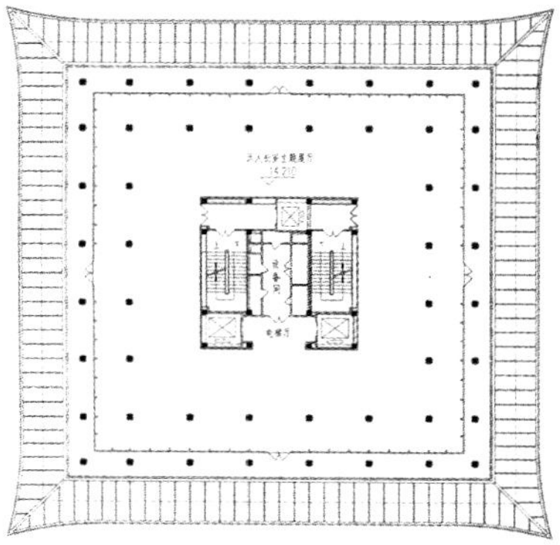

二层平面

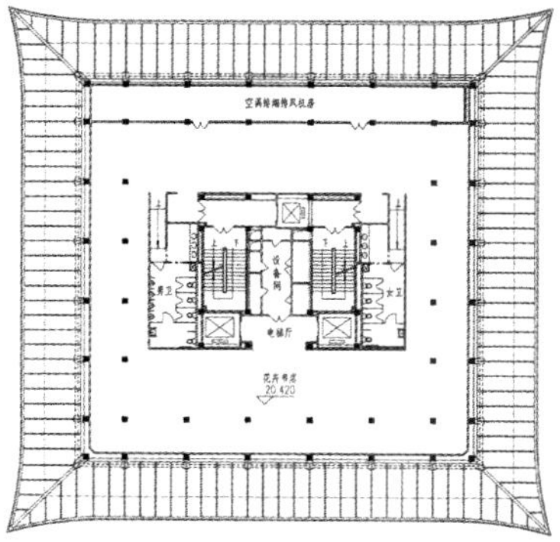

二层暗层平面

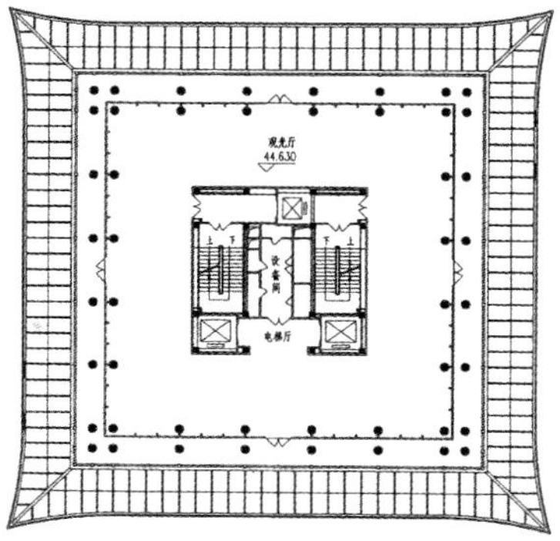

五层平面

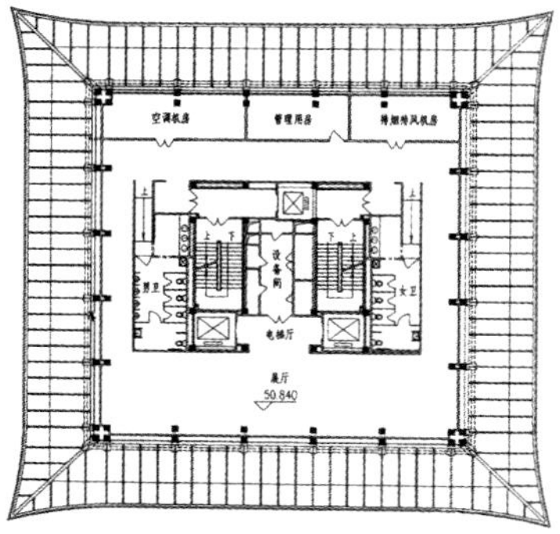

五层暗层平面

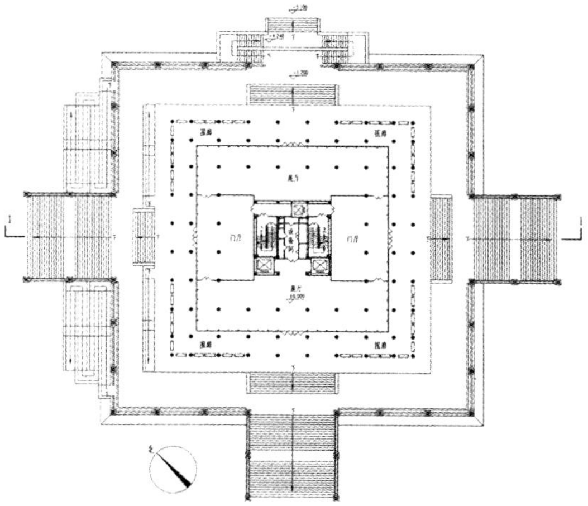

一层平面

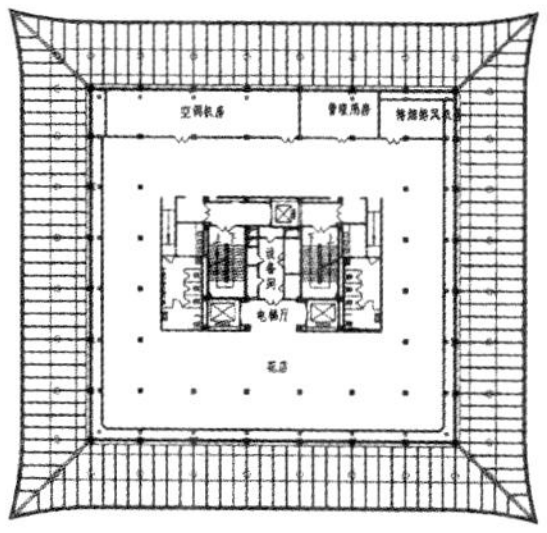

一层暗层平面

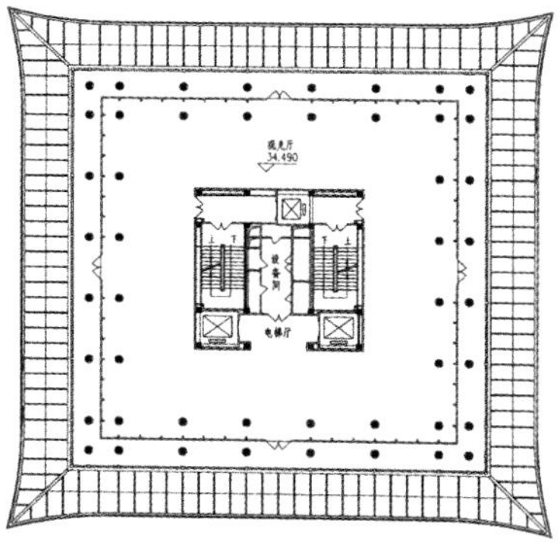

四层平面

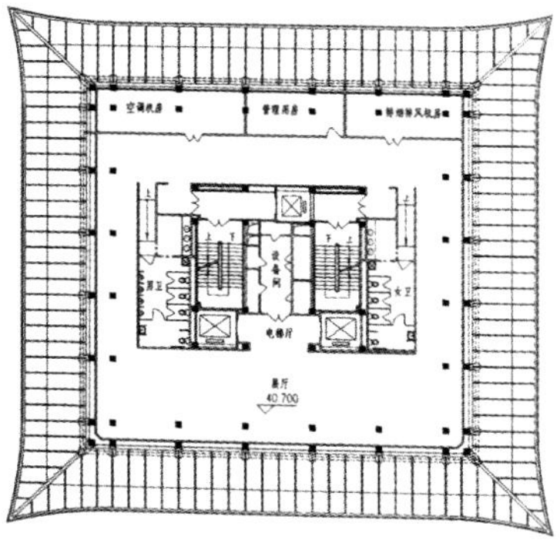

四层暗层平面

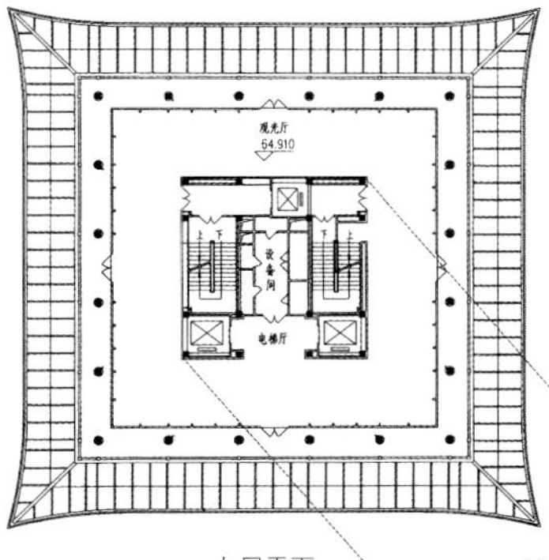

七层平面

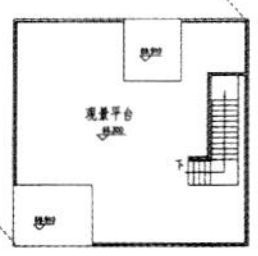

七层跃层平面

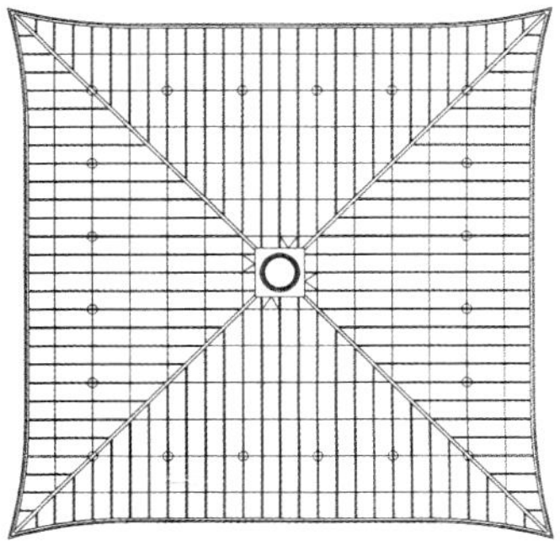

屋顶平面

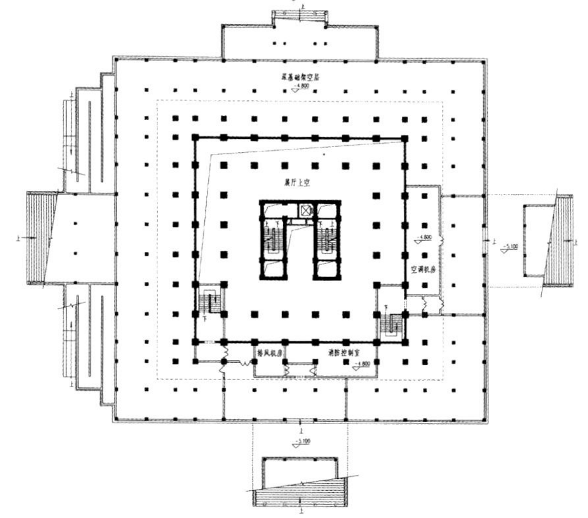

地下一层夹层平面

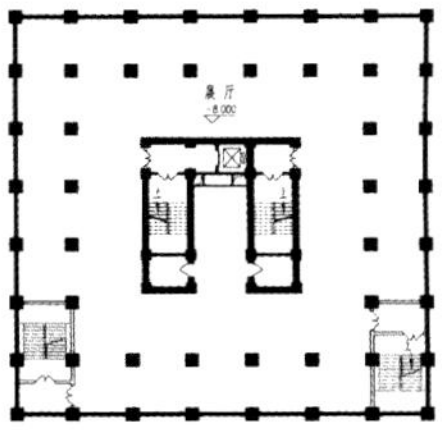

地下一层平面

Furthermore, the steel is recyclable and environmentally friendly. The roof and roof overhang are made of safety glass, creating an effect of a crystal tower with the glass curtain wall of the tower' s wall body.

Symbol of Scale: According to the detailed rules of implementation, Chang' an Tower is located on the Young Zhongnan Mountain. It is the highest building in the scenic zone, and also the meeting point of the main and auxiliary axes of the landscape.

The exterior styling of the Chang' an Tower has the characteristics of traditional timber-structure pagodas of the Tang Dynasty: every layer of roof overhang has a layer of supporting pedestals with a patterned layer-by-layer structure. Roof overhangs of different layers are sloped upward and their dimensions are spacious, reflecting the characteristics and charms of timber-structure architectures in the Tang Dynasty- the roof overhang seems to point out to the far distance.

Roof overhangs and neckings are jointed by metal parts. This is the abstraction and conclusion to the traditional construction techniques of brackets under roof overhangs. A glass curtain wall is behind outer channel columns and fixed with pillars and girders by glass ribs. All these techniques add a modern touch to traditional Tang-style architecture.

Steel structures and related constructional elements are widely used in Chang' an Tower, which have many advantages, such as light weight, high strength, good seismic resistance, convenient industrial production, energy and land saving, recyclable, and fewer pollution during construction or demolishment. In view of the above, Chang' an Tower is regarded as an example of 21st century "green architecture".

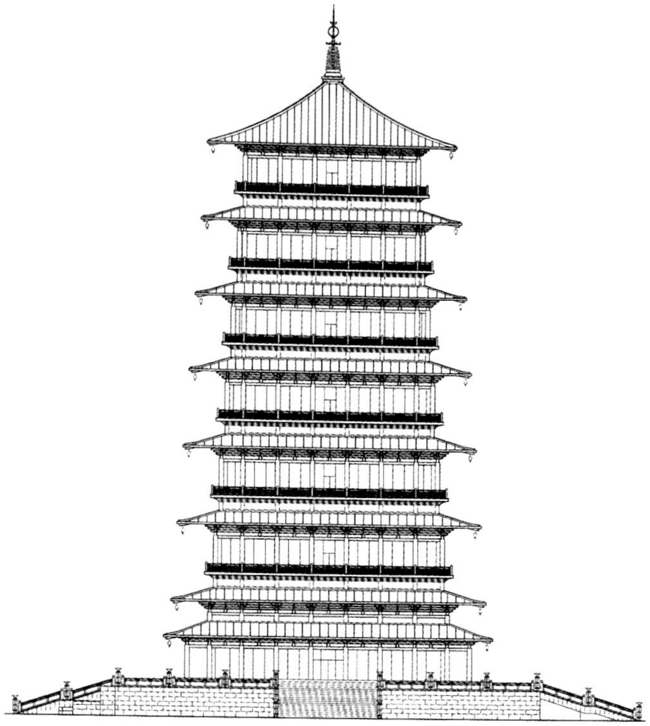

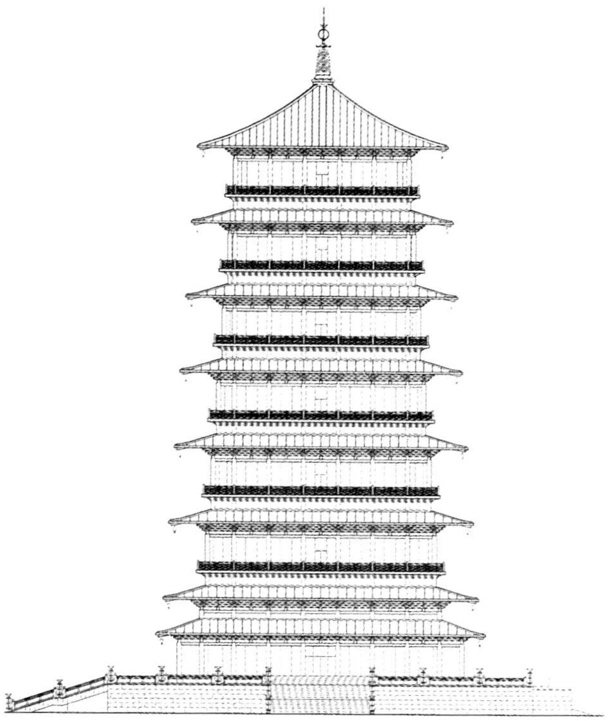

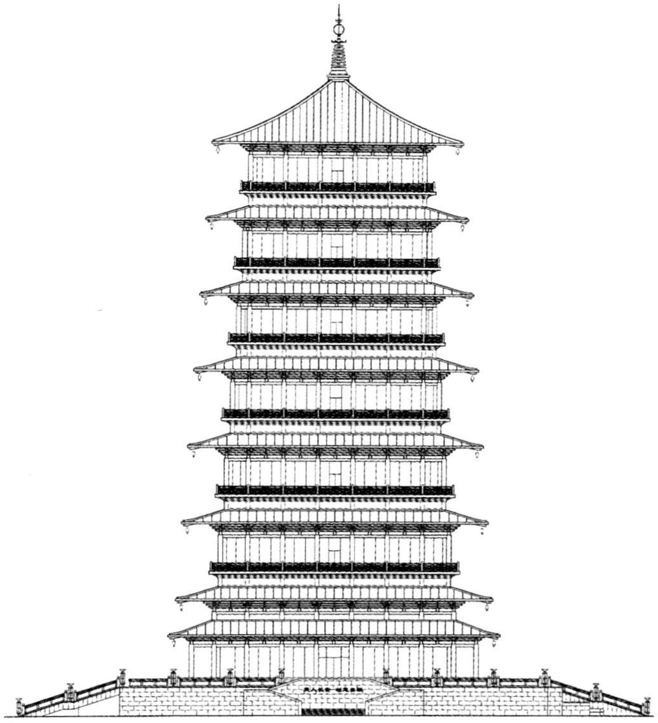

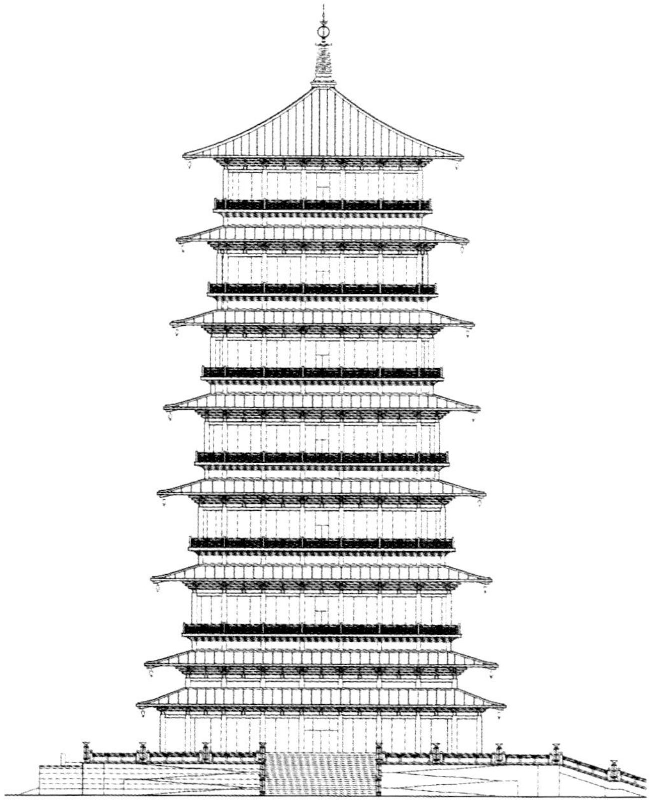

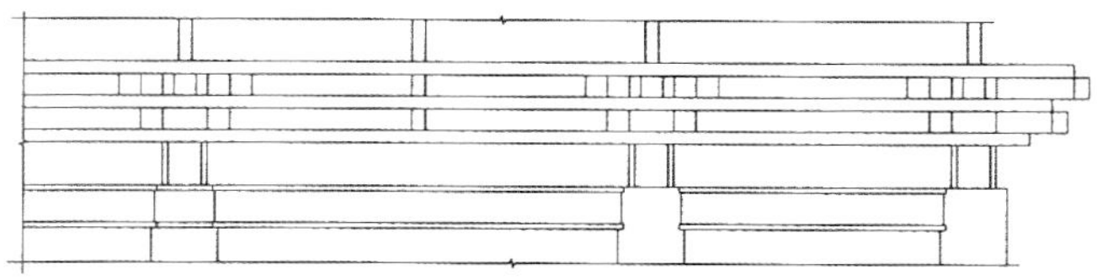

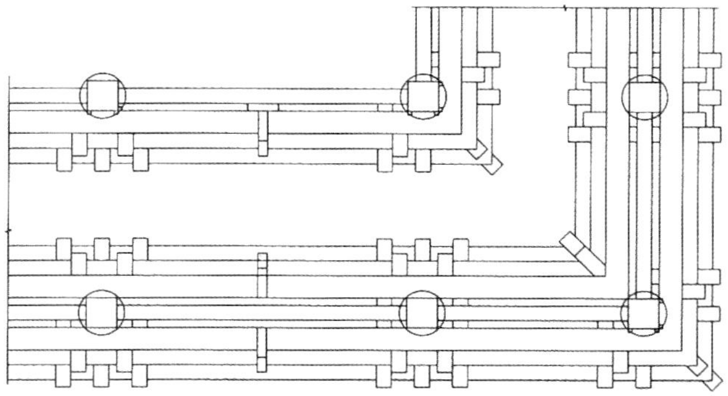

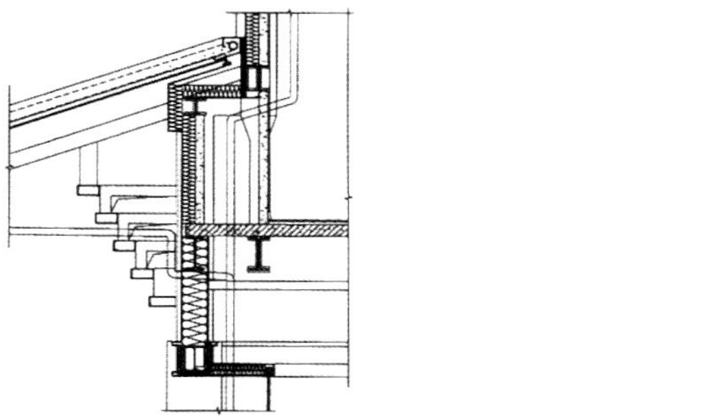

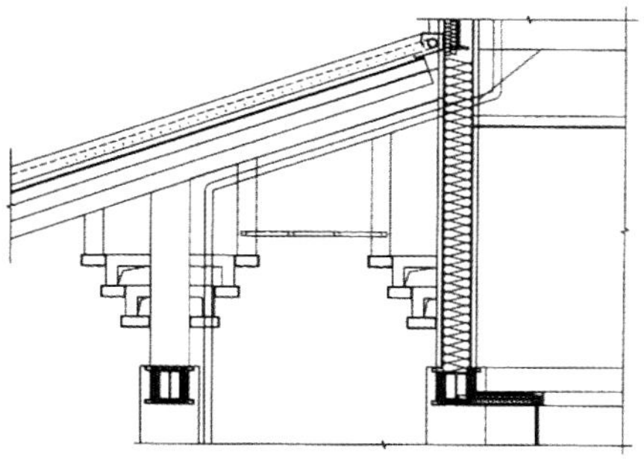

展区介绍:

长安塔位于2011西安世园会园区制高点小终南山上，是2011西安世园会的标志，也是园区的观景塔，是俯瞰全园景观最好的景点。

在设计上保持了隋唐时期方形古塔的神韵，同时增加了现代元素，既体现了中国建筑文化的内涵，又彰显出时尚现代的都市风貌，是生态建筑的实践和示范。

此塔具有“天人长安”文化的标志性，其形象带有“长安”的特色。此塔根据唐长安宝塔的特点设计成方形，塔外观造型具有唐代木结构塔的造型特点。反映历史上长安人与自然的共生共荣，此塔具有文化的标志性。

全塔采用先进的内钢框架支撑结构，自重轻、施工快，且钢材为可循环材料，节能环保。屋顶及所有挑檐均采用安全玻璃，与墙体的玻璃幕墙共同构成水晶塔的效果。

长安塔还具有尺度的标志性：根据实施性详规，长安塔坐落在小终南山上，是景区的制高点建筑，亦是主、副两条景观轴的交汇点。

长安塔外形具有唐代传统木塔的特点：一层挑檐上面有一层凭座，逐层收分，充满韵律。各层挑檐尺寸开阔上扬，体现唐代木结构建筑出檐深远的特色和风采。檐下与柱头之间用金属构件组合，是传统建筑檐下斗拱系统的抽象和概括。玻璃幕墙退在外槽柱内侧，通过玻璃肋与柱、梁固定。一系列处理使唐风唐韵的建筑充满了现代感。

长安塔还大量采用钢结构和相关构件，具有自重轻、强度高、抗震性能好、便于工业化生产等优势，而且节能省地、可循环利用，建造和拆除时对环境污染较少，被誉为21世纪的“绿色建筑”。

Theme Pavilion

创意馆

Designed by: Plasma Studio and Groundlab

Team: Eva Castro, Alfredo Ramírez, Eduardo Rico, Holger Kehne, Jorge Ayala, Nicoletta Gerevini, Evan Greenberg, Hossein Kachabi, Nadia Kloster, Steve de Micoli, Filippo Nassetti, Mehran Garlegui.

Design Theme: the Pavilion is aimed at representing the profound culture and splendid history of Xi' an. More importantly, the creative design also foreshadows a bright future for this city.

设计单位：Plasma 建筑事务所和 Groundlab 景观事务所

设计团队：伊娃·卡斯特罗、阿弗莱多·拉米雷兹、爱德华多·黎哥、豪格·凯南、尼可·杰列维尼、伊凡·格林伯格、侯赛因·可查比 娜迪亚·克劳斯特、史蒂夫·米科利、菲利普·娜塞提、梅兰·格列格

设计主题：体现西安的深厚的文化积淀和辉煌的历史，通过富有创意的设计展现西安光明的未来。

Introduction:

The Theme Pavilion, one of the four landmarks of the International Horticultural Exposition 2011 Xi' an, is centrally located and was designed to be integrated with the surrounding water and land terrains. Its architectural layout is in the shape of the Chinese character " 王 " consisting of irregular tri-wing geometrical objects. External wall faces made of bronze, wood, or shrubbery function as irregular interfaces, create a well-arrange artistic effect which has rich connotation.

The building is naturally divided into three separate display areas extending to the waters, with the cross aisle connecting all of them. Visitors may reach the observation platform on the roof of the greenhouse through a series of inner display ramps. The main body of the architecture is made of reinforced concrete and in steel structure. The folded plate concrete roof was adopted to support the large span space structure of the architecture. This elegant construction demonstrates the highest level of the integration of function and form in architecture.

Highlight of this project: The architecture with bronze and glass facade is unique in shape, like a bronze work of art encrusted with diamonds standing by the lake.

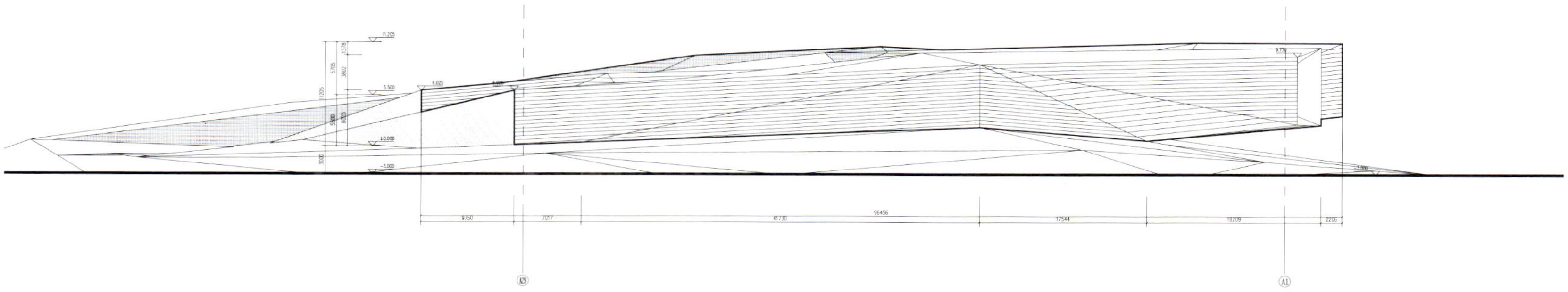

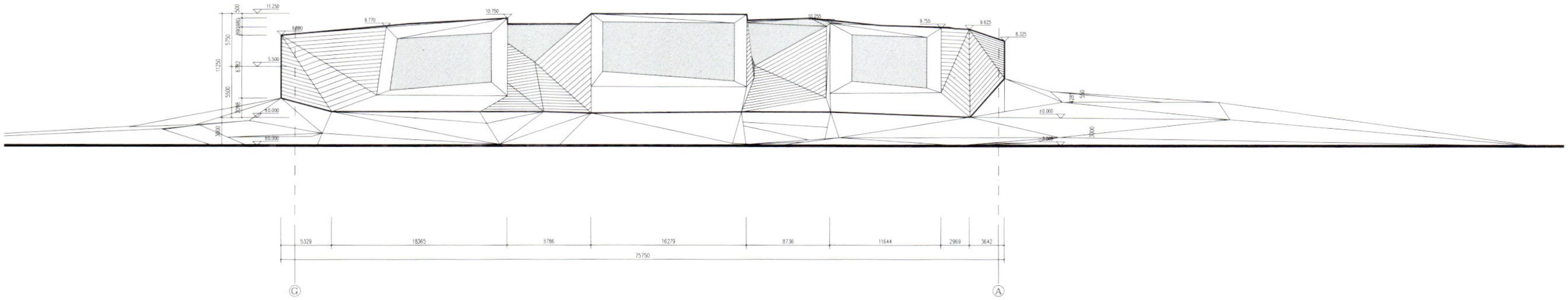

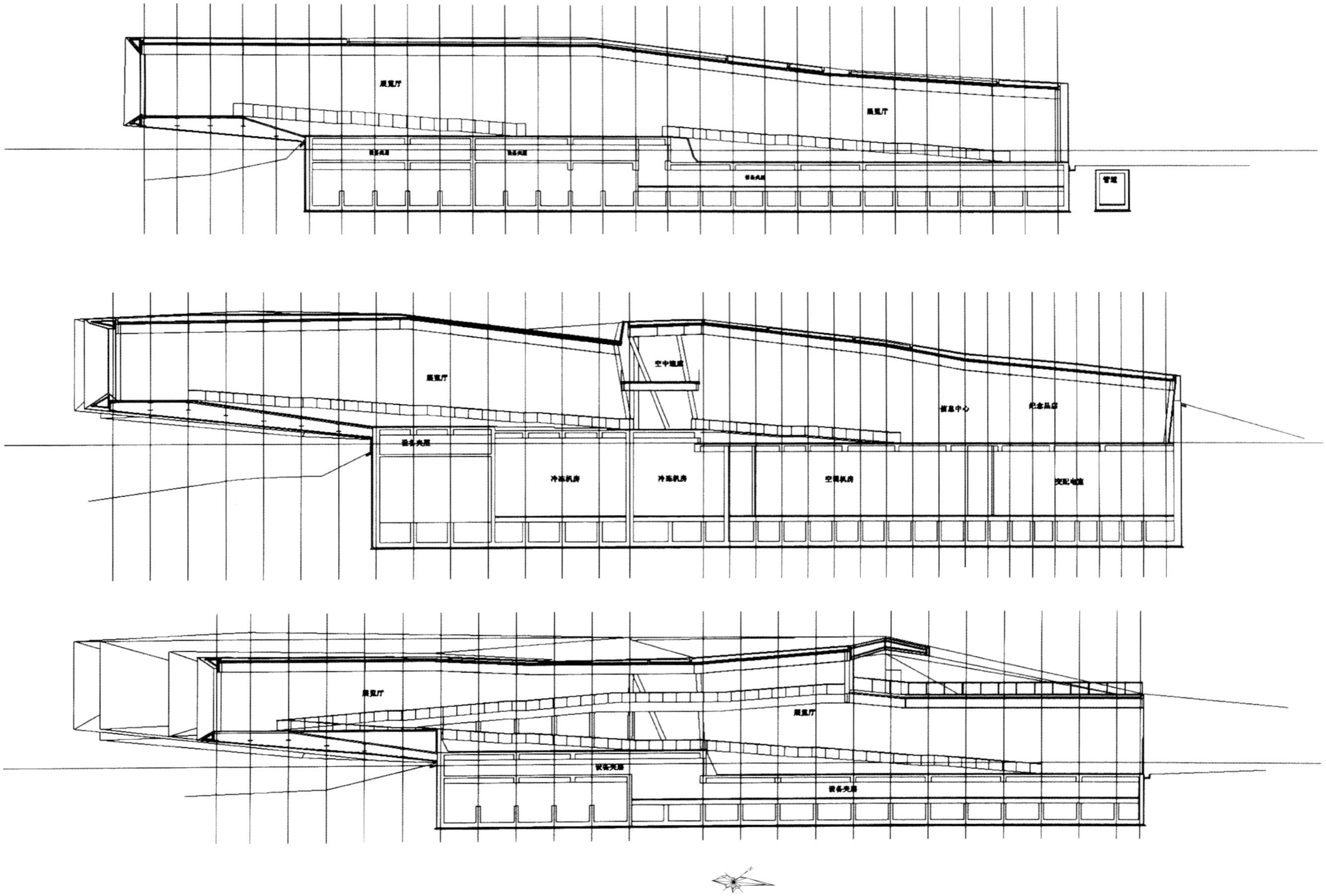

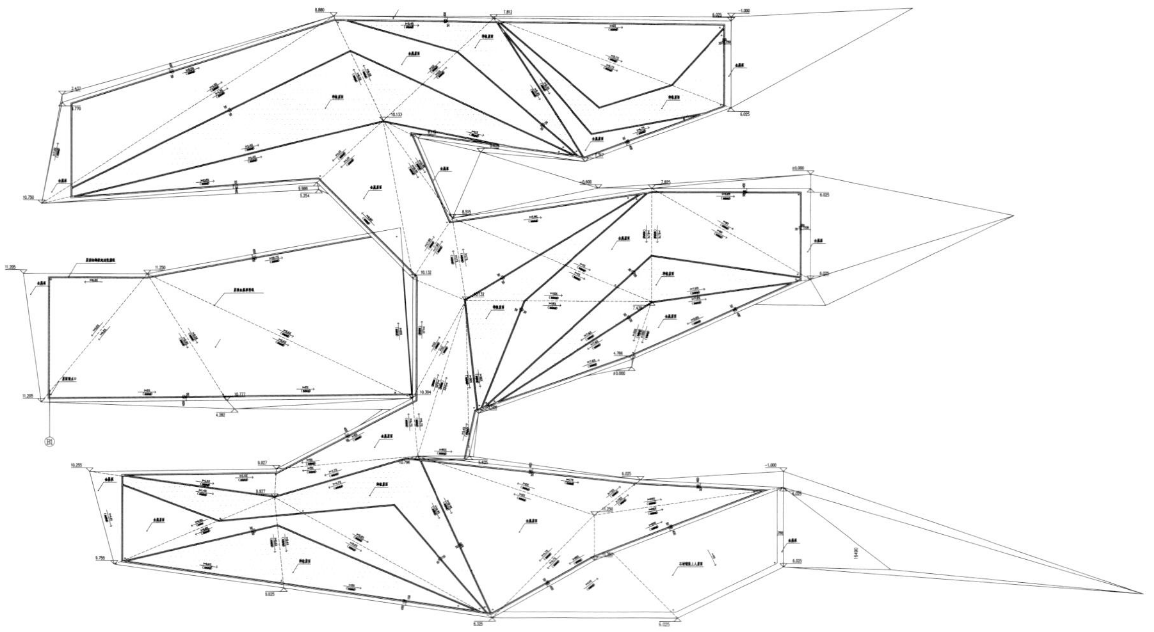

展区介绍：

创意馆位于2011西安世园会主轴线上，作为东道主展馆，整个展馆结合码头和周边场地进行设计，建筑布局呈“王”字型，由三翼不规则几何体组成，青铜金属、石材及花园式种植屋面等不同饰面的无规则衔接处理，形成了错落有致、内涵丰富的艺术效果。

建筑自然分割成三个相对独立、并向水面延展的展览展示区域，横向通道将其贯穿连接。游客可以经由一系列的室内展示坡道上达屋顶观景平台。结构主体采用钢筋混凝土结合钢结构的形式，屋顶使用了混凝土褶皱板结构，以满足建筑对大跨度空间的需求，优美的结构形式体现了功能与形式相结合的建筑艺术最高境界。

本项目设计亮点：建筑的外立面采用青铜与玻璃，建筑体型独特，犹如一件镶嵌钻石的青铜艺术品矗立在湖边，优雅而壮丽。

Greenhouse

自然馆

Design Theme:

Greenhouse is situated at the intersection of many characteristic landscapes, presenting various plants, ecological landscapes and typical plant landscapes of various climate zones.

设计主题：

位于许多特色景观的交汇点，展示多种类植物及其生态景观以及，不同气候带下的典型植物景观。

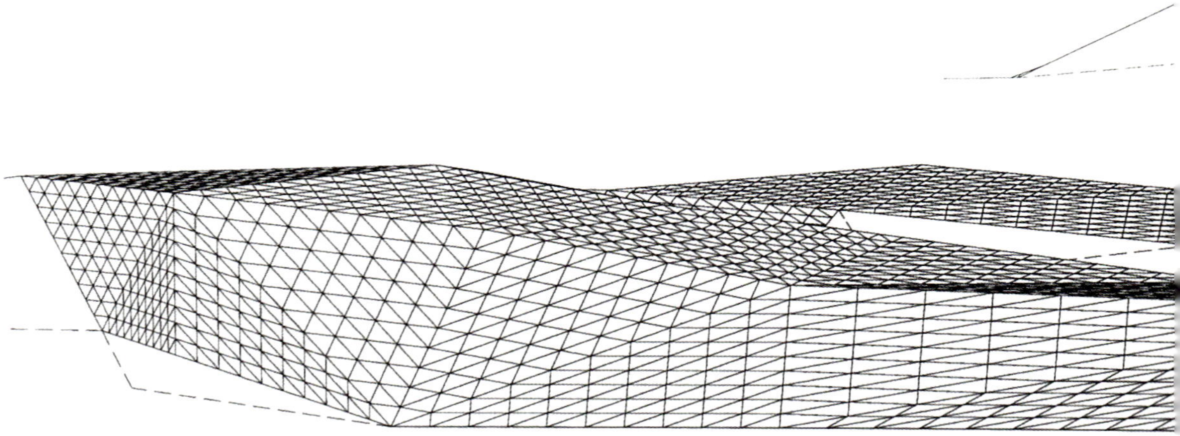

Introduction:

Greenhouse is the botanic greenhouse of the 2011 Xi' an International Horticultural Exposition that located at the Splendid Lake, it is an exquisitely building set in Qinling Garden in shape U. According to the requirements that the interior plants make to the environment, it changes the proportion of overground and underground of the building constantly, to reveal the rare plants and ecological landscapes in different districts and climatic zones. Greenhouse contains exhibition areas of tropical rain forest plants, cold plants and endemic plants. Tropical rain forest area illustrates various peculiar natural phenomena, while the cold plants area and the endemic plants area demonstrate curare, owl butterfly and Ficus plants with strangling phenomenon.

Greenhouse integrates glass, wood and a little concrete in the facade of the underground part. It is built against mountains in folds, perfectly matching the topography. In other words, the size of the Pavilion is reduced from the perspective of height, thus ensuring that beautiful scenery of the lake and the opposite garden can be enjoyed from various height levels at the Pavilion.

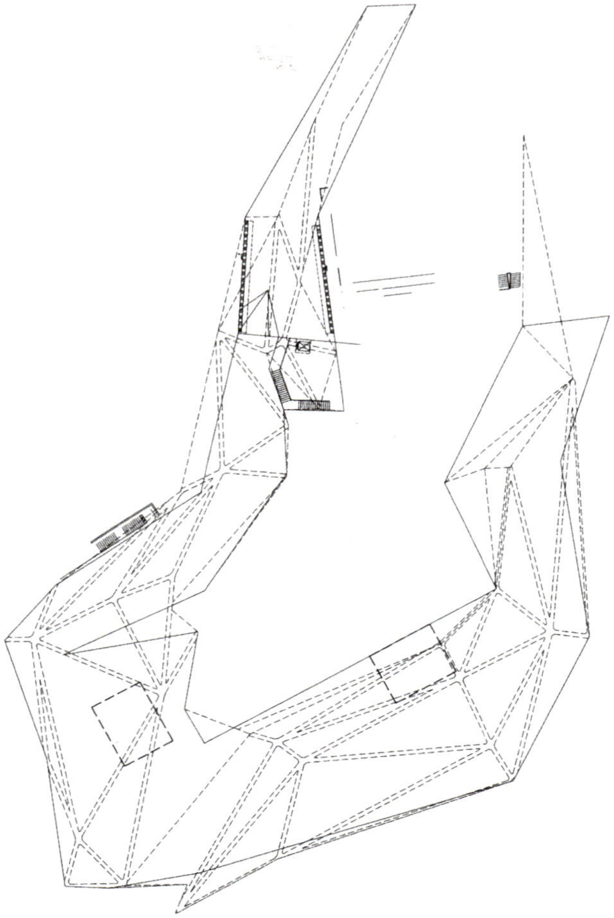

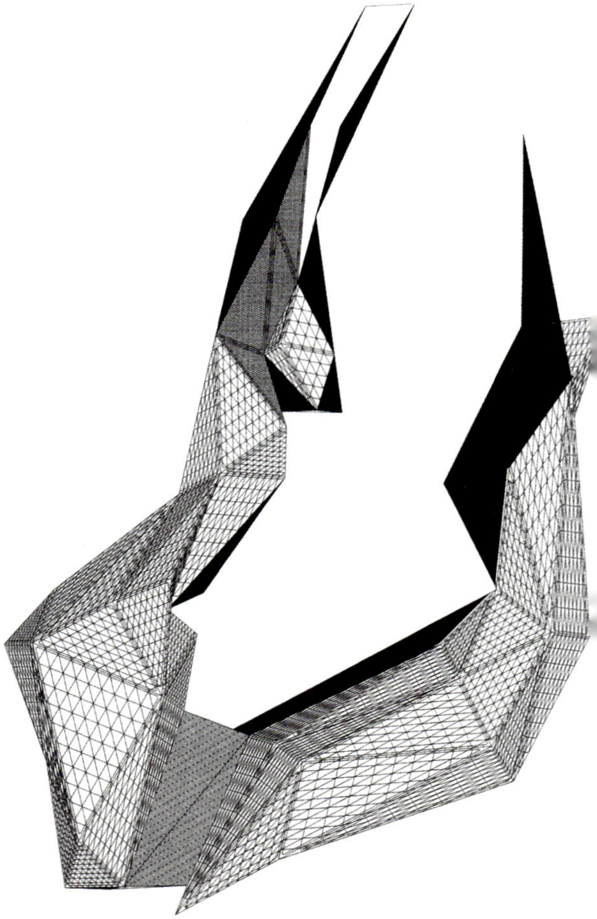

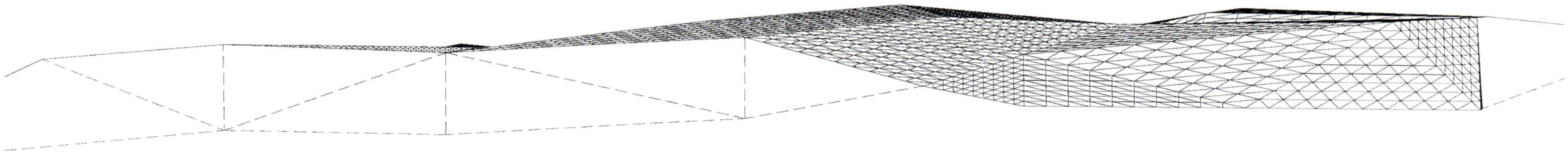

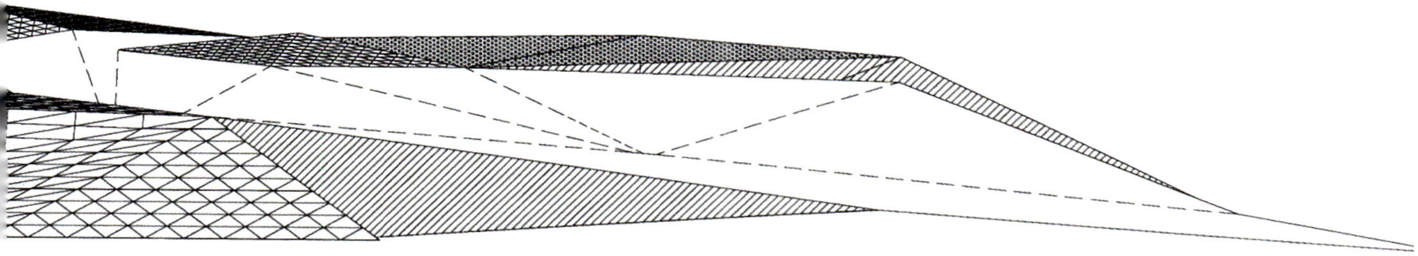

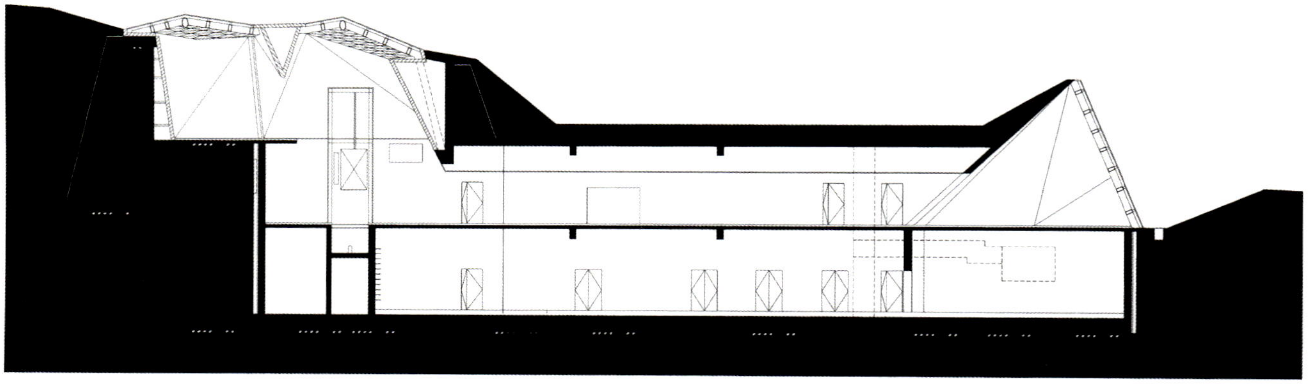

展区介绍：

自然馆是 2011 西安世园会的植物温室，位于锦绣湖畔，是一个镶嵌在秦岭园中、U 字型玲珑剔透的建筑。这个根据室内植物对环境的要求不断改变建筑物地上和地下部分的比例，主要展示地球上不同地域、不同气候带的珍稀植物及生态景观。自然馆包含热带雨林植物展区、极冷植物展区和特有植物展区，热带雨林展现了独木成林、空中花园、老茎生花、老干结果等奇特的大自然现象，而极冷植物展区和特有植物展区展示了箭毒木、猫头鹰蝶、有绞杀现象的榕属植物等植物，奇妙万千。

建筑半埋地下立面材料选用玻璃、木材与少量混凝土结合，倚山而建，层层叠叠，与地形完美结合，即从高度上、视觉上弱化了建筑的体量，保证在建筑室内可以从不同标高领略湖面和对面花园的美景。

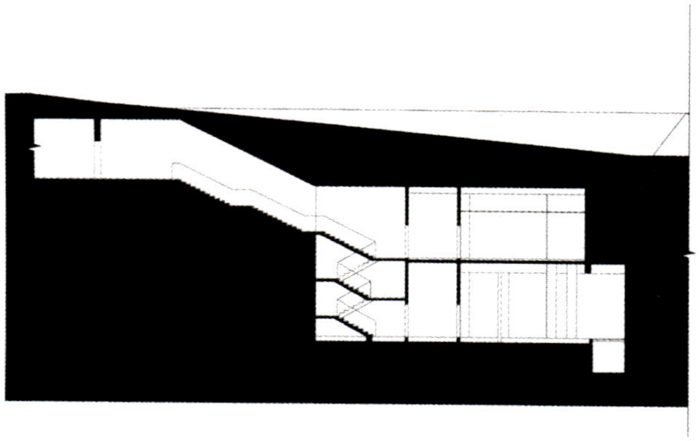

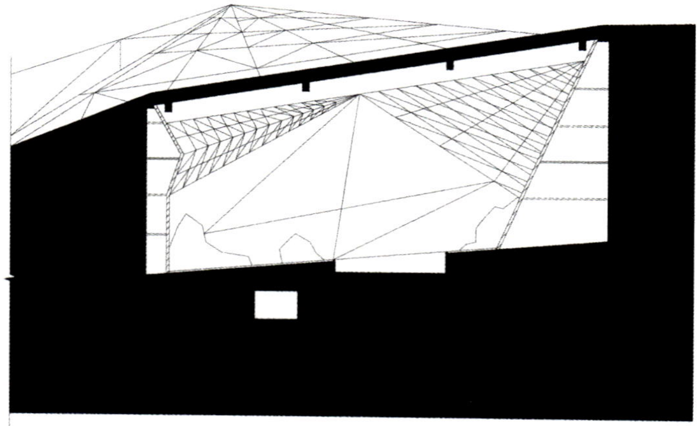

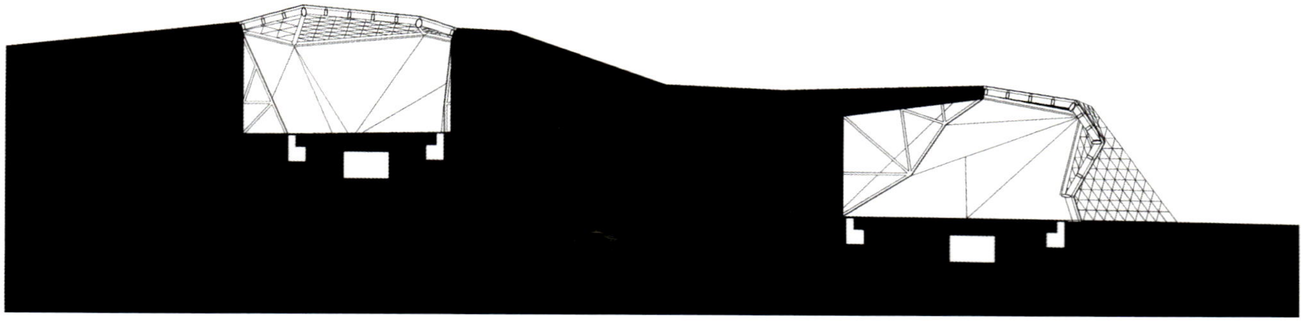

Guangyun Entrance

广运门

Design theme:

The design of the pedestrian bridge across the Expo Avenue is based on the theme of the Xi' an International Horticultural Exposition.

设计主题:

步行大桥横跨世博大道

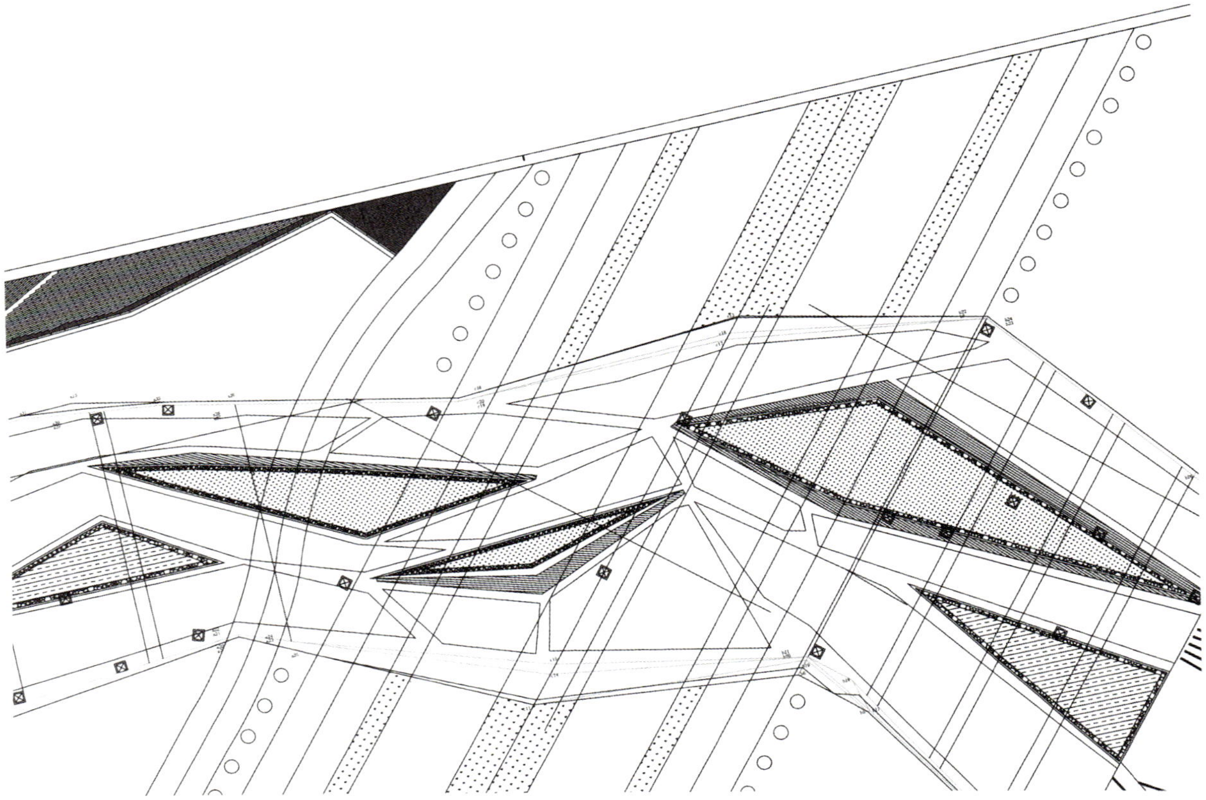

Introduction:

Guangyun Entrance, located at the northeast part of the exposition site, is the main entrance of the International Horticultural Exposition 2011 Xi' an, and it crosses the 60-meter wide Expo Avenue. The avenue can hold as many as 20,000 passengers during peak hours. The ramp is made up of water features and square flowers gardens, with horticultural features paving the way from the highest end to the lowest end, well-integrating with the Chang' an Flower Valley. This magnificent design makes a great impression.

The design is based on the theme of the Xi' an International Horticultural Exposition. Taking advantage of the natural topography of Chang' an City, artificial buildings are combined to natural environment perfectly to showing the historical culture of China as well as the modernity and green charm of the city. The architectural structure is reinforced concrete, but the materials such as the natural wood and granite from different regions have been chosen as the ornament featuring the Chinese elements.

As the starting point of the whole Exposition site, the entrance integrates the design elements representing the color of green horticulture such as plants and waterscape. There is a plant rack on the bridge, so the green plants are able to creep along the rack which functions as a shade and frame for the vertical greening. In addition to meeting the use requirement, the architectural design also fully considers the requirements of the disabled for passing the bridge.

In designing the general planar layout, the first step is to design the moving line of people from the parking lot, with the two separate routes for people in and out of the Exposition site.

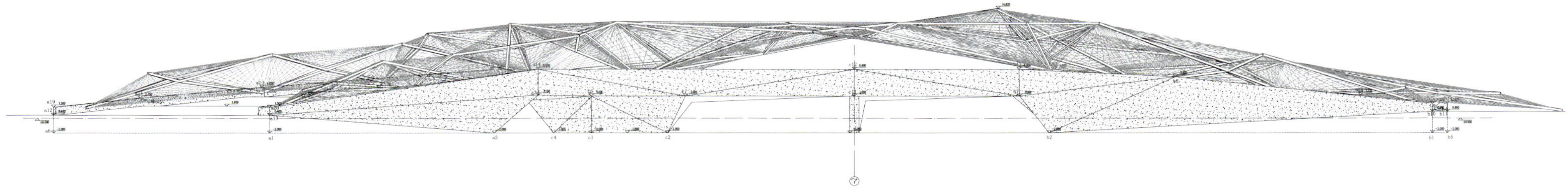

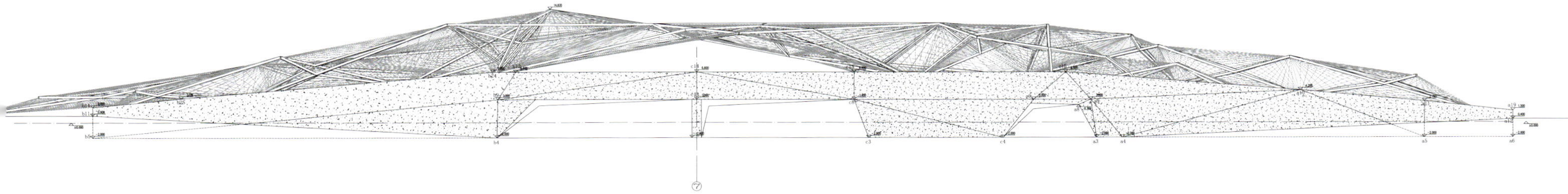

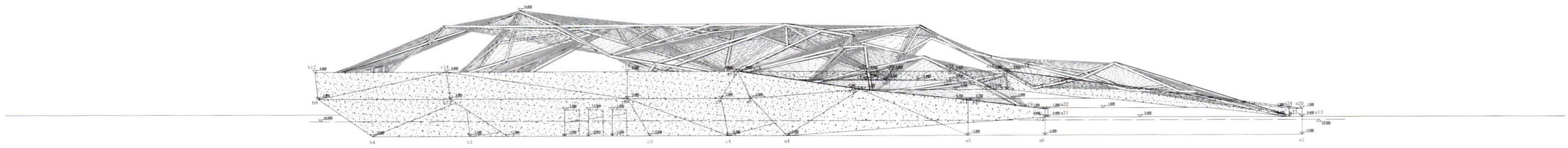

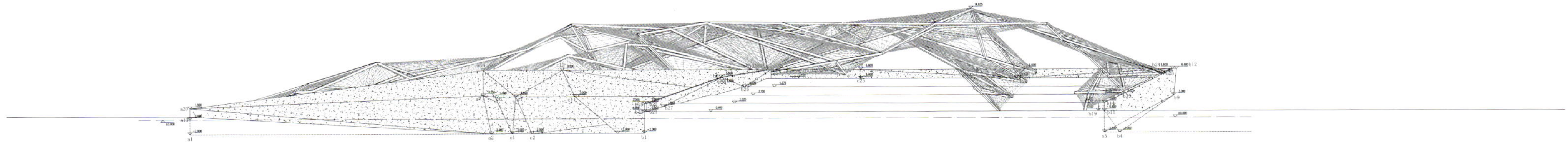

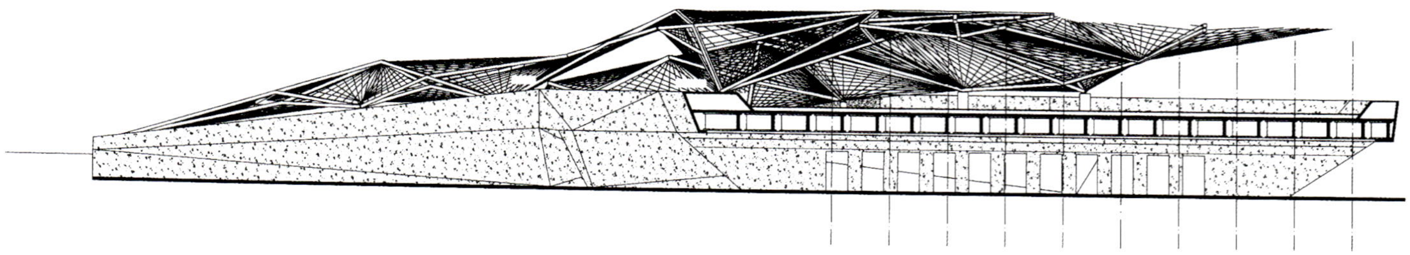

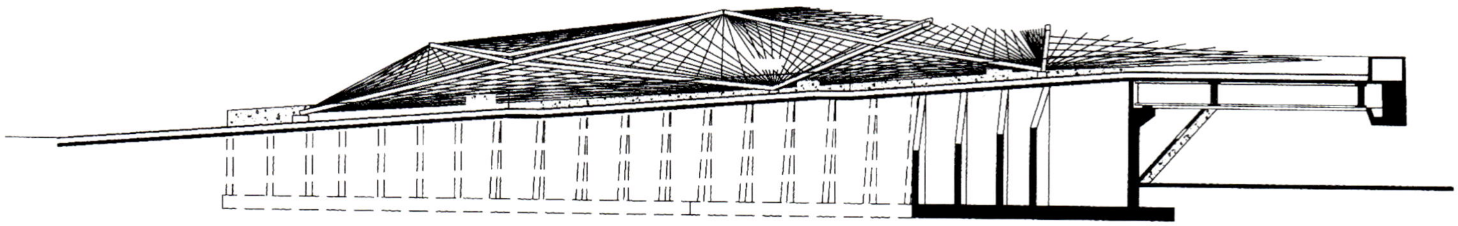

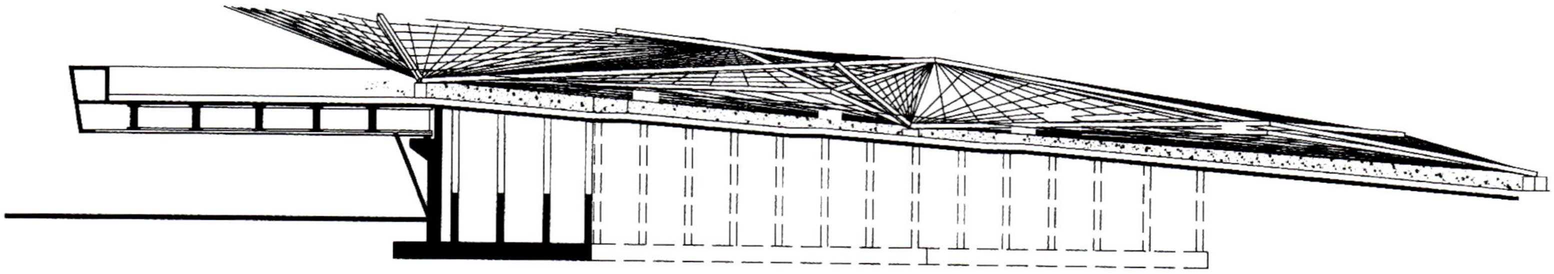

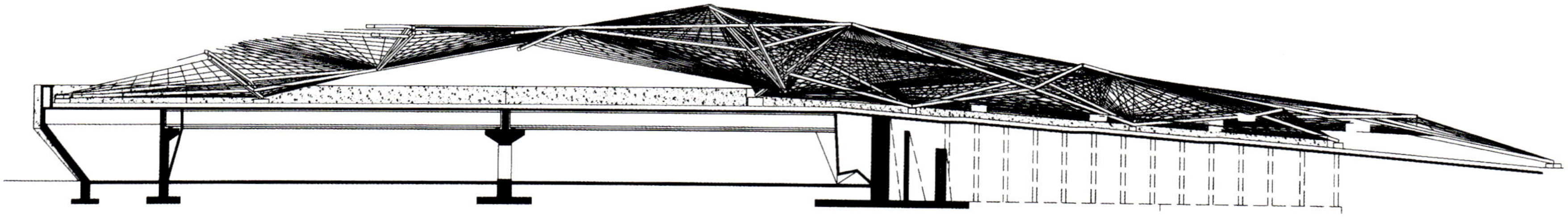

展园介绍：

广运门是 2011 西安世园会主入口，位于园区东北部，横跨 60m 宽的世博大道，高峰期每小时可通行 2 万多人。由踏步、水景、方块式园艺花卉造型组成的坡道把上下之间联系起来，与长安花谷浑然一体，以其恢宏的气势形成强烈的震撼效果。

其设计概念紧扣西安园艺博览会的主题，利用了古都长安的自然地形，使人工建筑与自然环境完美结合，既彰显出了中国历史文化，又表达了中国时尚、绿色的魅力。建筑结构为钢筋混凝土结构，但是材料选择上以自然的木材及地域性的花岗岩为点缀，中国元素亦巧妙地穿插其中。

大门作为整个景观园区的起点，运用植栽，水景等反映绿色园艺色彩的设计元素贯穿其中，桥上设植物廊架，有绿色植物攀爬其上，垂直绿化并提供遮阳。根据建筑的使用要求，也充分考虑了残疾人的通行需求。

总平面布局源自停车场人流的动线，进出人流各自分开。

Special Service Zone

特色服务区

Special Service Zone shows impression on South China and exotic features. It is comprised of three parts, such as Aqua South China, Southeast Asian Street and European Avenue, integrates catering, shopping, entertainment, leisure and consumption as a whole. It provides thoughtful and personalized service for the visitors, and serves as beautiful characteristic landscape in the Xi' an Expo.

Aqua South is an architectural complex with white walls and black tiles that full of southern charm of "small bridge, flowing water and people", waterfront section is for visitors to rest and exercise, boats on the water are moving catering point; architectural modeling and gardening in Southeast Asian Street reflect the typical Southeast Asia style; European Avenue takes European small town in contracted style as main buildings.

Romance by the Ba River is an architectural complex full of simple Shaanxi rural charm and bold Qinchuan style, you can rest and shopping in the beautiful scenery, and enjoy your meals on the floating boat; Yefeng Waterfront is in delicate and artistic Southeast Asia style, assorted with graceful and exotic songs and dances, make you feel yourself in overseas; Continental Style is concise and fashionable European town, a waterfront rest space that set in the middle and the fancy car parade make you enjoy yourself and forget going home; Ba Liu Yi Hotel is a window that the exposition opens to the world to welcome visitors all over the world.

特色服务区展现了江南印象与异国风情，西安世园会的服务区将由“水韵江南”、“东南亚风情街”和“欧洲街”等部分组成，集餐饮、购物、娱乐、休闲和消费为一体，既为园内游客提供周到而个性的服务，又成为西安世园会亮丽的特色风景线。

水韵江南为粉墙黛瓦充满“小桥、流水、人家”江南韵味的服务建筑群，滨水段可供游人休憩、活动，水上小船是流动的餐饮点；东南亚风情街建筑造型与园林园艺体现典型的东南亚风格；欧洲街以风格简约的欧式小镇为主建筑。

灞上人家是极具淳朴陕西乡韵、豪迈秦川风情的建筑群，既可在波光摇橹、月影沉醉的美景中休憩、购物、活动，又可在水上小舟用餐；椰风水岸是精致唯美的东南亚风格建筑群，辅以曼妙的异域歌舞活动，使人恍若置身域外；欧陆风情是简约时尚的欧陆小镇，中间布置临水休憩空间，结合花车巡游等欧洲文化表演，令人乐而忘返；灞柳驿酒店则是世园会开向世界的一扇窗口，来迎接世界各地的游客。

Romance by the Ba River

灞上人家

Design Theme:

to create a model of ecological construction, to display the humanity and natural landscape of Guanzhong Region of Shaanxi Province.

设计主题：

打造生态建筑典范，展现陕西关中的人文精神和自然风貌。

Introduction:

"Bashang", where Emperor Xuanzong held a large-scale water transport exposition, was the site of "Banquet at Hongmen" in the war between Liubang and Xiangyu, known as Chu-Han War. As one of the three major service zones for the International Horticultural Exposition 2011 Xi'an, the project aims to display special architecture and landscapes for visitors, who will experience more about the humanity and natural landscape of the Guanzhong Region of Shaanxi Province.

The architectural style takes its inspiration from the Chinese traditional "Siheyuan" (quadrangle courtyard) and borrows ideas from the "Sishuiguitang" (Huizhou Architectural Style), which is showcased in one episode of the TV program "Hui Merchants". In addition, the design focused on the extending range of architectural styles, the variation of surface and the flexibility of function, turning the architecture itself into a treasure trove containing new life and catalyzing new stories by removing the simple cultural connotation of the architecture. In terms of group organization, the "cluster" layout was adopted so that the architecture could get close to nature as soon as possible.

Four Design Highlights:

In construction: In fact, the abundant complex came into being through rollover and mirror combination of one unit. Architectural form in details is repetitive, that means the architectural nodes are

灞上

the same.

In Architecture Form: The architectural style stems from "Sishuiguitang" (which means the rain flows along the side slope of the roof to the courtyard from four directions), the major architectural feature of Chinese traditional "Siheyuan" (quadrangle courtyard). "Sishuiguitang" symbolizes that the wealth from all the four corners is just like the rain from God, which continuously flows into one' s own home. The design scheme adopts the architectural style of "Sishuiguitang". The courtyard is regarded as a natural absorber, implying a natural concept of "the unity of man and nature".

In general Layout: The architectural complex, in a "cluster" layout, consists of each independent architectural unit. In this way, the overall architectural shape rises and falls with the terrain, matching with the surroundings. And the original topography will be well protected as far as possible. The architectural complex constitutes a natural habitation among the dwellings.

In Landscape Design: The landscape planting in the "Romance by the Ba River" Service Zone, based on the architectural design, displays a village environment full of fields and gardens as well as fun. After appreciating the varied expo landscapes, the visitors will feel calm as if they came back to nature. In terms of the landscape planting, the Exposition boasts authentic plant varieties with definite themes, mainly growing the indigenous plants. Planting trees simulates natural planting method for piles and clusters of trees in front of village houses. And the shrubs are planted to imitate a beautiful natural scene on the slopes on the both sides of a river bank.

展园介绍：

“灞上”乃刘邦、项羽楚汉争霸中鸿门宴的所在地，唐玄宗曾在此举办规模盛大的水运博览会。作为 2011 西安世界园艺博览会的服务区之一，本项目试图提供给人以特别的建筑与景观体验，让游人更多的体验到陕西关中的人文精神和自然风貌。

建筑以中国传统四合院为蓝本，采取《四水归堂》的形式，并以建筑尺度的延展，表皮的变异与功能的变通作为切入点，力图使建筑摆脱简单的文化标识。在群体组织上，采取簇群式的布局方式，让建筑最大程度地亲近自然。

其具有四点生态景观设计亮点：

建造上：丰富的群体实际上由一个单元，通过翻转、镜像组合而成。细部很丰富的建筑形体都是单元性重复的，这意味着建筑的节点都是相同的。

形态上：形式来源于中国传统四合院建筑的主要特征之“四水归堂”。四水归堂，四方之财如同天上之水，源源不断地流入自己的家中之意。本方案中采用《四水归堂》的形制，天井作为人对于自然的接纳，暗喻了一种天人合一的自然观。

总体布局上：建筑采取簇群式布局方式，由各自独立的单元散落而成。这样，建筑总体的形态随着自然的地势而起伏，与周围环境形成了良好的呼应；同时尽可能地保存原来的地形地貌，形成了如同民居中的自然聚落。

景观设计上：灞上人家服务区景观种植配合建筑设计，体现田园、野趣的村落环境。景观种植品种纯净、主题明确，并以地方乡土植物为主。乔木种植方式模仿村落宅前以及路旁成片、成簇的自然种植，灌木种植，还模仿河岸两侧慢坡优美的自然景象。

天下第一面
子农
Zi Nong
FIRST NOODLE UNDER THE SUN

European Avenue

欧陆风情

Design Theme:

Extracting the feature of European traditional royal garden and creating the feeling of European town.

设计主题：

提取欧洲传统皇家园林风格特点，营造欧洲小镇感觉。

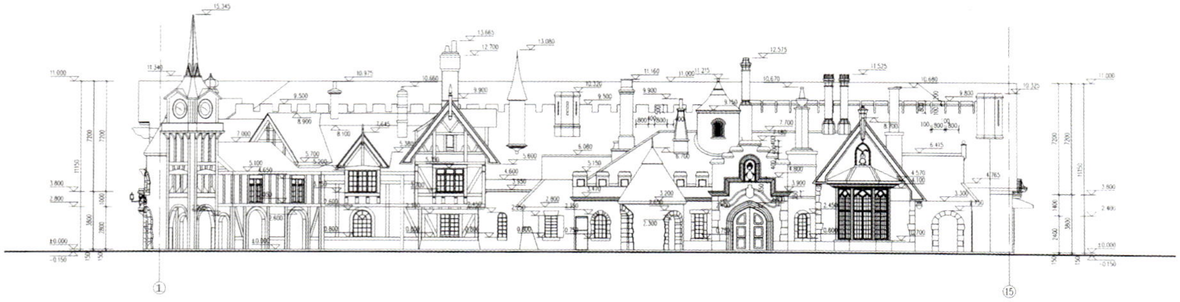

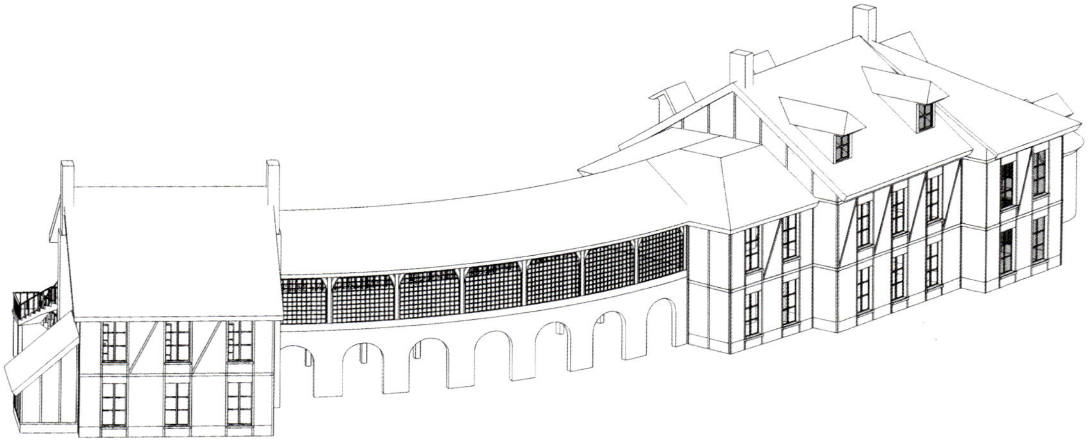

Introduction:

The project tries to extract the feature of European traditional royal garden (such as geometrical-pattern landscape planning) combining the typical planning feature of European urban residences as well as countries and towns (the combination of axes, square, obelisk, oblique and linear blocks; typical urban texture like foundation, harbor, and inside courtyard) to reflect more mixture and combination of European traditional symbols in the architectural pattern and form the friendly architectural space. A European town construction group with prominent style and typical feature has been built.

Design Highlights: 1. Building 1 of the European Avenue stands across the street from Building 3. The chimney, the round tower, the bell tower and the mermaid statue far away have added rich color of fairy tale. The water platform on the bank is the best position to appreciate the panoramic of the whole garden as well as the exhibition units surrounded by the green bushes behind the platform. 2. The two buildings near Building 3 are connected by the outdoor corridor. The colorful parterres are adorned in the front yard and back yard. The Chinese Canteen and the Western-style pastry Café are on each side of the corridor, from which you could overlook the gardens of various types in the tranquil atmosphere. 3. The musical fountain square is the main passage of connecting the domestic exhibition zone and the international exhibition zone. The ground with classical texture and the surrounding constructions bring the best of each other. The fountain in the center would express its own perfect with the beautiful music like a dancing fairy. 4. As one of the constructions with exotic style, the Wishing Pool is also called Happy Fountain. The perfect combination of the green fence and the vivid stone sculptures, together with the evening lamplight and the sky in dawn, presents fantastic landscape.

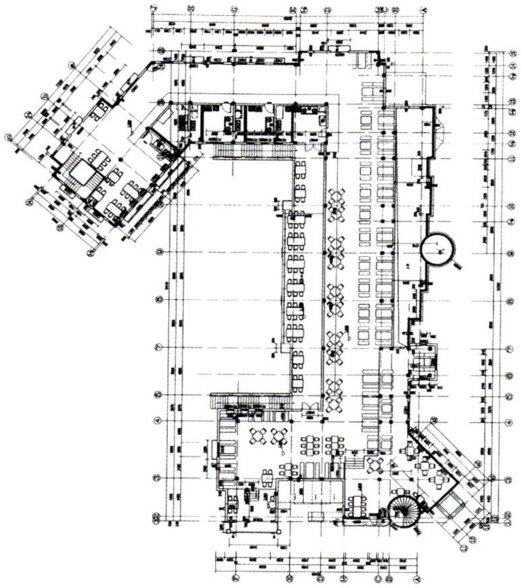

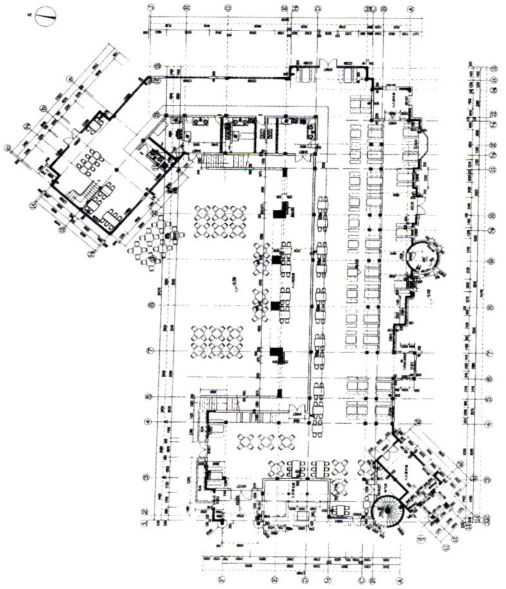

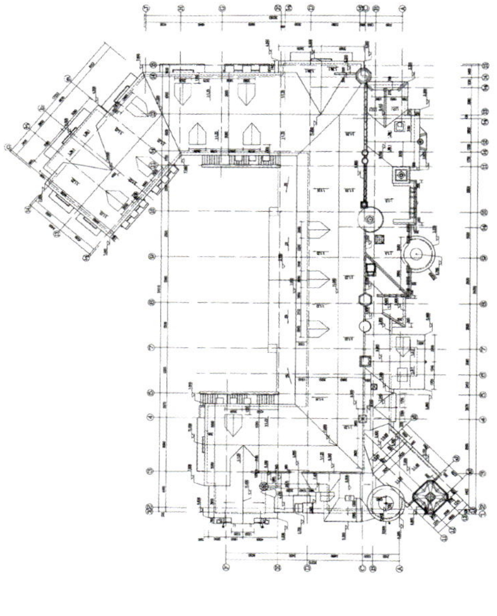

一点味餐厅

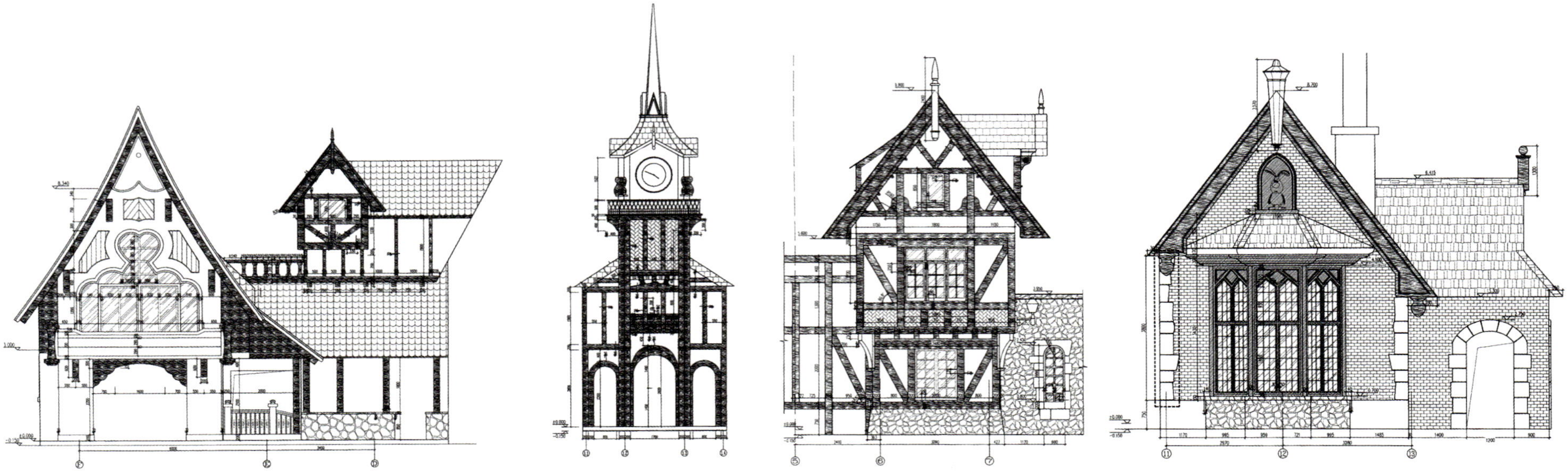

展园介绍：

本项目尝试提取欧洲传统皇家园林风格特点（如几何形状景观规划形式），结合欧洲城市住宅及乡村城镇的规划典型特征（轴线、广场、方尖碑，斜向与直线街区组合。喷水池，港口，内庭院空间等典型城市肌理），反映更多将欧洲传统建筑符号以建筑形式进行拼贴混搭、组合，形成亲近人的建筑空间尺度，塑造出一个风格特征突出、具备典型特征代表的欧洲小镇建筑群组。

设计有三大亮点 :1、欧陆风情 1 号楼与 3 号楼隔街相对，烟囱、圆塔、钟楼、远处的美人鱼雕塑为湖岸增添了浓郁的童话色彩。岸边的亲水平台可欣赏整个园林全景的最佳位置，背后是被绿色包围的各个展元，展元的景观都能一览无余。2、与 3 号楼毗邻的两座建筑由室外游廊连接。姹紫嫣红的花圃点缀在前庭后院，别有一番景致。以游廊为界一分为二，前为中餐厅，后为西点咖啡屋。在此眺望远方形色各异的园林，恬静、美味。3、音乐喷泉广场是连接国内展园和国际展园的主干道。古典纹饰的地面造型与周边的建筑相得益彰。中心的喷泉会随着音乐的高低起伏来诠释自己的完美，犹如舞动的精灵。4、许愿池又名幸福喷泉，是感受异国风情的建筑之一。修剪绿篱和栩栩如生石材雕塑的完美结合，傍晚加之灯光和昏黄天色的映衬，景象异常迷人。

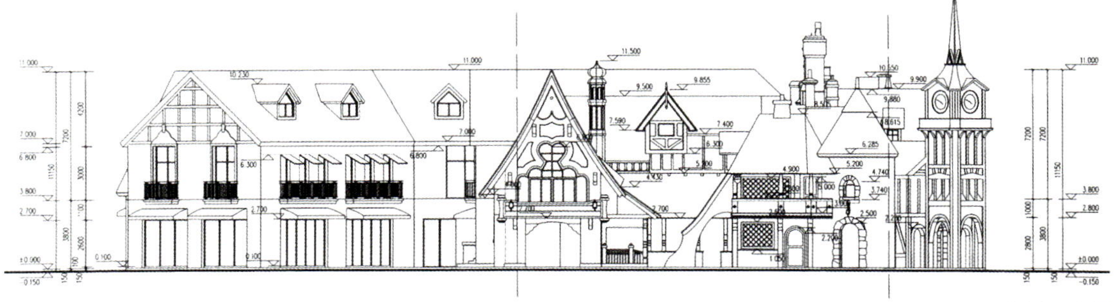

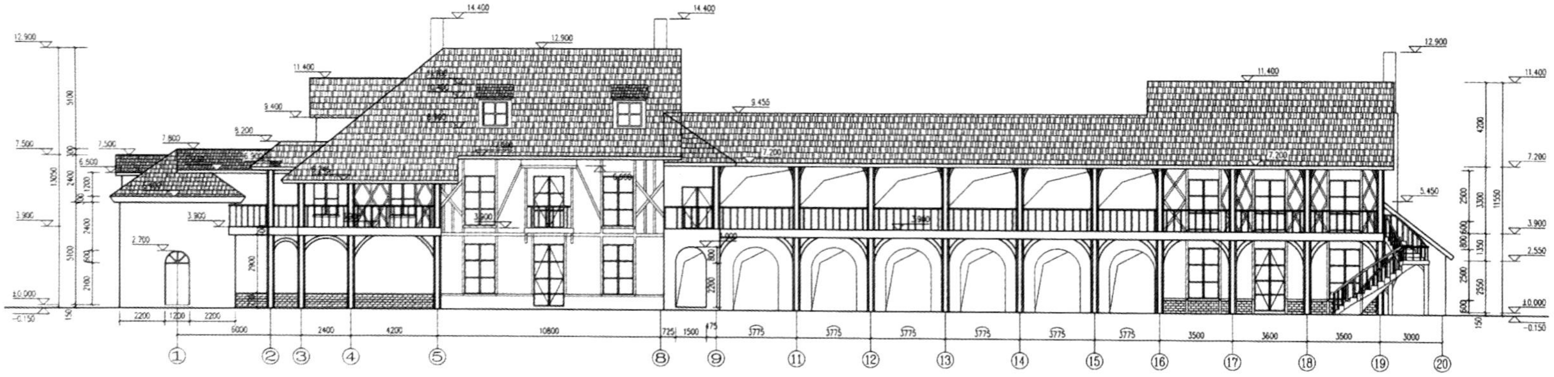

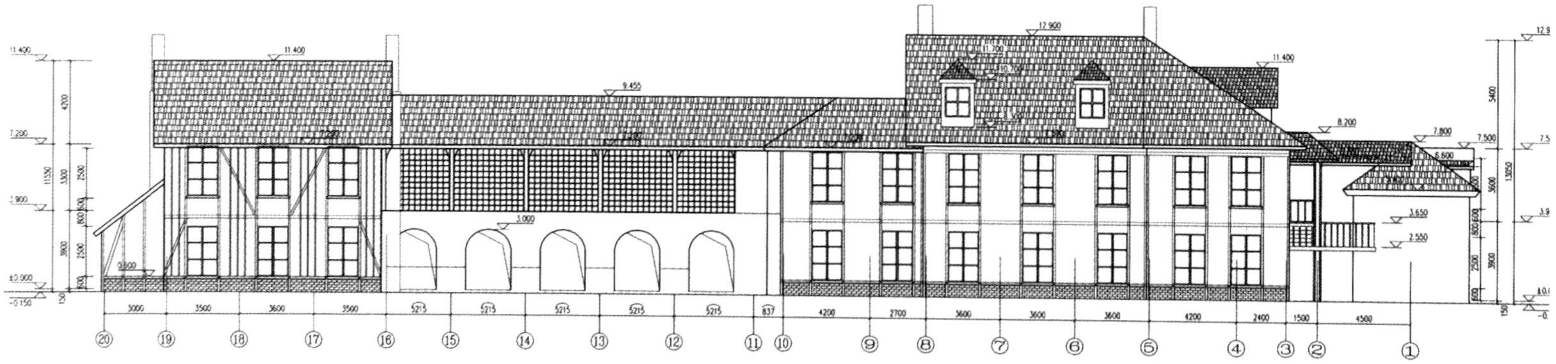

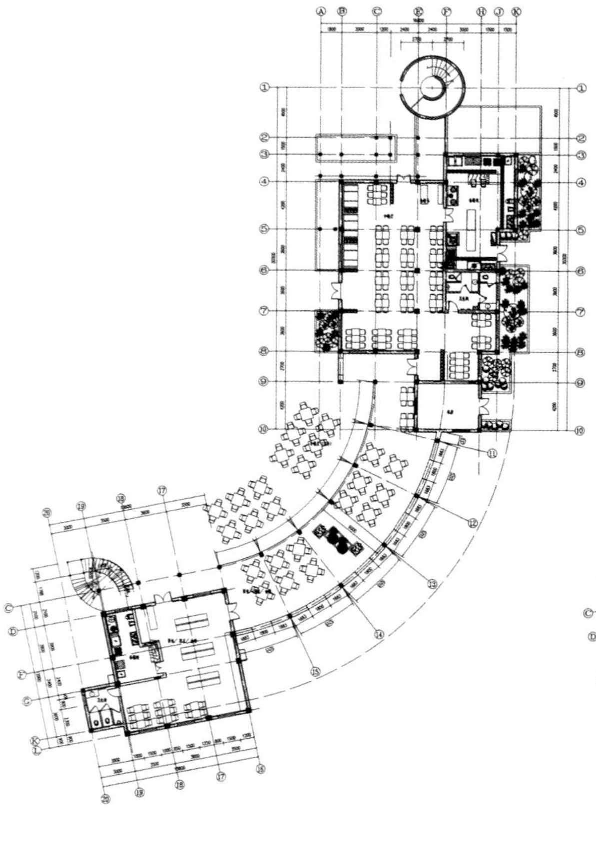

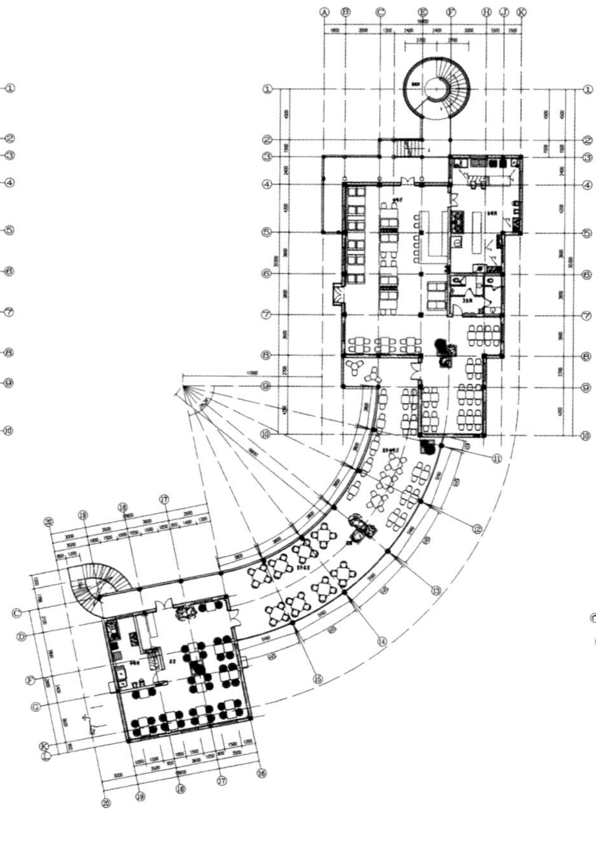

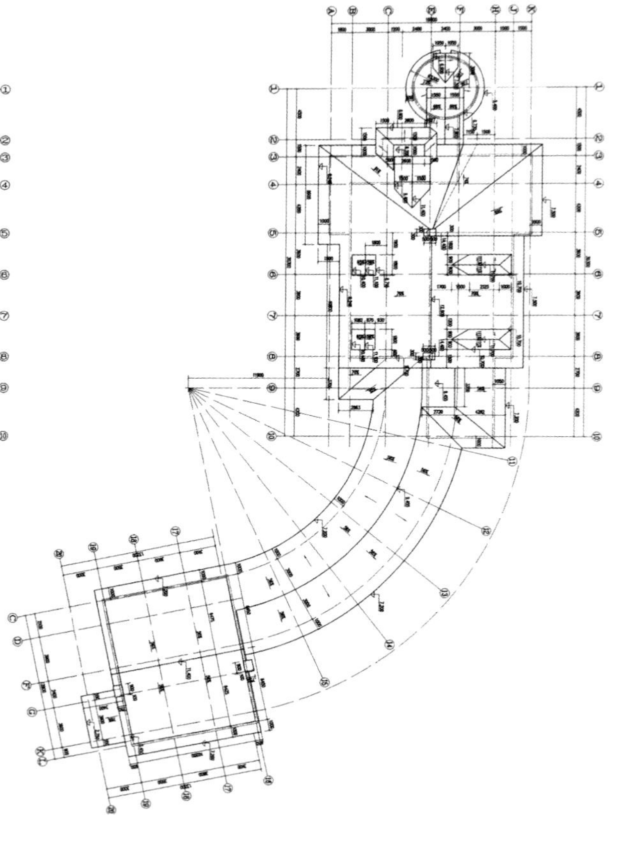

体验店

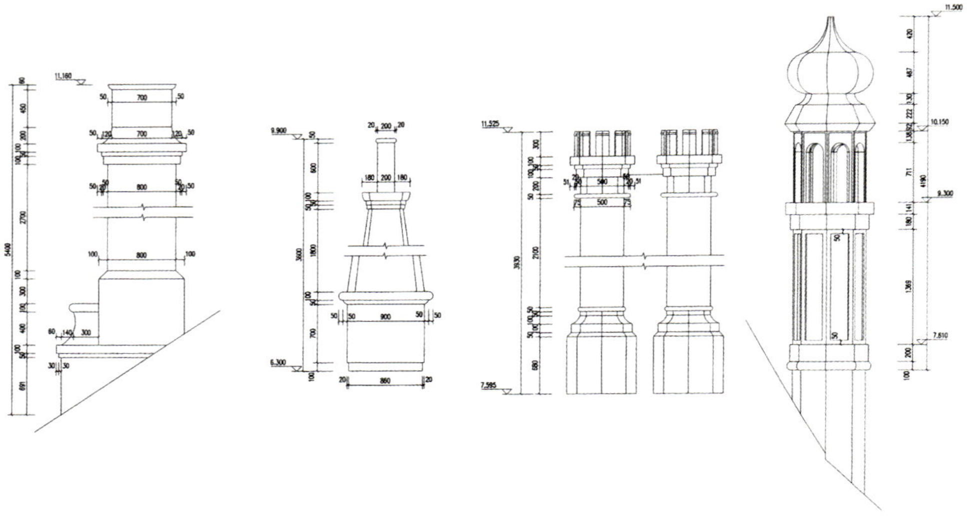

120
西安急救中心
EMERGENCY CENTER

EXPO

Coconut Wind and River Bank

椰风水岸

Design Theme:

Extracting the essence of Southeast Asian landscape and experiencing Southeast Asian style.

设计主题：

提取东南亚地区园林风格，感受东南亚风情。

Introduction:

Coconut Wind and River Bank is an enthusiastic and natural exotic area full of life and charm. Building 3, 4, 5 and 6 which look like royal constructions enjoy a great view. They are designed with magnificent Thai-style multi-storey roofs. Building 5 with towering spire is the landmark of this area. Building 2 is designed in typical Southeast Asian style to be simple and romantic with regional characteristics.

The project has tried to extract the essence of the gardens in Southeast Asia, such as the free-style landscape layout, to combine with the typical characteristics of the urban planning (axis, square, block formed by curved lines and straight lines, port, internal courtyard, etc.). Different buildings are combined with suitable space sizes to create a stylish and typical Southeast Asian town.

Lightspot of the design: the Southeast Asian garden features the characteristics of naturalness, health and leisure. Towering subtropical trees such as coconut trees and palm trees are combined with the lower bushes and shrubs to create a lively atmosphere of tropical garden. Quiet pavilion, winding bridge, zigzag water streets surrounding the buildings, beautiful green plants and the buildings are combined together to showcase Southeast Asian style.

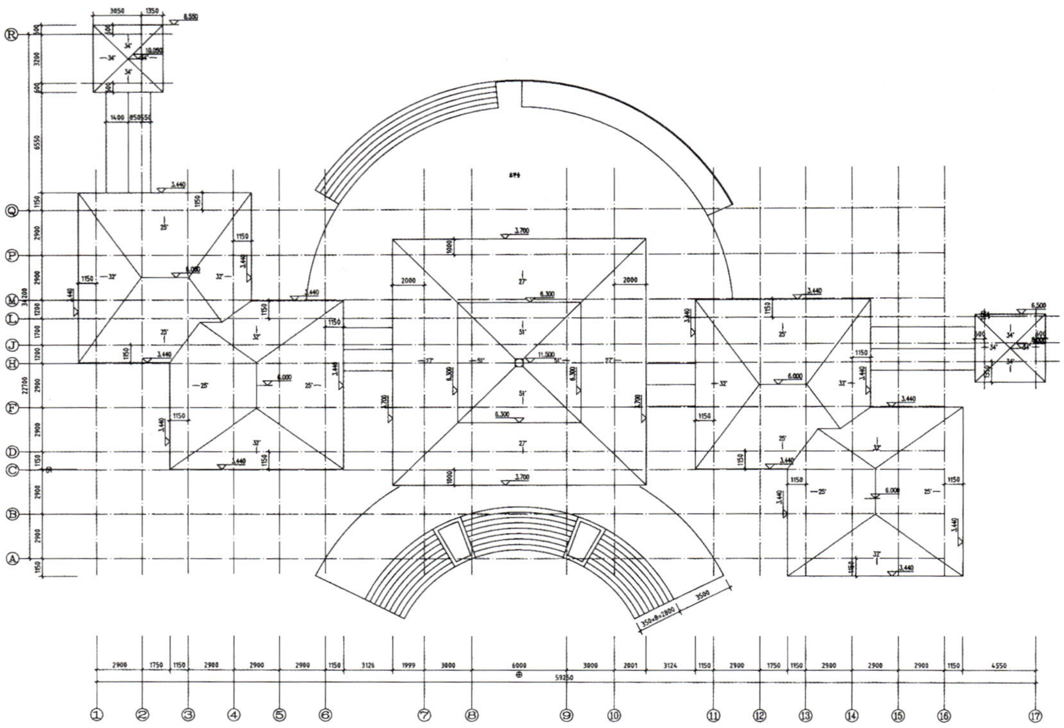

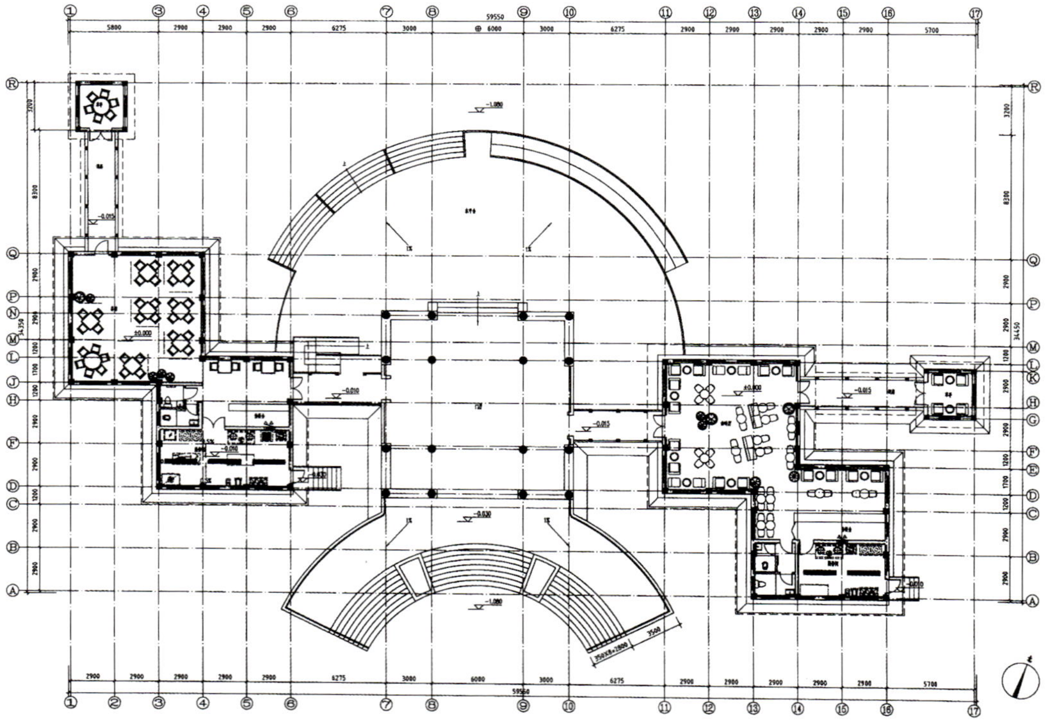

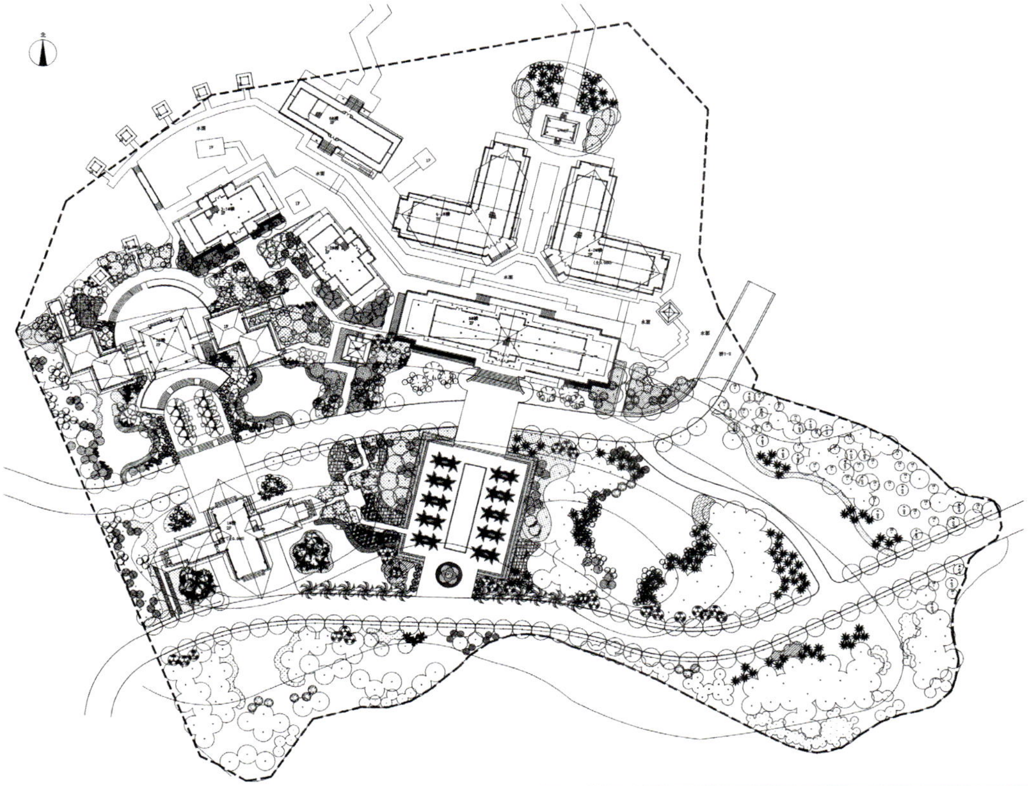

展园介绍：

椰风水岸是一个热情、自然、充满活力和风情的异国区域。以类似皇家建筑的3号楼、4号楼、5号楼、6号楼为主要观景地，采用华丽的泰国传统——多层屋顶作为建筑表达形式，其中塔尖高耸的5号楼为景区内标志型建筑；而2号楼则为典型的东南亚建筑风格，简朴、浪漫、尽显其特有的地域性符号。

项目尝试提取东南亚地区园林风格特点（如自由的景观规划形式），结合东南亚乡村城镇的规划典型特征（轴线、广场，曲线与直线街区组合，港口，内庭院空间等典型城市肌理），以建筑形式拼贴混搭、组合，形成亲近人的建筑空间尺度，塑造出一个风格特征突出、具备典型特征代表的东南亚小镇建筑群组。

项目设计亮点：东南亚展园秉承了自然，健康和休闲的特质，椰子树、棕榈树等高大的热带植物配以低矮的灌木丛，营造出充满生机的热带园林氛围。静谧的凉亭、曲折的栈桥、围绕着楼体的蜿蜒水街，人影绰绰，轻声笑语，婀娜多姿的绿色植物与建筑在荡漾的碧波中交相辉映，东南亚风情一览无遗。

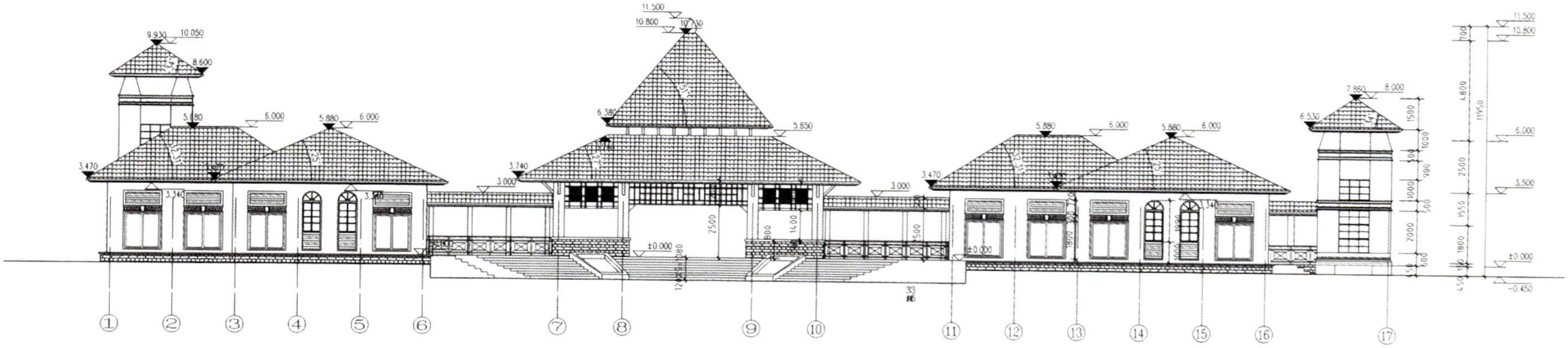

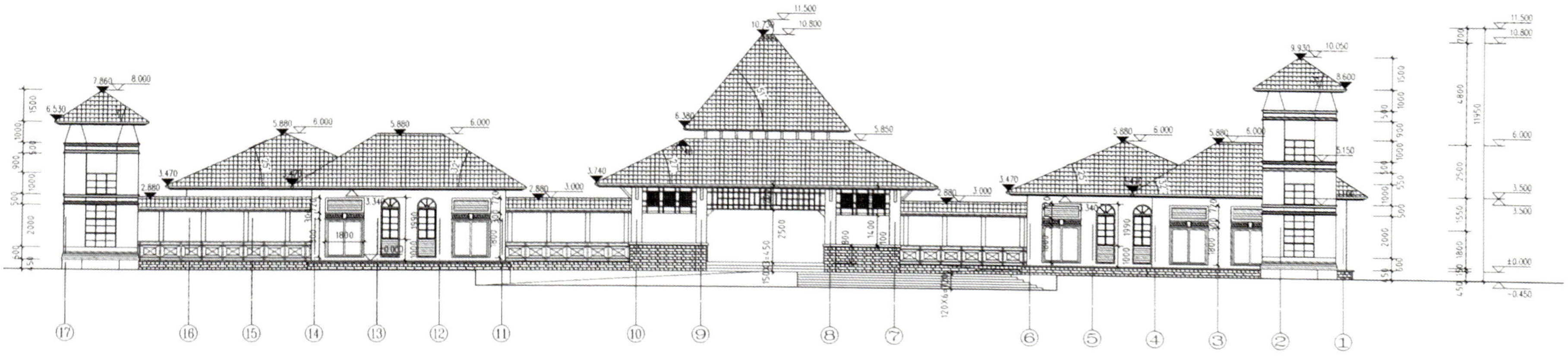

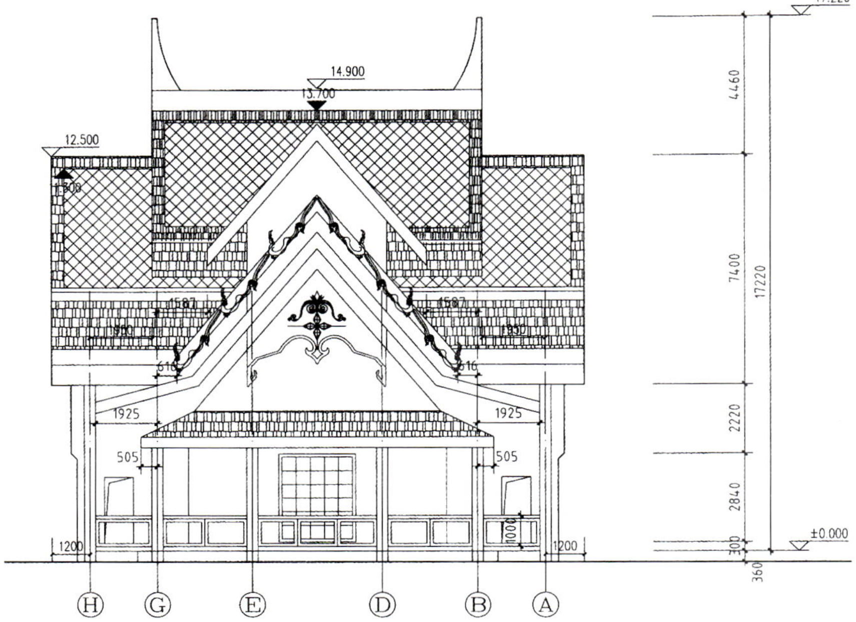
14.900
13.700
12.500
17.220
4460
7400
17220
2220
2840
300
360
±0.000
1925
1925
505
505
1200
1200
H
G
E
D
B
A

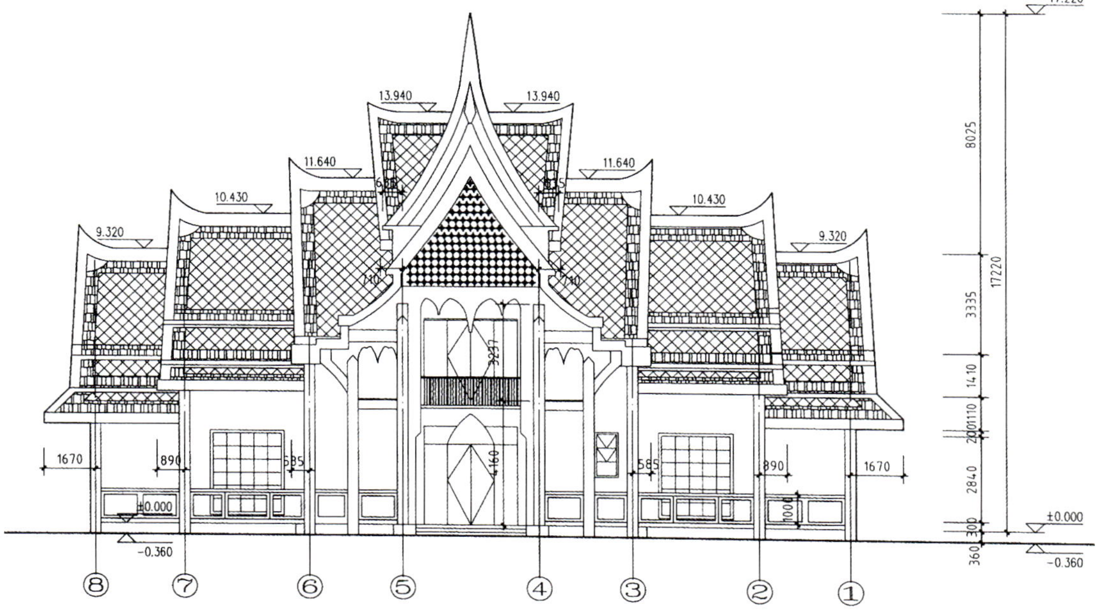
17.220
13.940
13.940
11.640
11.640
10.430
10.430
9.320
9.320
8025
3335
17220
1410
2840
300
360
±0.000
-0.360
1670
890
585
4160
585
890
1670
±0.000
−0.360
8
7
6
5
4
3
2
1

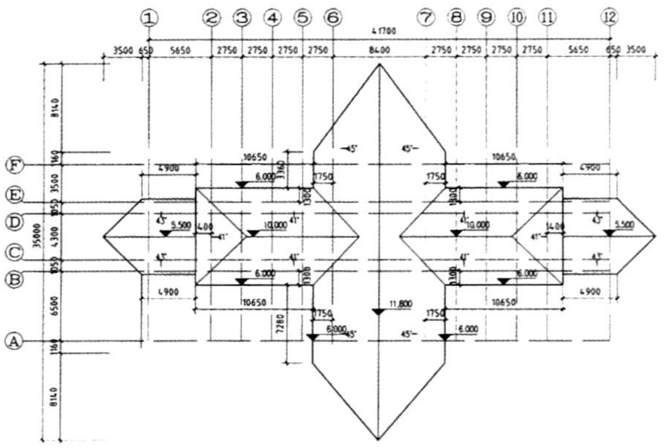

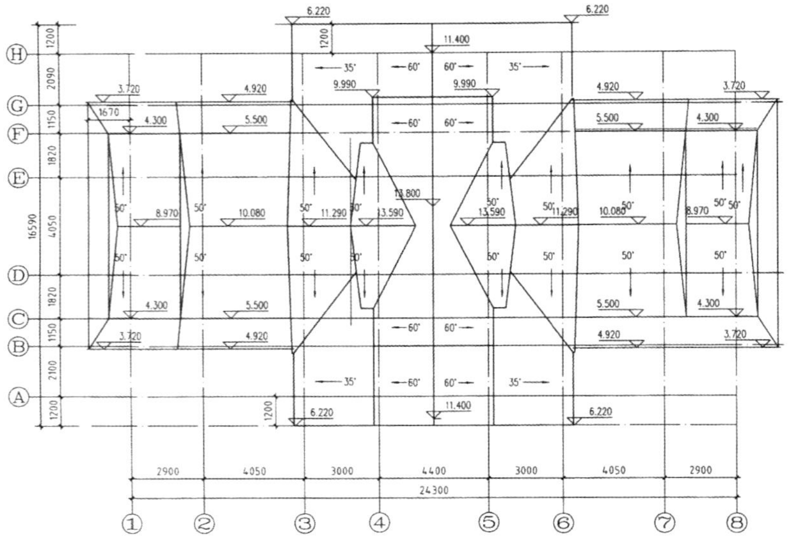

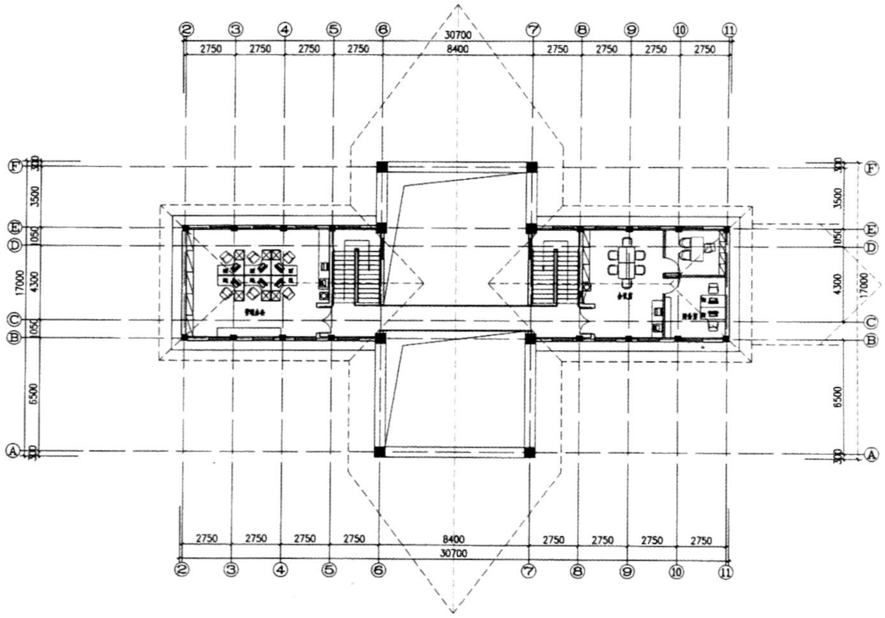

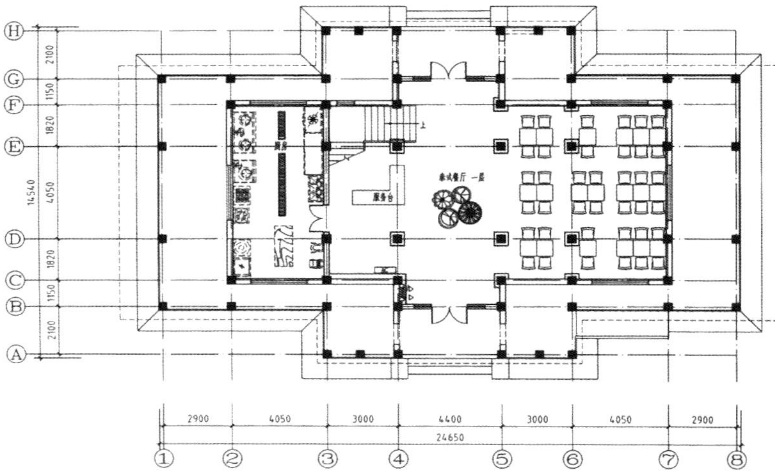

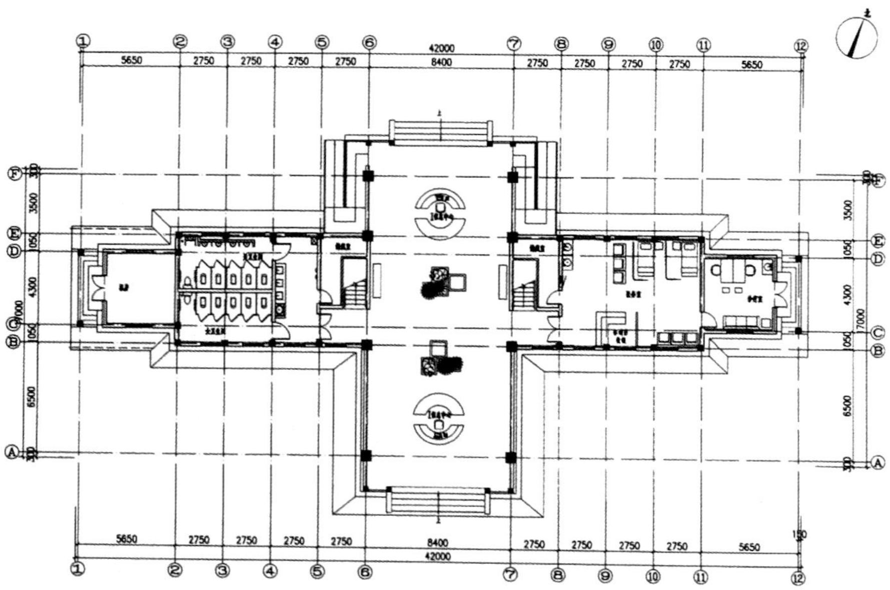

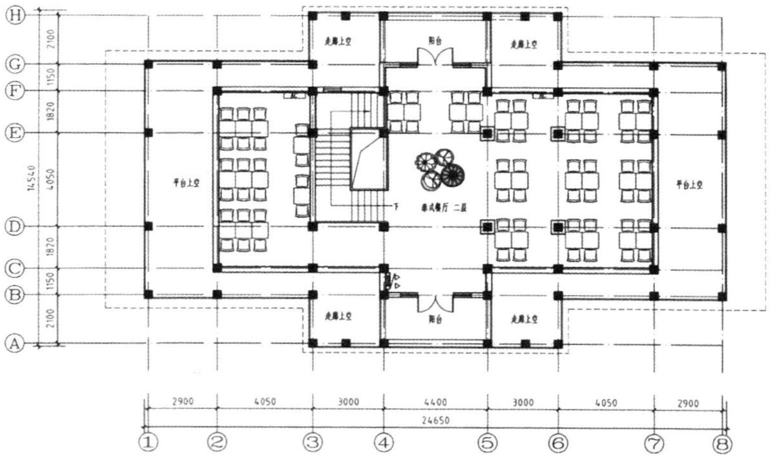

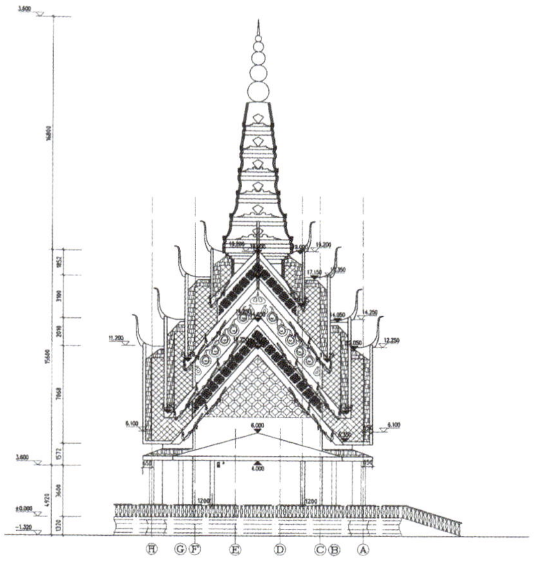

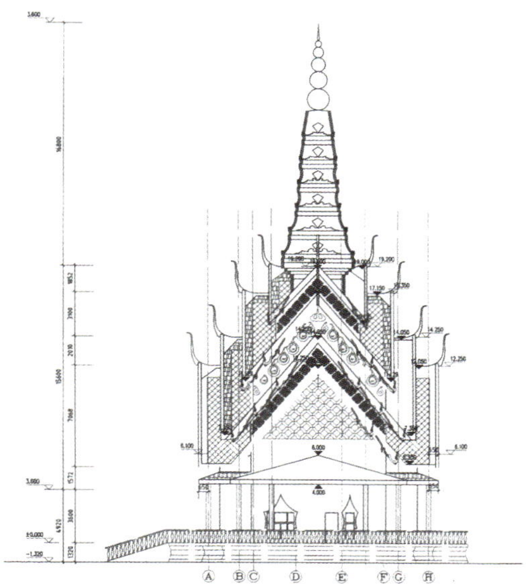

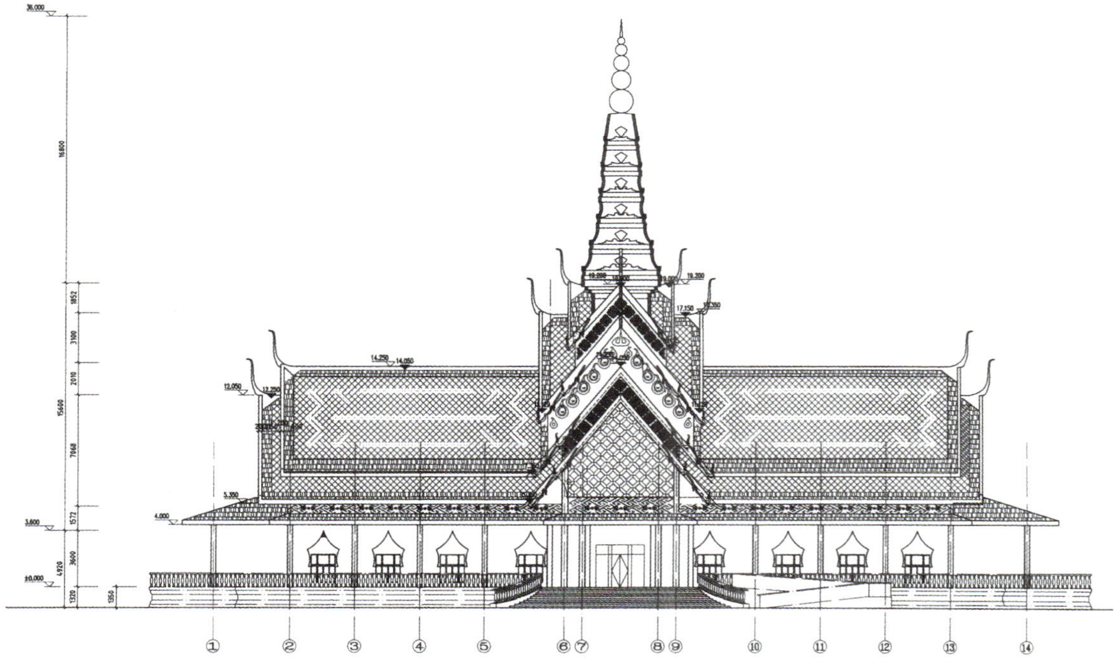

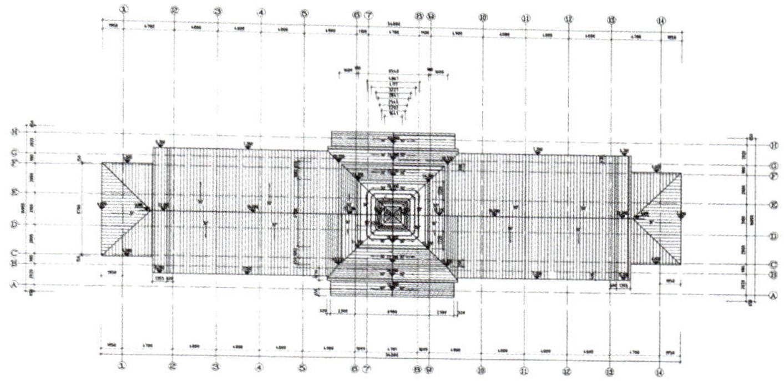

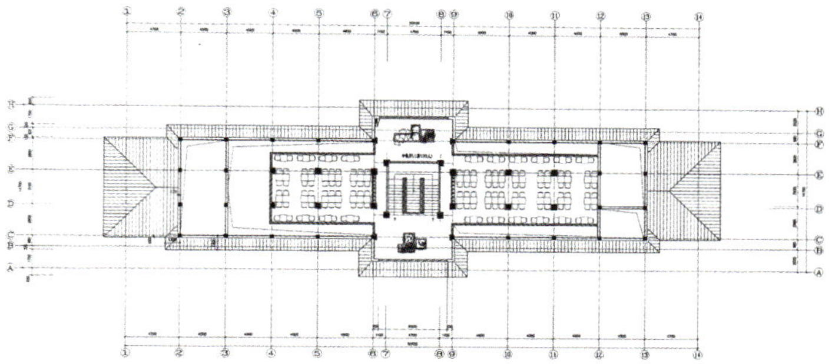

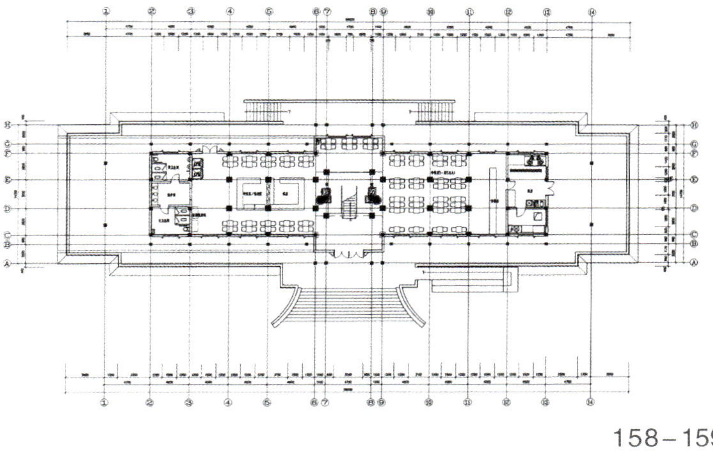

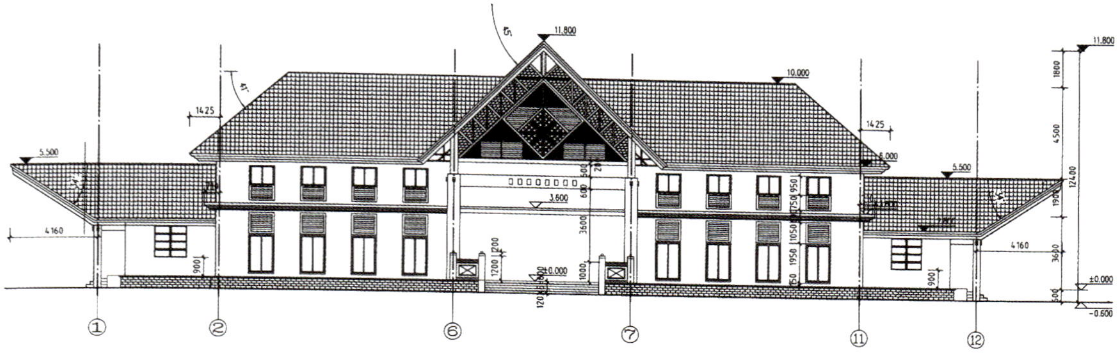
11.800
10.000
5.500
4.160
3.600
±0.000
-0.600

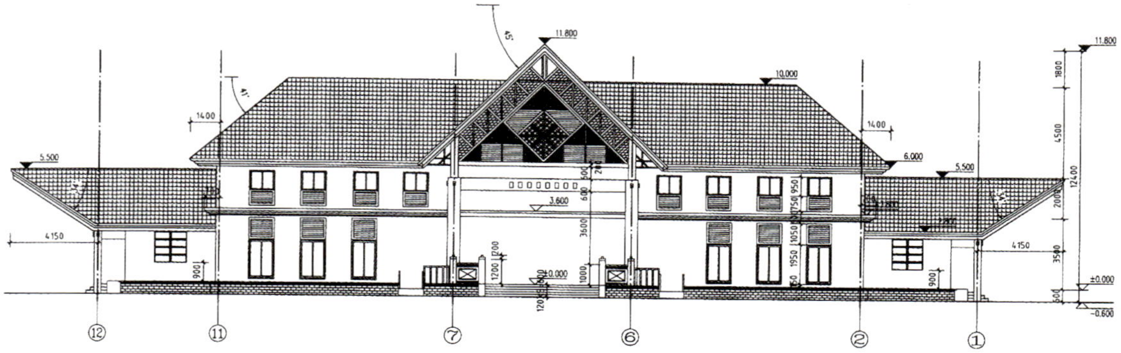
11.800
10.000
6.000
5.500
4.150
3.600
±0.000
-0.600

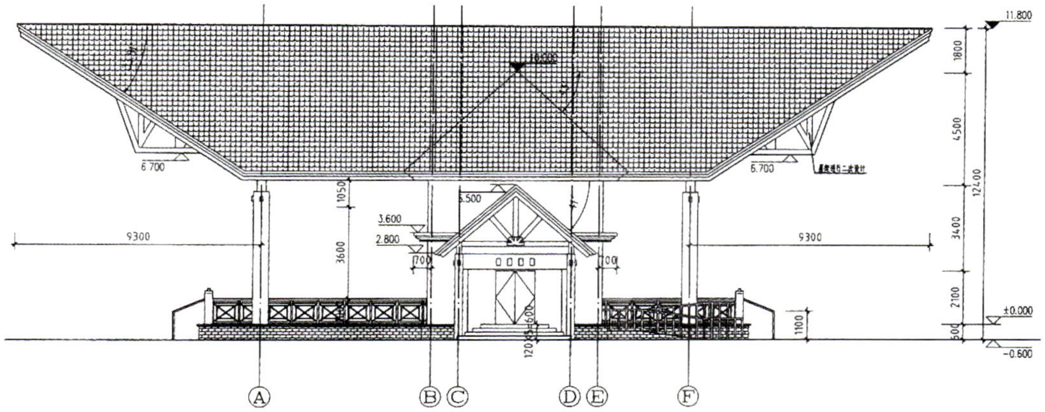

11.800
1800
4500
12400
3400
2100
600
6.700
6.700
6.500
3.600
2.800
9300
9300
1050
3600
700
700
1100
±0.000
-0.600
A
B
C
D
E
F

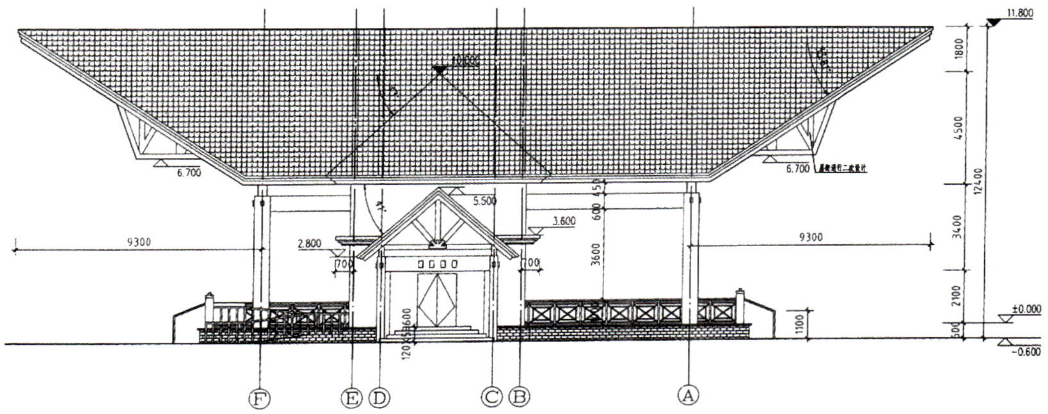

11.800
1800
4500
12400
3400
2100
600
6.700
6.700
5.500
3.600
2.800
9300
9300
3600
700
700
1100
±0.000
-0.600
F
E
D
C
B
A

Grand Palace

灞柳驿

This service zone is located in the southwest corner of central area of Xi' an Garden Show Park, covers an area of 93,770m², it is a window that the exposition opens to the world.

该服务区位于西安园博园中心区的西南角，占地面积为 93770m²，是世园会开向世界的一扇窗口。

Introduction:

Ba Liu Yi Hotel is located in the southwest corner of central area of Xi' an Garden Show Park. It is an independent island that surrounded by water, covers an area of 93,770 m², designed in accordance with five-star standard, scale of construction is 27,000 m². It is a comprehensive hotel that integrated ecology, leisure and entertainment as a whole.

The characteristic Chanba district (established in September 2004, one of the "four districts, one port, two bases" that Xi' an prioritizes), the surrounding rivers, islands, mountains' natural landscape and the unique climate of Northwest deeply rooted in here. Closed curve that running along the island' s terrain draws the outline of the hotel' s boundary through brick wall, inside there are ever-changing batten side walls, two walls form a private and interesting corridor space. The hotel keeps the island' s original landform as much as possible, large-scale cultivated woods shapes the ecological oxygen bar, realizes the service zone' s graceful and grand natural scenery.

The interior is divided into five different functional areas, including excutive reception area, VIP reception area, Ambassador Area and two executive room areas. During the World Horticultural Exposition, this hotel is an important reception center.

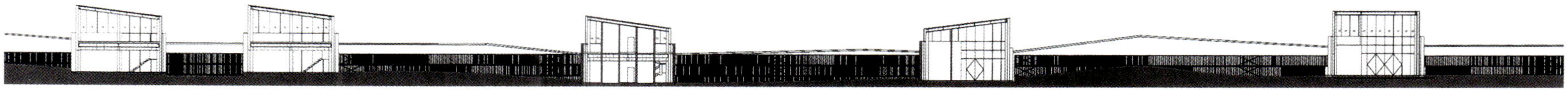

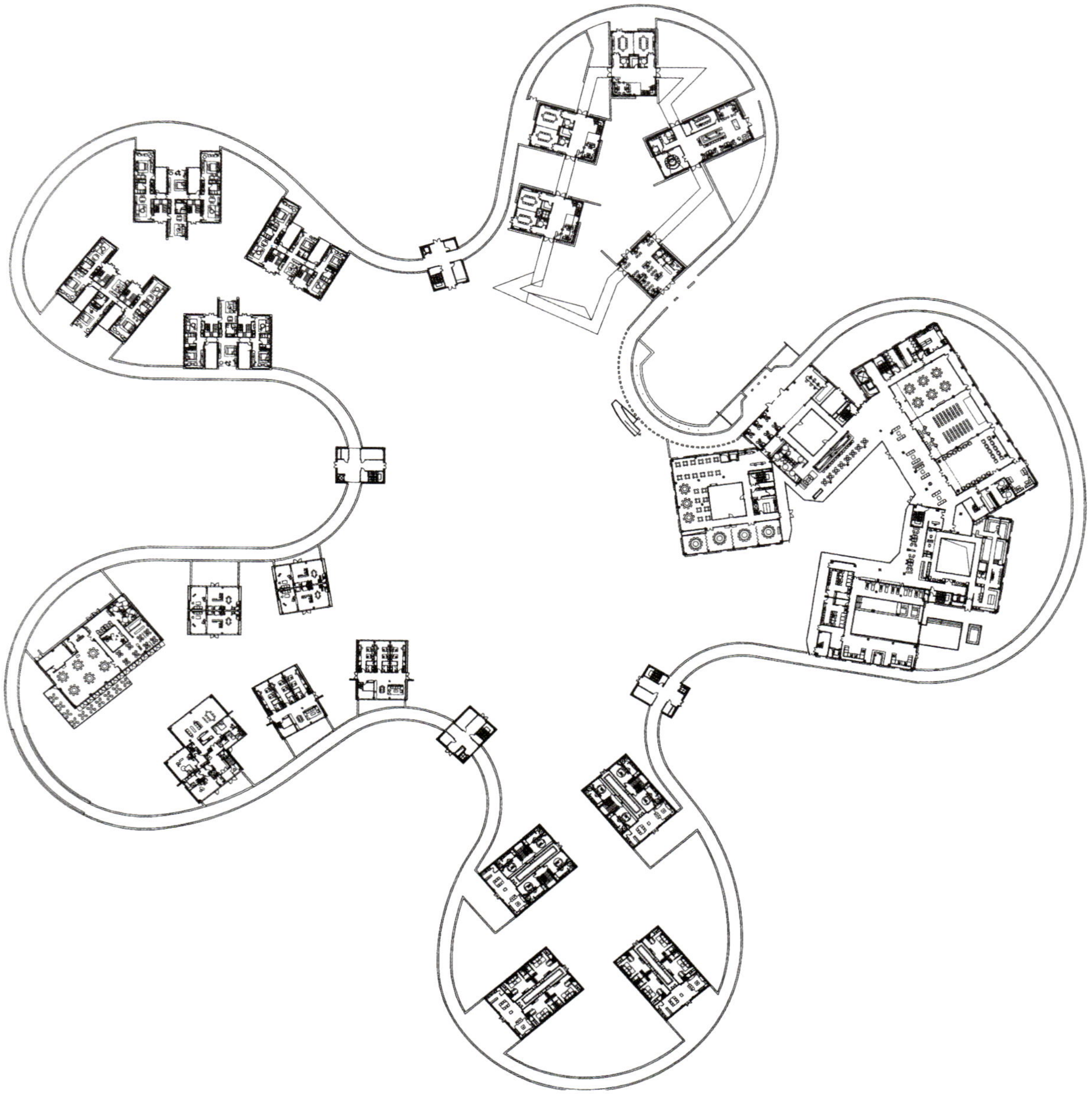

This hotel takes the stone wall as the center to form a five-petal area, each petal represents a unique landscape feature, five areas extend direct to the water, pointing to five service areas respectively. These five areas occupy the five most beautiful waterfronts, relatively independent, but also connect with each other through the central courtyard. Central courtyard forms a large public gathering space for cultural activities and art exhibition. According to the different terrain, five areas adopts five different architectural forms, thus produce characteristic spaces no matter for interior use or for exterior observation.

Preparatory Office of the World Horticultural Exposition also invites contemporary Chinese landscape panting master—Mr. Cui Zhenkuan create a painting, known as "the boundless Qinling" (cang mang qin ling tu), two meters high, six meters long, permanently hanging in the lobby of Ba Liu Yi Hotel. This painting themed as Qinling, which represents the theme "Eternal peace & harmony between nature & mankind, nurturing the future earth---a city for nature, co-existing in peace" of the exposition the most, and highlights the architectural feature of Chanba Ecological Zone.

The exposition is the window that Ba Liu Yi opens to the world. This hotel with special significance, whether the building itself or the interior decoration, will represent Xi' an, Shaanxi even China to meet the tourists from all over the world.

展园介绍：

西安世园会灞柳驿酒店位于西安园博园中心区的西南角，为一座四面环水的独立小岛。酒店占地约 93 333 m^2，按五星级标准设计，建设规模约 2.7 万 m^2，是集生态、休闲、娱乐为一体的综合性酒店。

浐灞新区的独特区位（成立于 2004 年 9 月，是西安市重点发展的“四区一港两基地”之一）和周边河流、岛屿、山脉的自然景观以及西北大地独特的气候植于其中，随小岛地形流动的闭合曲线通过青砖墙勾勒出酒店的边界，内侧则是连续变化的木条边墙，两道围墙形成了私密又富于趣味变化的廊道空间。酒店最大可能地保留原有岛屿地貌，大面积种植树林形成城市的生态氧吧，实现园区服务区的优美宏大的自然之景。

酒店内部分为五个不同的功能区，包括行政接待区、VIP 接待区、国宾区和两个行政客房区。

世界园艺博览会期间，该酒店是重要的接待服务中心。

酒店以石墙为中心，形成五个“花瓣”区，每片花瓣代表了独特的景观特色，并且五个区面向水面的方向延伸开去，分别指向五个服务区。五个区分据小岛景色最为优美的五处水岸，既相对独立，又能通过中心庭院互相联系，中心庭院形成了一个大的公众聚集的文化活动和公共艺术展示空间，根据岛屿地形环境差异，五个区采用五种不同的建筑形态，并由此产生了无论是内部使用体验，还是外部视觉观察来看都各有特色的系列空间。

世园会筹备办还特邀当代中国山水画大家——崔振宽先生担纲创作水墨画，现定名为《苍莽秦岭图》，高 2m、长 6m，永久悬挂在灞柳驿酒店大堂内。该画以秦岭为题材，最能代表世园会“天人长安、创意自然”的主题，也最能突出浐灞生态区的建设特色。

西安世界园艺博览会灞柳驿酒店是世园会开向世界的一扇窗口。这蕴含特殊意义的酒店，无论是它的建筑本身，还是内部装饰，都将代表西安、陕西，乃至于中国来迎接世界各地的游客。

International Garden

国际园

By advertising International Horticultural Exposition in more than forty countries and thirty domestic cities, there are 43 international gardens in the 105 outdoor exhibition gardens which includes 27 international gardens of the U.S.A, Canada, the U.K, France, the Netherlands, Germany, Argentina, Australia, India, South Korean and so on, as well as three gardens of the International Network for Bamboo and Rattan.

The ecological garden of the Netherlands, the delicate garden of Italy, the lotus pavilion of South Korean, the Yokohama Garden of Japan, the natural landscape of the U.K, the classical garden of French …. When entering into the international gardens of the International Horticultural Exposition 2011 Xi' an, one can experience the exotic flavors of varied countries.

通过在全球 40 多个国家、国内 30 余个城市对世园会的特色与魅力进行宣传，室外展园 105 个中国际园占 43 个，国际园涉及美国、加拿大、英国、法国、荷兰、德国、阿根廷、澳大利亚、印度、韩国等 27 个国家和国际竹藤组织等 3 个国际组织。

荷兰的生态花园、意大利的精致园林、韩国的爱莲亭、日本奈良的枯山水、英国的自然风景、法国的古典园林……走进西安世园会国际园，可以感受到精彩纷呈的“万国风情”。

Paris Garden

巴黎园

Area: 1,508 m^2

Paris Garden will be a beautiful "Castle Garden in Capital of Fashion" with the characteristics of traditional French garden.

展园面积：1 508 m^2

该园力求打造时尚之都，城堡花园，整洁优美，传统的法国花园。

Introduction:

Located in a south-north direction, Paris Garden highlights European style to be a beautiful "Castle Garden in Capital of Fashion" with the characteristics of traditional French garden". The original atmosphere of Paris is created with arrangement of fountains, gardens and castles. The theme of the garden is a castle amidst garden, showing the magnificence and majesty of European castles. Paris, the capital of fashion, has its charm brought to the Expo with panes leading to the castle, carefully arranged trees, and neat fountain design. Plantation in the garden consists mainly of bush, large flowerbeds, and lawn. It is a perfect spot to gain classy experience under the bright sun.

展园介绍：

展园以典型的欧洲风格为主线，"时尚之都，城堡花园，整洁优美，传统的法国花园特色"的构思由南向北贯穿全园。在景点的设置上，通过喷泉、花园、城堡的安排，完美地展现巴黎的欧洲风情。园区主题是被花园环绕的城堡，突显了欧洲城堡大气、庄严的感觉。通往城堡的小径、精心修剪的树木、整洁美观的喷泉，处处彰显时尚之都巴黎的风采。园区植物以灌木为主，大面积的花卉，草坪和灌木丛，阳光照耀，是度过高品位生活的好地方。

Garden of INBAR

国际竹藤园

Area: 1,450 m^2

Through the bamboo application in gardening construction and human manufacturing and living, the close relation between bamboo and human are presented.

展园面积：1 450 m^2

该园通过竹藤在生活中的广泛应用，生动地展示竹藤与人类密不可分的关系。

Introduction:

The Garden of INBAR consists of Bamboo Array Square and Concentric Square, covering 1 450 m^2 land area. The garden's main construction is the Bamboo Exhibition Hall, given from German Goethe Institute as a gift. Through the bamboo application in gardening construction and human manufacturing and living, the close relation between bamboo and human would be presented and the core theme of harmony between human and nature would be expressed. Inside the trapezoid exhibition, three wide exhibition halls filled with various kinds of bamboo are arranged. The lamp light adds more mysterious atmosphere in the quiet night through the glass wall.

展园介绍：

国际竹藤展园占地面积 1450 m^2，由竹阵广场、同心广场两部分组成。其是以德国文化机构歌德学院赠送的艺术品——竹子展厅为主要建筑物的花园，将通过竹藤在园林建设和人类生产生活中的广泛应用，生动展示竹藤与人类密不可分的关系，诠释人与自然和谐共生这一核心主题。梯形的园区内，三个布满各类竹子的开阔展厅等边分布，灯光透过玻璃幕墙，在静谧的夜晚更添大自然的神秘气息。

竹藤組織

Netherlands Garden

荷兰园

Area: 2,714m^2

The Garden full of bamboos and rattans has shown the landscape of Netherlands as the "Country of Pastures and Flowers".

展园面积：2 714m^2

该园用布满竹藤的展园展现了荷兰"牧场之国"和"花之国"的景观特色。

Introduction:

The Netherlands is a world famous country for gardening works and flowers exporting. Varied flowers cover 40% of the Garden. Though there is not the famous windmill and sabot of the Netherlands, the garden full of bamboos and rattans has shown the harmony between human and nature perfectly.

The Netherlands Garden is built in triangular shape. It features the landscape of Netherlands as the "Country of Pastures and Flowers". The main part of the Garden consists of a steel Calla lily and a greenhouse made of glass. Carefully arranged water flows and vegetation show the advanced technology and unique concept of the horticultural and garden art in the Netherlands.

展园介绍：

荷兰是世界闻名的园艺和花卉出口大国，展园里各色花卉占据了展园将近到40%的面积，虽然少了荷兰著名的风车和木鞋等元素，但布满竹藤的展园依然完美呈现出了人与自然的和谐统一。

荷兰展园整个地块呈三角形分布，秉承了荷兰"牧场之国"和"花之国"的景观特色。展园主体建筑由一个钢结构的马蹄莲造型和玻璃花房构成，精心布置的水流与植被体现出荷兰先进的园林园艺技术与独特的园林园艺理念。

Umea Garden, Sweden

瑞典于默奥园

Area: 1,062 m^2

The Sweden Garden features plantation coverage and represents the characteristics of Sweden.

展园面积：1 062 m^2

该园以植物覆盖、发光石头为主，体现着瑞典国家的特色。

Introduction:

The Sweden Garden features plantation coverage, especially the European plants such as Italian clematis, honeysuckle, and silver birch. The unique exhibit of the Garden is a luminous stone, which represents the characteristics of Sweden. It can emit light by itself and light up the nearby plants in the evening. The specially arranged seats in the small forest invite you to stay for a while and enjoy the peace in such a European garden.

展园介绍：

瑞典于默奥展园主要是以植物覆盖为主，尤以欧洲特色植物意大利铁线莲、忍冬及白桦树为主。该展园有一特色展品发光的石头灯，石头体现着瑞典国家的特色，而且该石头能自动发光，周围的植物在夜晚灯光的照应下光彩夺目。树林中独特座椅让人忍不住坐下静静地享受欧洲庭院的惬意。

UMEÅ
SWEDEN
于默奥
瑞典

Sri Lanka Garden

斯里兰卡园

Area: 489 m²

It is a perfect combination of gardening with South Asian culture.

展园面积：489 m²

该园可称作南亚文化与园林园艺的完美结合。

Introduction:

This is a traditional building of South Asia. The double-floor steel structured exhibition hall has red gridiron pattern on the exterior, conveying the theme of environmental protection and nature. The style of the hall is a perfect combination of gardening with South Asian culture. As a country in South Asia, Sri Lanka shed South Asian flavor and Buddhist atmosphere as well, therefore, the garden open up gradually with the theme of Buddhism.

展园介绍：

传统南亚建筑。钢结构二层展馆，外部为暗红色方格装饰，体现环保与原生态主题。可称作是南亚文化与园林园艺的完美结合。作为南亚国家的斯里兰卡，展园散发着南亚异域风情，同时斯里兰卡的佛教气息浓厚，因此展园也紧扣了佛教的主题来展现。

Kazakhstan Garden

哈萨克斯坦园

Area: 1,123 m^2

The garden has told the story about the harmonious relationship between the Kazakhstan people and the nature.

展园面积：1123 m^2

该园显示出哈萨克斯坦人与自然密切结合、协调发展的关系。

Introduction:

Kazakhstan Garden features two main exhibition halls, which draw inspiration from felt tents, showing the historical background of Kazakhstan and casting new light on the ancient culture. The Donggan exhibition hall in Kazakhstan Garden will display the existing wedding garment and custom of Donggan people carried down from the Qing Dynasty. Plantation in the Garden includes cedar, poplar, ginkgo tree, midget crabapple, apple tree, hawthorn tree, tall fescue, blue grass, and ryegrass, telling the story about the harmonious relationship between the Kazakhstan people and the nature.

展园介绍：

哈萨克斯坦展园以两大展厅为特色，展厅以毛毡帐篷为基调，寓以展示哈萨克斯坦的历史文化背景，给古代文化赋予新的光彩。哈萨克斯坦东干族展示厅内将展现有东干人现仍保留的清朝时期的婚嫁服饰，习俗等。展园内布置有雪松、白杨、银杏、西府海棠、苹果树、山楂树、高羊茅、早熟禾、黑麦草等植物品种，充分显示出哈萨克斯坦人与自然的密切结合、协调发展的关系。

New Zealand Garden

新西兰园

Area: 1,167 m^2

The design idea of "technological garden" has penetrated into New Zealand Garden; perfectly integrating elements such as deck, plants, marsh and wetland.

展园面积：1 167 m^2

该园以科技的花园为主线，将甲板、植物、沼泽、湿地等元素完美融合。

Introduction:

The design idea of "technological garden" has penetrated into New Zealand Garden from south to north. It brings together the modern technology, architecture and garden art by carefully arranged water and land. The theme of the garden is to perfectly integrate elements such as deck, plants, marsh and wetland, using scientific design and practical design concept to solve the problem of building damage caused by draining and plant roots. All plantations in the garden are imported from New Zealand with high ornamental value.

展园介绍：

新西兰展园以科技的花园为主线，由南向北贯穿全园。在景点的设置上，通过水陆交错的精心设计，表达现代科技，建筑和花园的融合。景区主题是通过科学的设计方法和实用的设计理念，将甲板、植物、沼泽、湿地等元素完美融合，解决排水和植物根系对建筑物伤害的问题。园区内所有植物均从新西兰当地进口，有着非常高的观赏价值。

Italy Garden

意大利园

Area: 1,091 m²

With advanced and elaborate layout, the garden is also called "Garden of Tuscany".

展园面积：1091 m²

该园进行了先进和精巧的布局，又被称为"托斯卡纳花园"。

Introduction:

The "Garden of Tuscany" presented by Italy is designed by Politecnico di Torino. The design is closely related to the name of the Garden. Advanced garden technology is applied to the elaborate layout in the garden. Flowing light spots on the lawn and tree-wall cast by the sunshine will bring you to the sunny Tuscany.

展园介绍：

意大利展园"托斯卡纳花园"由意大利都灵理工学院主持设计。展园设计与名称紧密呼应，先进的园林园艺技术与精致巧妙的布局结构相结合，别致的树墙和草坪中洒下点点阳光，让人仿佛置身于明媚的托斯卡纳。

Burma Garden

缅甸园

Area: 1,680m^2

Burma Garden adopts architectural style of Southeast Asia, showing the romance of Southeast Asia and the Buddhist style of Burma.

展园面积：1680m^2

该园采用的东南亚建筑风格体现了东南亚的浪漫、凸显了缅甸的佛教风格。

Introduction:

Featuring 153 nationalities, 80% of Burmese have dressed in national costumes. Most of the customs are closely relative to the precept and doctrine of Buddhism.

Burma Garden of International Horticultural Exposition 2011 Xi' an adopts architectural style of Southeast Asia, showing the openness of Southeast Asia in terms of style, color and concept. The Buddhist temple in the rear of the Garden has elephant-shaped sculptures, which expresses the love for elephant in Southeast Asian countries. The open design of the space decorated with unique subtropical banana trees and palm trees integrates openness, flexibilities and functions.

展园介绍：

缅甸是一个有153个民族的多民族的国家，超过80%的人还穿着民族服装。缅甸人许多风俗习惯都离不开佛教的教规、教义。

此次西安世园会缅甸园采用东南亚建筑风格，从色彩、理念等方面充分体现了东南亚的浪漫，

展园前方的两只白象雕塑体现出东南亚国家对大象的“钟情”，后方红、白、黄相间的寺庙建筑则凸显了缅甸的佛教风格。同时展园中亚热带特有的芭蕉树、棕榈树郁郁葱葱，开阔的园林空间体现出通透性、流动性与功能性的融合。

缅甸 MYANMAR

缅甸 MYANMAR

Nepal Garden

尼泊尔园

Area: 1,087m²

Tasteful design and exquisite craft have told the urban development of Katmandu, showing the strong cultural atmosphere of Nepal.

展园面积：1087m²

该园以竹苞松茂的设计、精美的工艺讲述加德满都城市发展，充分展现尼泊尔浓烈的文化风情。

Introduction:

Tasteful design and exquisite craft have told the urban development of Katmandu, showing the strong cultural atmosphere of Nepal. Traditional Nepal Buddhist garden, delicate wood and stone sculpture, and cultural exhibits from Nepal create a rich culture flavor for the Nepal Garden. Exhibits featuring Asoka and the unique architecture allow the visitors to gain insight of the neighboring state that shares boarder with Tibet and seems at once familiar and strange.

展园介绍：

展园以竹苞松茂的设计、精美的工艺讲述加德满都城市发展，充分展现尼泊尔浓烈的文化风情。传统尼泊尔佛教园林，精细的木雕、石雕与大量来自于尼泊尔国内的文化展品赋予尼泊尔展园浓郁的文化气息，以古尼泊尔阿育王为主题的展品与建筑风格更使游人可以深入的了解这个与西藏接壤、熟悉又陌生、富有神秘色彩的邻邦。

India Garden

印度园

Area: 502 m²

India Garden features temple style of ancient India and the flavor of Buddhism.

展园面积：502 m²

该园设计极具古印度寺庙风格的同时蕴含佛教色彩。

Introduction:

India Garden features temple style of ancient India. The overall layout of the Garden adopts geometric symmetry with two gates on each side and main building sitting in the middle of the backside. Stone seats are placed in the Garden for visitors to have a rest. Arborvitae and hibiscus surrounding the building creates a peaceful atmosphere.

As the birthplace of Buddhism, India is world renowned because of the rich and mysterious culture of Buddhism. There are sculptures of Buddha of varied gestures with spring water spouting out behind. In addition, the Buddha, beast, grass made of bronze and the horse made of camel leather have shown the cultural flavor.

展园介绍：

印度展园依照古印度寺庙风格建造。整个花园布局体现出完美的几何对称性，两座大门分列两侧，主体建筑位于园区正后方。展园内放置有供游客休息的石质座椅，金钟柏和木槿围绕展台四周，营造出静谧的氛围。

作为佛教的发源地，印度以其丰富而又神秘的佛文化闻名于世，展园里就有姿态各异的群佛雕塑，背后则是喷涌而出的泉水。此外还有青铜制作的佛、兽、草、木、骆驼皮制作的骏马等，展现其文化色彩。

India Garden

Bolivia Garden

玻利维亚园

Area: 1,350 m²

The garden has exhibits the American Indian culture. And the Gateway of Sun, a symbol of early Bolivia civilization established 1,000 years ago is greatly highlighted.

展园面积：1 350 m²

该园展示了印第安土著居民文化，其中一千多年前的太阳门文化的展现尤为突出。

Introduction:

The Garden mainly exhibits the culture of the Native American Indians. The first thing in sight when visiting the Garden dedicated to the American Indian culture in Bolivia is the Gateway of Sun, a symbol of early Bolivian civilization established 1,000 years ago. As you walk in the Garden, you will find the Lake Titicaca landscape on your side where stone-tree sculptures of Bolivia saline lake lie around and live shows of Indian costume, dance, and music will be presented. It will bring to the visitors the history, culture, and folkways of Bolivia with liveliness and agility.

展园介绍：

世园会玻利维亚展园主要展示印第安土著居民文化。展园以一千多年前的太阳门文化为起点，太阳门是美洲古代最著名的古迹之一，是印第安蒂亚瓦纳科文化的杰出范例。继太阳门后，展园一直延伸到世界上海拔最高的淡水湖的的喀喀湖景观，并穿插以玻利维亚盐湖景观石树雕塑及印第安传统服饰、舞蹈及音乐现场演出，把玻利维亚历史文化及民风民俗以鲜活灵动的方式展现出来。

安第斯园今日活动
SISAY 乐队
Andes Sound 安第斯音乐
SISAY

Russia Garden

俄罗斯园

Area: 2,288m²

The Russia Garden features serenity of countryside life, illustrating the original appearance of good old farms of Izborsk in Pskov perfectly.

展园面积：2288 m²

该园以传统恬静的乡村风格为主调，完美呈现了普斯科夫洲 Izborsk 小镇风格。

Introduction:

Unfolding in east-west direction, the Russia Garden features serenity of countryside life. The original appearance of good old farms of Izborsk in Pskov is perfectly illustrated by the artfully arranged traditional wooden houses, small wells, and outdoor bathing place.

Russian people were living around the forests in the early times. The traditional architectures were made of wood from the forests. Since the end of the 10th century, after the reception of Christianity, stone architectures have appeared under the influence of Byzantium. However, traditional wooden architectural craft is reserved to be used in the stone architectures. In International Horticultural Exposition 2011 Xi' an, the most representative wooden architectures are brought to the visitors of Xi' an Horticultural Expo as a complete set. The theme of the Garden is the wooden house surrounded by orchard. Houses built with one-piece wood are a traditional building craft in Russia. The major plantation in Russia Garden is orchard, including apple trees and cherry trees. Fruits hang heavily on the trees, lighting the Garden up as scattering stars.

展园介绍：

俄罗斯展园以传统恬静的乡村风格为主调，由东向西贯穿整个花园。在景点的设置上，通过传统原木木屋、小取水井，露天浴场的安排，完美呈现了普斯科夫洲 Izborsk 小镇，古老而悠久的农场原貌。

俄罗斯民族早先居住在森林周围，就地取材搭建居所，传统建筑是以木造为主。自 10 世纪末接受基督教后受了拜占庭影响开始出现石造建筑。然而，俄罗斯人传统的木造建筑技术仍保留下来，混用于石造建筑上。在这次世园会上，俄罗斯人把最能代表民族风格的木质建筑原封不动地带来。园区主题正是被果园环绕的木屋，使用整根木料加工建筑的房屋，在俄罗斯是古老而传统的建筑工艺。俄罗斯展园主要植物为果树，苹果，樱桃，等果实累累挂在枝头，点缀整个花园。

俄罗斯

Nigeria Garden

尼日利亚园

Area: 2,149 m²

With primeval and natural atmosphere, it has shown the beauty of simplicity and uniqueness.

展园面积：2 149 m²

该园以原始的自然气息，呈现出一种简单而独特的美。

Introduction:

The main building of Nigeria Garden is in "L" shape with thatch roof and ambulatories at the gate and the front door which connect with a round granary. Two column-shaped African totems are standing on the two sides of the gate. Complicated pattern and bright colors have shown a kind of primeval and mysterious atmosphere.

Nigeria Garden outlines a "desert-savanna-tropical forest" landscape structure by rendering the different plants in desert, water and tropical forest. The totem sculptures give expression to the folkways and unique landscape of Africa. The well arranged Garden creates special landscape by using various plants, curved path, and unique architecture.

展园介绍：

尼日利亚园主体建筑呈 L 型，茅草屋顶，大门口及后门处建有回廊，旁边连接着一座圆形的谷仓。大门两侧矗立着两座圆柱形的非洲图腾柱，复杂的图案和鲜艳的色彩透出一股原始而神秘的气息。

尼日利亚园分区种植着棕榈树、仙人掌等植物，通过沙生植物——水生植物——热带雨林植物的变化，勾勒出"沙漠景观——稀树草原景观 -- 热带雨林景观"的设计层级，用图腾雕塑体现了非洲的民风民俗和景观特色。植物配置多元，园路自然流畅，建筑风格独特，达到了层次分明，特点突出的景观效果。

Philippines Garden

菲律宾园

Area: 574 m^2

Arranged from east to west, the Philippines Garden features enthusiastic tropical views.

展园面积：574 m^2

该园以热情洋溢的热带风光为主题，由东向西贯穿整个花园。

Introduction:

Arranged from east to west, the Philippines Garden features enthusiastic tropical views. Special thatched sheds are built in the Garden to create an atmosphere of tropical rain forest. The theme of the Garden is the peaceful rural Philippines with highlights such as lanes and lawn. The major plantation of the Garden is tropical plants that can paint your summer experience with the color of ocean.

The handiworks of Philippines are varied with the local color. There are handmade sculptures and icons, thin bobbinet lamps, gold, silver and pearl jewellery, handmade baskets, and household utensils, all of which are boutique. Inside the Garden, pearls of varied colors and colorful shell ornaments are full of tropical flavor and the national style of Philippines.

展园介绍：

菲律宾展园以热情洋溢的热带风光为主题，由东向西贯穿整个花园。在景点的设置上，通过热带风情的茅草凉亭的设置，呈现出热带雨林浓厚的气息。园区主题是菲律宾农家风格，通过小径、草坪的装点，展现菲律宾农家温馨环境。菲律宾展园主要植物为传统的热带植物，为夏季装点一丝海岸的色彩。

菲律宾的手工艺品种类繁多极富地方色彩，有手工塑像、圣像，以及家居用的薄罗纱灯、金银珍珠首饰，手工篮子、器皿等，无一不是精品。在展园内，各色珍珠、彩贝饰品等均充满着热带风情与菲律宾的民族风格。

Thailand Garden

泰国园

Area: 1,012m²

It has integrated the classical elements of traditional Thai architecture, art and lifestyle, giving expression to the Thai garden style.

展园面积：1 012m²

该园融入泰国传统建筑、艺术以及泰式生活方式中的经典元素，展现出浓郁的“泰式”风格。

Introduction:

There are two exhibition halls in Thailand Garden. No.1 exhibition hall is designed in the architectural style of Lanna Thai of north Thailand 700 years ago. No.2 exhibition hall is in architectural style of Ayutthaya Period of 650 years ago. The design of two halls has highlighted Thailand view and Thai style with red and gold as the main colors. It has integrated the classical elements of traditional Thai architecture, art and lifestyle.

Thailand Garden has two classical Thai-style buildings, a great and a small. They look like grand palace in natural landscape. Rich art, culture and modern architecture of Thailand are integrated into the overall design of the exhibition hall by detailed engravings. The special stone steps and flowerbed in the yard give expression to the Thai garden style.

展园介绍：

泰国园共有两个展馆，1 号馆是 700 年前泰国北部兰纳王朝建筑的风格，2 号馆为 650 年前大城王朝风格，两馆的设计均突出了泰国视角和泰式风格，以红色和金色为主色调，融入泰国传统建筑、艺术以及泰式生活方式中的经典元素。

泰国展园中一大一小极具古典风格的泰式建筑就像田园中升起的光辉的宫殿，通过细节雕刻将泰国多元化的艺术文化和现代化的建筑物浓缩在展馆的整体设计中。庭院中配以特色石质台阶和花圃，展现出浓郁的“泰式”庭院景观风格。

North Korea Garden

朝鲜园

Area: 770m²

The Garden is a theme park featuring the long history of North Korea. It has three themes including "country of long history", "country of natural landscape" and "country of freedom and peace".

展园面积：770m²

展园以朝鲜历史悠久的主题公园为主，主要由“历史悠久之国”、“山清水秀之国”和“自由和平之国”构成。

Introduction:

The Garden is a theme park featuring the long history of North Korea. It has three themes including "country of long history", "country of natural landscape" and "country of freedom and peace". The exhibition building with white walls is covered with grey tiles in the "mountain-top style", showing prominent characteristics of North Korean architecture. Inside the building are pictures of North Korea reflecting the combination of history and modernity. Bucket arches, cloud columns, green tiles, and brown doors in the building fully convey the architectural style. The famous Kimjongilia from North Korea represents the nation's rich experience and skill in horticultural field. It also plays an important role in the whole Garden.

展园介绍：

展园特色以朝鲜历史悠久的主题公园为主，主要由3个部分构成，即“历史悠久之国”，“山清水秀之国”以及“自由和平之国”的展示。展馆以朝鲜山顶式的青瓦白墙建筑为主，内部主要展示朝鲜图片反映历史与现代的结合，展馆的斗拱、云柱、绿色的瓦片、棕色的门，都充分展现了朝鲜的传统建筑风格。朝鲜闻名遐迩的金正日花也无处不体现着花卉领域的经验和技术，对整个展园也起到了举足轻重的作用。

Sunchon Garden South Korea
韩国园

Area: 910m²

Aeryeonjeong, a Bower in Suncheon style, is the exhibit of South Korea Garden, showing the flavor of traditional Asian pavilions.

展园面积：910m²

该园主要展示韩国顺天特色亭子爱莲亭，充分展现了亚洲传统亭阁的韵味。

Introduction:

Aeryeonjeong (Aeryeon means "loving the lotus"), a Bower in Suncheon style, is the exhibit of South Korea Garden. Built in 1692, Aeryeonjeong is a masterpiece of traditional garden architecture of South Korean. It is mainly designed in red and cyan with delicate decorative patterns. Two of the four stone pillars are immersing into the lotus pond to be natural and elegant which makes the visitors experience the flavor of traditional Asian pavilions. The bower sits near a pond, giving expressing to the characteristics of traditional gardens in South Korea.

The exhibition Garden is surrounded by a ring of archaistic wall. The archaistic wall applies traditional South Korean skills to be piled up by the grid layers of long cornerstones. The materials of Aeryeonjeong and archaistic wall come from South Korea, and they are installed by the South Korean workers to fully show the features of the country.

展园介绍：

韩国展园主要展示韩国顺天特色亭子爱莲亭，此亭是以韩国传统亭阁昌德宫后苑的亭子为样本，原建于1692年，此亭是传统韩国庭院的代表。其以红、青色为主，绘满精致的花纹，四个基柱石中有两个基柱石浸入莲花池水中，形态自然、简洁，令参观者充分感受到亚洲传统亭阁韵味。亭子旁有一水池无处不体现着韩国传统的庭院特色。

展园周围也由韩国仿古墙所包围，作为韩国顺天园边界的瓦片围墙采用韩国传统技法、由长基石方格层堆积建造而成。爱莲亭和古老墙砖所用的原材料全部来自韩国，并由远道而来的韩国工人现场自行安装，完全体现了韩国的本土特色。

Germany Garden

德国园

Area: 2,680m^2

Special lanes are paved in the Germany Garden representing the history of Oldenburg.

展园面积：2 680m^2

该园内铺装特色道路，代表奥尔登堡历史踪迹。

Introduction:

Special lanes are paved in the Germany Garden representing the history of Oldenburg. Five distinctive sub-exhibition zones in the Garden are designated for exhibits with different features. An invisible flower belt in the center of the Garden grows special flowers from Germany. The beach and water affinity platform near the lake has the artistic conception of sea-shore view.

展园介绍：

德国展园内铺装特色道路，代表奥尔登堡历史踪迹。展园内设有5个小展区，每个展区布置不同的展品，体现不同的特色。中央有一隐形花带，布置不同体现德国特色的花卉。靠近湖面有一沙滩及亲水平台，给人一种亲临海边眺望大海的意境。

STADT OLDENBURG

HORST JANSSEN
UND DIE ROMANTIKER

Japan Hokkaido Garden

日本北海道园

Area: $844m^2$

Hokkaido Garden Japan features natural flowerbeds from south to north to be a sea of flowers.

展园面积：$844m^2$

该园以自然的花朵为主线，由南向北贯穿整个花园，是真正的花的海洋。

Introduction:

Hokkaido Garden Japan features natural flowerbeds from south to north. The Garden is a sea of flowers without complication of buildings, highlighting the beauty of flowers. The interweaving flowerbeds in different colors look like a rainbow carpet. The theme of the Garden is purple lavender, which represents the beautiful and romantic Hokkaido. All the flowers exhibited in Hokkaido Garden get their seeds from Japan, bred in Xi' an, and then transplanted in the Garden.

展园介绍：

日本北海道园以自然的花朵为主线，由南向北贯穿整个花园。北海道花园完全是花的海洋，去掉繁杂的建筑，突显自然花卉的美丽。多种色彩的花朵交织，如同彩虹般绚烂。园区主题是浪漫的薰衣草，紫色醇香的薰衣草，代表着北海道的美丽和浪漫。北海道园所展出的鲜花都是从日本带来的种子，在西安育种，然后再将苗木移至园区的。

Yokohama Garden Japan

日本横滨风园

Area: 804m²

The Garden adopts the design concept of "Harmonious Environment – New in Old" with the promenading mode around pond and fountain originated from the Edo Time.

展园面积：804m²

该园主要采用"和谐的环境 – 新旧融合"的设计理念、江户时代的池泉漫步模式来展示。

Introduction:

The Japanese Yokohama Garden adopts the design concept of "Harmonious Environment – New in Old" with the promenading mode around pond and fountain originated from the Edo Time. The exhibition hall adopts the most advanced Japanese technologies such as spray system, green-roof, solar and wind power, as well as the traditional environmental technology gained through the long history of Japan (reed screen, watering, and wind-bell). The garden-path, fountain, and plants around the exhibition hall freshen the hall and together create an elaborate picture of Japanese Garden.

展园介绍：

日本横滨风园采用"和谐的环境 – 新旧融合"的设计理念，主要采取江户时代的池泉漫步模式。在展厅上，除了采用日本最先进的迅速蒸发喷雾系统、绿化屋顶、太阳能风能等技术外，还融合进凝集日本古代的历史和文化的环境技术（芦苇屏，洒水，风铃）等。展厅周边与花园小路、喷泉、池塘、植栽等相结合，在淡化展厅的同时，使之与日本庭园达到完美的融合。

横滨风

日本
横滨风园

Nara Garden Japan

日本奈良园

Area: 501m²

The Garden has interpreted the leisurely life state of Nara. The theme of the design is to integrate this kind of life state into garden.

展园面积：501m²

该园诠释了奈良当地的一种闲适的生活状态，设计主题正是将生活状态融入园林。

Introduction:

This is a Japanese-style garden, with a dry landscape showing the core content of Japanese gardening and horticulture. There are no buildings in the Garden, interpreting the life state of Nara: life is like a journey, and the scenery along the way and the mood with which you travel are more important than the journey destination. The plants, rocks and their arrangement reflect the Japanese cultural understanding of garden art. The overall design is elegant and harmonious. The periphery of the foundation is the magnificent mainland of China, and the three-stone group represents Xi' an. Under the mainland is the sea. In the sea, there is an island, on which Mount Mikasa of Nara City rests. The characteristic plants are black pine and yew. In the stone lantern showcase beside Nara Garden, there is a precious cultural relic – Epigraphy of Shinsei, envoy to Tang court from Japan, which is discovered in Xi' an in 2004.

展园介绍：

日式园林，枯山水展示了日本园林园艺最精髓的最核心的内容，全展园没有任何建筑物，诠释了奈良当地的一种生活状态"生活就像一场旅行，不在乎目的地，在乎的是沿途的风景和看风景的心情"，其中的植物，石头以及布局等体现了日本文化对园林艺术的理解。整体造型素雅协调，地基的外围是雄伟的中国大陆，中间的立石、添石、控石三尊石组代表西安，下面是海，奈良的三笠山便坐落在海中的岛上。特色绿植有黑松，红豆杉。奈良园旁的一个石灯笼展示柜中，还展有一件弥足珍贵的文物，这就是2004年在西安发现的，日本遣唐使"井真成"的墓志铭。

斯 里 兰 卡
Sri

Turkey Garden

土耳其园

Area: 715m²

Turkey Garden aims at creating a grass-root Turkish garden style, creating a strong flavor of Turkish style, standing opposite the European street at a distance.

展园面积：715m²

该园景观设计旨在营造一种平民化的土耳其园林风格，使整个园区与北部的欧陆风情街形成对景，营造一种浓郁的土耳其式风格。

Introduction:

Turkey Garden aims at creating a grass-root Turkish garden style. The whole landscape area is divided into two connected yards. A European fountain, a wishing tree and a Byzantines pavilion are in the center of these two yards. Artistic landscape creations, paths and traditional Turkish dwellings create a strong flavor of Turkish style, standing opposite the European street at a distance.

展园介绍：

土耳其园的景观设计旨在营造一种平民化的土耳其园林风格。整个景观区域分为两个院子，既独立又相互联系，分别以欧式喷泉，许愿树及拜占庭风格亭子为中心，分配景观小品及流通路线，结合传统土耳其民居建筑，使整个园区与北部的欧陆风情街形成对景，营造一种浓郁的土耳其式风格。

土耳其园
Turkey Garden

长冰淇淋

Bangladesh Garden

孟加拉国园

Area: 1,163m²

Bangladesh Garden is a structured garden with symmetrical layout. With the design concept of simplicity, it has illustrated the theme of low-carbon lifestyle.

展园面积：1 163m²

该园是一座极其规整的对称式分布园林，力求简约的设计理念，表现低碳环保的主题。

Introduction:

Bangladesh Garden is a structured garden with symmetrical layout. With the fountain pool in center as the axis, buildings are arranged on two sides to echo each other. The vegetation in the Garden consists mainly of Ligustrum quihoui and pear tree. The open space in Bangladesh Garden gives not only a panoramic view but also the expression of the design concept of simplicity, which illustrates the theme of low-carbon lifestyle.

The resting area in the rear of the Garden shows human oriented concern. The performance area allows the visitors to appreciate traditional Bengal culture closely. Five giant photos, namely Bengal Tiger, National Memorial of Bengal, and Sunrise on The Sea, are very popular among the visitors.

展园介绍：

孟加拉国是一座极其规整的对称式分布园林，以中间的喷泉水池为轴线、两侧的建筑相互呼应，绿植主要有小叶女贞，梨树等。园区体现出简约、清新的设计理念，开阔的设计令内部景物一览无余。展园力求简约的设计理念，表现低碳环保的主题。

展园中不乏人性化设计，后部设有休息区域。表演区域则使游客可以近距离接触极富传统孟加拉风情的文化活动。5 幅巨型的照片，分别是孟加拉虎、国家纪念馆、海上日出等也受游客的极大关注。

Kalamata Garden Greece

希腊园

Area: 500m²

Greece Garden is also called "Kalamata Garden of Greece", aiming at showing the tourist attraction, culture and architecture of Mediterranean region using modern technology.

展园面积：500m²

希腊园又名"希腊卡拉马塔园"，其旨在用现代科技表现希腊地中海地区的旅游观光、文化和建筑特色。

Introduction:

Greece Garden is also called "Kalamata Garden of Greece". Kalamata is the famous historical city of Greece with a long history of 2,700 years, which is world famous because of the International Dance Festival held in July annually. Kalamata Garden of Greece aims at showing the tourist attraction, culture and architecture of Mediterranean region using modern technology. The Garden is divided into five parts. Each part has a central element, a pictorial banner and a kinescope recording, introducing the overall information of Greece, as well as its tourism, culture, architecture, and cuisine.

In addition, the Garden also applies special metal material, video, audio, and special light effects to present a new image of Greece. Such a system featuring videos and audios creates a feast of Greek culture and information.

展园介绍：

希腊园又名"希腊卡拉马塔园"，卡拉马塔市是希腊的历史名城，建市已有2700年历史，因每年7月举办的国际舞蹈节而享誉世界。希腊园旨在用现代科技表现希腊地中海地区的旅游观光、文化和建筑特色。展园分为五个部分，各有一个中心元素及一个画报横幅与屏幕录像，分别介绍希腊地区的整体印象、旅游、文化、建筑和特色美食。

此外，通过运用特殊金属材质、视频、音频与特效灯光表现希腊现代化的一面，通过这样的视频与音频系统创造一场希腊文化信息的盛宴。

EXPO XI'AN
希腊园
Greece Garden
面　积：500平方米
设计主题：科技与文化的结合
概况介绍：
AREA: 500 m²
DESIGN THEME: Science, Technology and Culture
INTRODUCTION
EXHIBITOR: Municipal Government of Kalamata, Greece

Pakistan Garden

巴基斯坦展园

Area: $747m^2$

Pakistan Garden highlights the horticultural style of the Mughal Dynasty with reference to the Islamic architecture.

展园面积：$747m^2$

该园重在体现莫卧儿王朝的园艺精神，并融合伊斯兰建筑理念。

Introduction:

Pakistan Garden highlights the horticultural style of the Mughal Dynasty with reference to the Islamic architecture.

The main body of the garden is a building with Islamic design features surrounded by four rose flowerbeds. The black onyx fountain at the center of the flowerbeds symbolizes that water is the source of life; the red, yellow, white, and pink roses blossoming in the flowerbeds embodies the fundamental elements essential for a harmonious world.

The main building inside is a "palace" integrating ancient Indian and Pakistani architectural style of 200 years ago. It will display the local specialties and handiworks of Pakistan. The landmark in the center of the garden is made of typical Pakistani stones.

展园介绍：

巴基斯坦展园重在体现莫卧儿王朝的园艺精神，并融合伊斯兰建筑理念。

展园主体是一栋伊斯林寺庙风格的建筑，分室外展和室内展两部分，室外展展出巴基斯坦的国花茉莉花等，园区中央有四座盛开玫瑰的花坛，花坛中心的黑玛瑙喷泉象征水是万物生命之源，花坛中红、黄、白、粉四色玫瑰分别象征了构成和谐世界的不同基本要素。

室内展的主体建筑为融合了 200 多年前古印度与巴基斯坦建筑风格的"王宫"。"王宫"内将展示巴基斯坦的土特产、手工艺品等。立于园区中央的标志建筑则用巴基斯坦特有的石头筑成。

Gabon Garden

加蓬园

Area: 623 m²

The theme of the Garden is "energy-saving, low-carbon, environment-protection and ecological Central and West African style".

展园面积：623 m²

该园主题是节能、低碳、环保与原生态的中西非风情。

Introduction:

The Garden is arranged with an exhibition hall in the rear and a pavilion in the front. The exterior decoration of the cube-shaped ecological building features gridiron pattern. Central Africa style can be seen in the pavement made of gravel and bluestones. The vegetation in the Garden consists mainly of local plants such as Ligustrum quihoui and palm plants. The unique appearance and architectural style is closely integrated with the slogan of International Horticultural Exposition 2011 Xi'an, "Green Leads Fashion", illustrating the theme of "energy saving, low carbon and environment protection". It has shown strong atmosphere of Central and West Africa.

展园介绍：

整个园区后部一展馆，前部一景亭的布局。方格造型的外装修，原生态方形景亭，碎石铺装与青石铺装均体现了中非风格。绿植以本地植物为主，有小叶女贞、棕榈科植物等。展园特色的建筑外观与风格与西安世园会口号"绿色引领时尚"紧密结合，体现节能，低碳，环保的主题，体现了浓郁的中西非风情。

Afghanistan Garden

阿富汗园

Area: 1,251m²

The design of the Garden fully reveals Afghanistan's long history and distinctive national features.

展园面积：1251m²

该园的设计充分表现了阿富汗古老的历史和丰富的民族特色。

Introduction:

Afghanistan Garden is designed with a framework consisting of the Herb Garden, Taxkorgan Garden, Paghman Garden, and Rose Garden. Within the framework, there are classical Central Asian elements such as pavilions, cobbled roads, Istalif pottery, carved wooden doors. The Garden fully reveals Afghanistan's long history and distinctive national features.

展园介绍：

阿富汗展园的设计以草药园、塔什库尔干园、帕格曼园和玫瑰园为骨架，配以凉亭、鹅卵石路、伊斯塔利夫陶器、雕刻木门等中亚典型庭院元素，充分表现了阿富汗古老的历史和丰富的民族特色。

Hong Kong Garden, Macau Garden and Taipei Garden

港澳台北园

Hong Kong Garden, Macau Garden and Taipei Garden have shown the integration, mutual cooperation, common development and progress between the Mainland, Taiwan, Hong Kong and Macau.

We can see that the flavor of Taiwan is strong in Xi' an Horticultural Expo. Taipei Garden, inspired by Taiwan snacks and orchids, has shown the hope for reunion of the people on both sides of the Taiwan Straits. Hong Kong Garden has chosen the representative plants of Xi' an, Hong Kong and South China to imply the integration of China. The landmark of Hong Kong Garden – "Wind Tree", a giant windmill, has become the classic work of environmental protection. They are the symbolic plants in Xi' an Horticultural Expo. Comparing to Taipei Garden and Hong Kong Garden, Macau Garden is designed to be freshing. Yellow garden has shown the special cultural background of Macau in unique way. It has fully integrated the features of Chinese and Portuguese architecture. In addition, the delegation of Macau participating in Xi' an Horticultural Expo hope that there will be more interactions and cooperation between the Mainland, Hong Kong, Macau and Taiwan to make full use of their advantages.

港澳台北园在不经意间透露着两岸四地的相互融合和合作、共同发展与进步。

可以看到，这次西安世园会中的台湾元素比较浓，由台湾小食、台湾盛产的兰花到以圆的发想设计的台北展园，都透露了海峡两岸人们期盼团团圆圆的心情。而香港园选择和搭配西安、香港和华南地区代表性植物正寓意全中国各地紧密融合，园中的主要标志物——风临树也成为整个世园会绿色环保的经典之作，是世园会具象征意义的植物！相对台北园和香港园来说，澳门园的设计相当清爽，黄色的园区以特有的方式展现澳门特殊的文化背景，充分地融合了中国和葡萄牙建筑的特色，此外，参加世园会的澳门代表希望今后两岸四地可以有更多互动，发挥优势，加强合作。

Hong Kong Garden

香港园

The theme "Hong · Kong · culture · charm" shows the unique natural environment of Hong Kong, and the concept of the harmony and virtue between man and nature.

该园主题是"香·港·风·情"，展述了香港特有的自然环境和天人合德的概念。

Introduction:

The theme "Hong · Kong · culture · charm" shows the unique natural environment of Hong Kong. The elements in the theme includes Wind (wind-energy tree), Mountain (geological park), Water (circulating water), Joy (touching the joyful mind), together with dancing butterflies, beautiful songs of birds, and the green belt along the third-ring road (local plants in Xi' an, plants in East China, and special plants in Hong Kong). The animals, plants, wind, mountain, water and other natural elements arouse the joyful feeling in our inner world. They also illustrate the concept of the unity of city, human and great nature (harmony and virtue between man and nature).

展园介绍：

主题"香·港·风·情"，展述香港特有的自然环境，以风（风能树）、山（地质公园）、水（循环不息之水）、喜（触动喜乐之心灵）配上蝴蝶起舞，雀鸟悦耳歌唱声及三环绿化带（西安本土植物、华东植物、香港本土特色植物）为主题，以动植物、风、山、水等自然元素牵引在人类内在世界心灵之喜乐，透过现代环保建筑与自然园林环境的配合，引领城市、人类与大自然合而为一（天人合德）概念。

香港蝴蝶
HEXAGONAL
VOLCANIC ROCKS

香港园主题
香·港·风·情

洋紫荊

Macau Garden

澳门园

With the theme of "Macau – an oriental city in South European style", the Garden has shown Macao's uniqueness as the meeting point of western and Chinese culture, where the two ran into each other for the first time and finally merged.

该园以“澳门——具有南欧风情的东方小城”作为主题，表示了澳门以深厚温婉的中华文化为底蕴，与西方的文化初识邂逅、相遇相知。

Introduction:

The theme of the Garden is "Macau – an oriental city in South European style". The front part of the Garden has a western style cloister and a mini square. The cloister has contracted decorative ribs and vases of classic orders. The black-and-white waved stone pave in Portuguese style in the mini square tells the story of the western culture coming from the sea 4 centuries ago. The building in South European style in the center of the Garden has colorful external walls, introducing the passion and joyful culture of South Europe. When looking at the Chinese cloister in the rear from the balcony of the building, one can see the red lacquer column, fan shaped window frames on the plant-wall, and the lattices with fret pattern forming a Chinese-style garden in implicit style. Such an arrangement also shows Macao's uniqueness as the meeting point of western and Chinese culture, where the two ran into each other for the first time and finally merged.

展园介绍：

该园的主题是“澳门——具有南欧风情的东方小城”。园区前景设有西式回廊及小广场，回廊上有简约西方古典柱式的装饰线及花瓶，小广场上的黑白波浪形葡国石铺面显示西方文化400多年前带着大海的气息踏浪而来。配合园区中央带有南欧风情的西式建筑，建筑物外墙强烈的色彩表现南欧热情开朗的风貌。由建筑物阳台眺望后方的中式回廊，以红漆圆柱，植栽墙上的扇形窗框，回纹窗棂，构成中华文化含蓄优雅的园景特色，亦表示澳门以深厚温婉的中华文化为底蕴，与西方的文化初识邂逅、相遇相知，建立独一无二的中西文化交汇点。

Taipei Garden

台北园

Taipei Garden is designed with the theme of "Flora Exhibition relay and the common development on both sides of the Taiwan Straits". The circles symbolize the completion of the successive Flora Exhibition in Taipei and the Horticultural Exhibition in Xi' an.

该园以"花博接力两岸共发展"为主题，以圆的发想设计代表着两岸花卉与园艺博览会的圆满与接力。

Introduction:

Taipei Garden is designed with the theme of "Flora Exhibition relay and the common development on both sides of the Taiwan Straits". There is an artificial hill with water flowing between and fragrant orchids blooming on it. The towering 101 building in Taipei Garden and the common ginkgoes in Xi' an form two intersecting circles in the heart-shaped lack in the center. The circles symbolize the solidarity, completion and one-mind, showing good will to the completion of the successive Flora Exhibition in Taipei and the Horticultural Exhibition in Xi' an. Such a design introduces the beauty of Taipei and Xi' an to the whole world. A cloister cuts through the Garden with a formative landscape made of three-dimensional glass fiber in the middle, symbolizing the snow mountain ridges of Taipei city.

展园介绍：

台北园以"花博接力两岸共发展"为主题，园内设有一座假山，山间有流水潺潺、兰香幽幽。台北的101大楼的耸立意象与西安市处处可见的茂密银杏，以两圆串联位于中央的心型湖泊；以圆的发想设计代表团结、圆满与同心，代表着两岸花卉与园艺博览会的圆满与接力，将台北与西安之美展现于全球。园中穿越一座回廊隧道，中心以立体玻璃纤维塑作造型山水，代表台北地形的雪山山脉。

Domestic Garden

内地园

The scales of the exhibitions in Domestic Garden are in various ranges, as well as the elaboration. Beijing Garden is a Yan pedestal good for enjoying a distant view, adopting the earliest artistic base; Yangzhou Garden has applied the featured constructions of Yangzhou to present the neighborhood landscape with bridge, running water and houses, absorbing the marrow of Chinese traditional garden; Shanghai Garden is the fanciest one, bringing the large area of fog by the fog effect technology to strengthen the original figures of the mountain and forest and outline the feature of southern Chinese riverside town.

Besides, the other exhibitions such as the Decorated Archway of Shanxi Garden, the Confucian Culture of Shandong Garden, the Qin–Tang Charm has shown the beautiful landscape and special culture of all over China.

国内展园的规模有大有小，精细程度也不一样。北京园是一个燕墩，这采取了园林最早的艺术形式台，是登高望远佳地；而扬州园则是一番小家碧玉，除了将扬州的特色建筑搬到这里，有种小桥流水人家的感觉，深得中国传统园林精髓；上海园最梦幻，它采用雾效处理技术，能够带来大面积雾效的独特感受，能很好地渲染山、林的原生态面貌，突显上海的江南水乡特色。

除此之外，山西园的大牌楼、山东园的儒家文化、陕西园的秦风唐韵……中国各地的秀丽风景和特色文化在谈笑之间能够尽收眼底。

Beijing Garden

北京园

Area: 1,598 m²

Beijing Garden has shown one stroke of Beijing gardens by integrating the environment with the culture.

展园面积：1 598 m²

该园采取开放的造园手法，使内外景观相互交融，展现了北京园林的一个侧面。

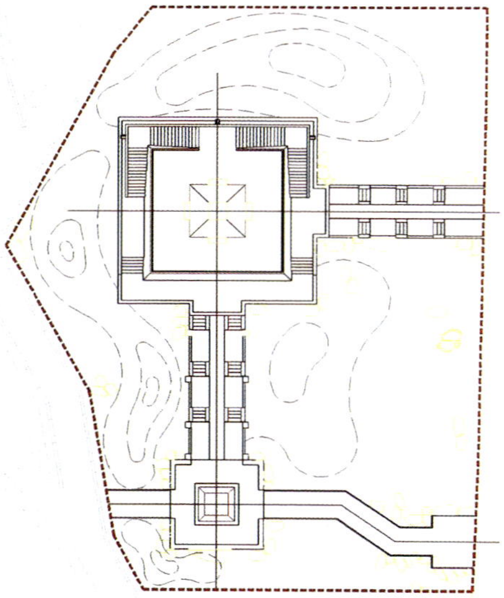

Introduction:

In combination with its long history and culture, Beijing Garden presents its geographic advantage of "facing Yellow River and Ji River in the south, Yan Mountain in the north, Taihang Mountain in the west, and Bohai Sea in the east". The idea of the layout scheme for the garden, "Yandun stele", is borrowed from the Yandun stele in Beijing, which was made upon the order of Qianlong Emperor. The image of stand —— a form of the earliest construction in Chinese garden, is used as the main elements of the garden. A stand is built near to the hill, which helps to raise the view point of sightseeing. Reaching the stand and looking around, one can command a panoramic view of the Garden, which will give full play to the advantage of the position surrounded by the whole garden; the layout of other elements of the garden is carried out with the stand as the main part, including paving the paths, planting the flowers and mounding the soil and allocating the stones. The fence-free layout may better integrate the garden into the surrounding, which presents the broad and tolerant mind of the capital of China.

The garden is designed in open style to combine the internal and external landscape together. Beijing Garden has shown one stroke of Beijing gardens by integrating the environment with the culture. It not only creates an image of garden but also reminds people of more things. People will find much in little, and complexity in simplicity.

展园介绍：

结合北京悠久的历史文化，展现“南襟河济，北枕燕山，西依太行，东临沧海”的地理优势。参照北京燕墩及其上面的乾隆御制碑为主题的“燕墩”设计方案。园中主体要素采用中国园林最早的建筑形式——台的形象，抬高游览视点，依山丘建筑。登台四望，全园美景尽收眼底，充分发挥地处佳园环簇的优势；园中以台为构图重心，展开其他园林要素的设计，铺路植花、堆土置石。北京园不设围墙，以便更好地融入周边大环境，以体现博大宽容的首都情怀。

全园采取开放的造园手法，使内外景观相互交融。北京园展现了北京园林的一个侧面，融入现场环境与文脉，不仅创造出一个园林形象，还使人产生像外之想，达到小中见大、寓繁于简、回味无穷的效果。

Shanghai Garden

上海园

Area: 1,126 m^2

Shanghai Garden has presented the new Shanghai of harmonious development between human and nature.

展园面积：1 126 m^2

该园展现了人与自然和谐发展的新上海。

Introduction:

Shanghai Garden has presented the new Shanghai of harmonious development between human and nature. Originated from the earliest county "Shenjiang", the main part "Shenyuan" is created for Shanghai Garden. The unique conical sunshine valley is set on the garden like a big umbrella. The sunken garden over water has made the natural and ecological wetland and given the satisfactory water experience through the water system, path system over water and aquatic plants system to realize the goal of returning to nature and purifying life. In the garden, the fog effect technique is used to bring the special feeling of large area, embellishing the original appearance of the mountain and forest and creating the healthy forest oxygen. The characteristic of southern Chinese riverside town has been finely displayed.

展园介绍：

上海园展现了人与自然和谐发展的新上海。上海市结合上海地区最早的县治"申江"，营建出主体为"申源"的上海展园。独具特色圆锥形阳光谷，如一顶大伞架设在上海园上。下沉式水上花园打造自然生态仿生湿地，反璞归真，结合水系统，水上步道系统，水生植物系统营造亲水体验，满足演绎和休憩。实现反璞归真，生命净化的目标。在展园区采用雾效处理技术，能够带来大面积无效的独特感受，能很好地渲染山、林的原生态面貌，同时营造出健康的森林氧吧，体现上海江南水乡特色。

Xiamen Garden

厦门园

Area: 1,100 m^2

The design concept of the Garden features the solemnity of traditional stone culture and openness of ocean culture, which spices the traditional setting with the beauty of simplicity of our time.

展园面积：1100 m^2

该园以内敛沉稳的厦门传统石文化和开放包容的海洋文化为出发点，为古典空间注入凝练唯美的现代情韵。

Introduction:

The design concept of the Garden features the solemnity of traditional stone culture and openness of ocean culture. It speaks the language of contemporary design, which spices the traditional setting with the beauty of simplicity of our time. Architectural style is traditional redbrick building in the south of Fujian Province. Sight spots are Stone-based Park with Water Scene, Hidden Bamboo Park, Spring Park of Blossom, and Fragrant Park of Archaic Chinese Rhymes.

展园介绍：

以内敛沉稳的厦门传统石文化和开放包容的海洋文化为出发点，融入现代设计语言，为古典空间注入凝练唯美的现代情韵。建筑风格是闽南传统红砖建筑，主要景点介绍有坐石临流所、梧竹幽居园、花海融春处、古韵流芳苑四个景点。

厦门园

Shanxi Garden

山西园

Area: 917 m^2

Shanxi Garden refines Shanxi' s history and culture, takes the main landscape in Royal Prime Minister' s Palace of Jincheng, Shanxi Province as matter.

展园面积：917m^2

该园提炼山西的历史文化和人文文化，以山西晋城皇城相府为主要景观和素材。

Introduction:

Shanxi Garden refines Shanxi' s history and culture, creates a exhibition park with pleasant atmosphere in the aspects of landscape environment and service facility. The entire distribution adopts the regular Chinese classical garden style, while the construction adopts Qing Dynasty architectural style of axial arrangement, to increase the depth and sequence of architectural complex. Small garden ornaments, stones, plants dotted naturally to highlight the relationship between axis and space, interior characteristic landscape in each scenic spot and the overall landscape.

Exhibition area displays Yushu Tower, Memorial Archway in honor

of merits and virtues, Character-carving, Land inscriptions and an "imperial edict" sculpture that located in Royal Prime Minister's Palace of Jincheng, Shanxi Province. Yushu Tower is originally located in Royal Prime Minister's Palace of Jincheng, Shanxi Province, boasting an inscribed tablet of "Wuting Village" and a couplet of "Arbors grow more lush in Spring while late-booming golden asters send out more aromatic fragrance in Autumn" written by Emperor Kangxi. Memorial Archway in honor of merits and virtues is a landmark archway in Royal Prime Minister's Palace. It is a four-pillar and three-story archway with stone carvings. The patterns of auspicious animals are carved around the base while the patterns of dragon and phoenix are carved at the top of the archway. Moreover, four Chinese characters of "Zhong (冢), Zai (宰), Zong (总), Xian (宪)" ("冢宰" refers to prime minister in feudal China and "总宪" refers to left censor-in-chief in Censorate) are engraved right above the archway. Character-carving: Six steles are inscribed with "Lian (廉 honest and clean)", "Jian (俭 thrifty)", "Zhong (忠 loyal)", "Qin (勤 hardworking)", "Xue (学 learning)" and "Xiao (孝 filial piety)" respectively, summarizing the noble morality of Chen Tingjing all his life long, the host of Royal Prime Minister's Palace. Land inscriptions: A couplet of "a benevolent family cultivated nine descendants with Jinshi degree (Jinshi imperial examination) and six out of three generations honorably became members of the Hanlin Academy" is engraved on the land. An "imperial edict" sculpture: In front of Yushu Tower is an imperial edict-shaped sculpture, on which "Royal Prime Minister's Palace" is engraved.

展园介绍：

山西园是提炼山西的历史文化和人文文化，从景观环境和服务设施方面创造舒适的宜人环境的展园。总体布局采用规则式中国古典园林风格，建筑采用中轴布置的清代建筑风格，以增加建筑群的纵深感和序列感，景观小品、置石、植物等都采用自然式景观布局，重点突出轴线和空间、各景点的内在特色景观和总体景观的的呼应关系。

展园以山西晋城皇城相府的主要景观为素材，展示了皇城相府的御书楼、功德牌坊、字刻、地雕、诏书雕塑等。御书楼源于山西晋城皇城相府，有康熙皇帝御赐的"午亭山村"石刻匾额及对联"春归乔木浓荫茂，秋到黄花晚节香"；功德牌坊则是皇城相府的标志性牌坊，四柱三楼，通体石雕而成，基周瑞兽相拥，坊额雕龙刻凤。正上方刻有"冢宰总宪"四个大字；字刻中六块刻有"廉""俭""忠""勤""学""孝"的石雕，概括了皇城相府的主人陈廷敬一生的高尚品德；地雕也雕刻着"德积一门九进士，恩荣三世六翰林"，御书楼前还放置一座造型为诏书的雕塑，上面刻有"皇城相府"。

Guangxi Garden

广西园

Area: 1,011 m^2

Stating the ecological concept of harmonious development in Guangxi, the expo has presented the green landscape and simple customs of local people.

展园面积：1 011 m^2

该园阐述广西人与自然和谐发展的生态理念，展现绿意盎然的山水风景以及纯朴的民俗风情。

Introduction:

The memorial arch at the gate of "Bagui Harmonious Court" is originated from Guangxi Dragon Boat, on the top of which crouch several stone frogs. Beside the arch, there is a huge stone of 1.5m high and 1.2m wide, on which carved three Chinese characters: 广西园 (Guangxi Garden). The design elements of Guangxi Garden are Hongshui River, Wooden House and Duet Pavilion, creating a natural landscape of Zhuang Ethnic.

Stating the ecological concept of harmonious development in Guangxi, the expo's theme of harmonious co-existing of city and nature" is outlined and the green landscape and simple customs are presented.

There is a cobbled road with pavements of Zhuang Ethnic pictures and a wooden long corridor. In the center of Guangxi Garden, there is a 200 m^2 artificial river expressing the power and culture of Hongshui River. In order to widely show the water culture of Guangxi, a 1.5m drop is set to lift up the waterside pavilion and give prominence to the high mountain and flowing water.

The major sight of Guangxi Garden – Zhuangxiang House is totally made of wood with the ethnic feature of long ridge and short eaves.

The plants in the garden are cultivated in sheet with over 60 species. They are arranged in random order and clear layers, forming the classical gardening style together with the mountain stones and flowing water. For example, a small landscape beside water is set with exquisite plants. The bottom layer is green grass, on which the boxwood, purple leaf barberry and Japan Euonymus are planted. These three little plants are of the color red, yellow, green, which livens up the space. Beyond the bottom layer, there are spherical plants such as purpus privet, schefflera octophylla and French holly. The higher pomegranate, red maple and osmanthus flowers are the main roles of the landscape with their own color, appearance and scent. The plants beside water are carefully selected by the designer; the leaves of philodendron, jasminum nudiflorum and dwarf lilyturf tuber are droping like slender white hands touching the water.

展园介绍：

“八桂谐苑”的大门牌坊以广西龙舟为原型，牌坊顶上蹲着几只石蛙。牌坊旁一块高约1.5m、宽1.2 m的巨石上镌刻着“广西园”三个大字。广西园是以广西母亲河——红水河、壮族特色木楼、对歌亭等为设计要素，营造出一个自然的壮乡小景。

阐述广西人与自然和谐发展的生态理念，突出“城市与自然和谐共生”的世园会主题，展现广西绿意盎然的山水风景以及纯朴的民俗风情。

园内有铺成壮锦图案的石子路、一条木质的长廊。广西园的中央是一条约200m² 的人工河，诠释着红水河是广西的能源之河，文化之河。为了在园内尽情展现广西的水文化，特别设置了1.5m的落差，抬高水榭，突出高山流水的意境。

广西园内的主景——壮乡楼是全木建构的，长脊短檐式的屋顶透露出浓郁的民族特色。

园区都是成片地种植同种植物与众不同，植物种类达60多种，高低错落有致，层次分明，配上山石流水，颇具“聚拳石为山，环斗水为池”的古典园林风范。以水边的一个小景为例，植物配置就很有讲究。最底层是绿草，草上种着豆瓣黄杨、紫叶小檗和大叶黄杨。这3种植物个头都不高，却有着红、黄、绿三色，一下子点活了空间。再往上一层是球形植物，有小叶女贞、鸭脚木、法国冬青。更高一些的石榴、红枫、桂花等乔木是小景的主角，颜色、形态、香味各有所长。水边的植物也是设计者精心挑选的，有春羽、迎春、麦冬等，它们的叶子都是向下垂的，仿佛青葱玉手在撩拨水面。

Hubei Garden

湖北园

Area: 1,091 m^2

With "water" as the subject, it is to create a modern garden full of varied garden landscapes that reflect a green philosophy.

展园面积：1091 m^2

该园以"水"为题材，营造一个园林景观多变、以植物造景为主、体现绿色时尚的现代园林。

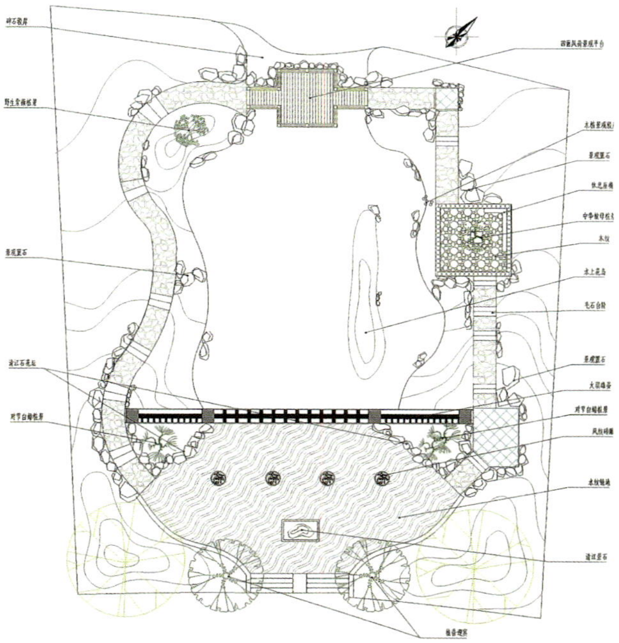

Introduction:

Located in the transition zone of North and South China, Hubei is one of the provinces having the richest resources and species of plants in China. With the promotion of "Three Gorges Dam Project" as the theme and "water" as the subject, Hubei Garden is committed to presenting the plant characteristics of Hubei and to creating a modern garden full of varied garden landscapes that reflect a green philosophy by focusing on the plant landscaping.

展园介绍：

湖北地处南北过渡带，是中国植物资源和植物种类最丰富的省区之一。湖北展园以弘扬三峡大坝工程为主线，以"水"为题材，力求展示湖北的植物特色，营造一个园林景观多变、以植物造景为主、体现绿色时尚的现代园林。

Hebei Garden

河北园

Area: 1 ,003 m^2

The garden builds the classical courtyard as the "Lotus Academy" representing Hebei Hebei culture.

展园面积：1 003 m^2

该园以"莲池书院"来塑造代表河北文化精髓的古典庭院。

Introduction:

Hebei Garden is a classical courtyard presenting the culture essence of Hebei. Originated from the Baoding Ancient Lotus Pond and built as the "Lotus Academy" representing Hebei culture, the courtyard contains Painting Collection Studio, Linyi Pavilion, the Gallery, the Lotus Pond and the Artificial Hill. The garden has fully expressed the original charm of the royal palace and garden and integrated the garden styles of southern and northern part of China. By the artistic expressions of painting, sculpture and penmanship, the time features of gardens in Yuan, Ming, Qing dynasties are displayed vividly.

展园介绍：

河北展园展现了河北文化精髓的古典庭院。其采用历史文化名城保定的保定古莲花池为原型，以"莲池书院"来塑造代表河北文化精髓的古典庭院，园内景点主要分布为：藏画斋、临猗亭、长廊、莲池、假山等。展园既充分表现出其原有的皇家行宫园林的风韵，又融合了我国南北园林建园艺术的风格与精髓，展园的建筑、山石、水体、道路及亭台构成园林的骨架，运用绘画、雕刻、书法为艺术表现形式，形象地表现元、明、清各时期园林的时代特征。

雪香书院

Xinjiang Garden

新疆园

Area: 977 m^2

In accordance with the theme, it creates a landscape garden with the characteristic of horticulture and the regional cultural connotation of Xinjiang .

展园面积：977 m^2

该园围绕主题，塑造出具有新疆园艺特色和地域文化内涵的景观展园。

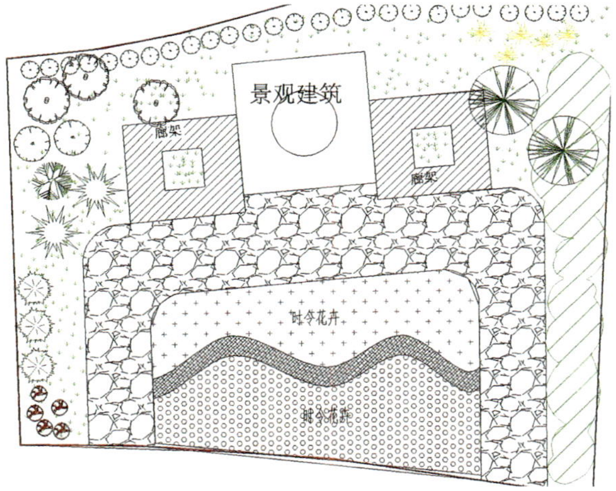

Introduction:

The design concept of Harmonious Xinjiang Garden focuses on the theme of the Exposition, "eternal peace & harmony between nature & mankind, nurturing the future earth" . It emphasizes the idea of people-oriented, cultural inheritance, innovative development and harmonious nature and follows the principal of "adjusting measures to local conditions" to build the Garden. Therefore, a landscape garden with the characteristic of horticulture and the regional cultural connotation of Xinjiang is created. Architectural features: Theme constructions of local features are built in the Garden with the ambulatory decorated around; besides, the Garden is embellished with the typical species, such as silicified wood and stones created by the wind erosion. These have all reflected the regional features of Xinjiang and multi-ethnic fusion. Characteristic plants: The characteristic plants are populus diversifolia, picea schrenkiana, ulmus densa litw, tamarix, etc.

展园介绍：

围绕"天人长安，创意自然"的主题，强调以人为本、传承文化、创新发展、和谐自然的思想，遵循因地制宜建园的原则，塑造出具有新疆园艺特色和地域文化内涵的景观展园。建筑特色是采用地方特色主题建筑，周围布置回廊，点缀硅化木、风砺石等特有物种，体现新疆地域特色和多民族融合的内容。特色植物主要以胡杨、雪岭云杉、圆冠榆、红柳等为主。

和谐新疆园

Heilongjiang Garden

黑龙江园

Area: 2,594 m^2

The theme of Heilongjiang Garden is "The Snowy Resort, The Ecological River". Natural and realistic gardening skills are applied in this Garden.

展园面积：2 594 m^2

该园设计主题为"冰雪盛地・生态龙江"，结合了自然与写实的现代造园手法进行设计。

Introduction:

The theme of Heilongjiang Garden is "The Snowy Resort, The Ecological River". The overall design adopts natural and realistic gardening skill to present the special landscape of four seasons in Helongjiang Province through the forest, snow and culture. Presented by the modern realistic style, the functions are divided into three sections: main construction section, ecological landscape section and snow culture section. The rendering is developed by the concept of one center and two main veins as well as the radiative lines. One center stands for the main construction; two veins means: green vein – forest landscape, blue vein – streams; radiative lines – the Heilongjiang featured cultural elements like snow landscape and other special small landscapes. Through the constructions with Dongbei local characteristics, the miniature of primitive forest and the peaceful humanity concept, a natural and ecological garden of north territory is created.

The main construction is named "The Crown of Golden Rooster". The color green symbolizes the forest; the charcoal wall stands for low carbon and environmental protection; the circle exhibition hall represents unity. In the garden, the smooth wooden lanes travel through the forest and form small spaces underneath the forest, creating a tranquil and pleasant space in combination with the tall and green trees.

展园介绍：

黑龙江展园设计主题为“冰雪盛地·生态龙江”，总体设计采用自然与写实相结合的现代造园手法，通过森林、冰雪、文化等元素将黑龙江省独特的四季风光展现给参观者。展园在景观风格上以现代写实来体现，在功能区域上分为三大区域：主体建筑区、生态景观区、冰雪文化区。设计上以一个中心两条主脉和辐射流线的理念来构图。一个中心是主体建筑，两个主脉分别为：绿脉——森林景观；蓝脉——溪流；辐射流线——冰雪景观及特色小品等黑龙江特色文化元素。通过东北地域特色的建筑、原始森林缩影和祥和的人文理念共同营造一个具有北疆冰雪、自然生态的黑龙江展园。

主体建筑命名为“金鸡之冠”。官与冠谐音，绿色则象征大森林；炭色墙体象征低碳、环保；圆形展厅象征着团结。展园中，自然流畅的木栈道穿梭在林内，在林下形成局部小场地，与高大苍翠的大树相结合，营造一个林下幽静惬意的空间。

Yangzhou Garden

扬州园

Area: 1,049m²

Focusing on the theme of "Green Spring of Poplar", the classical gardening elements of Yangzhou is displayed through the gardening skills of handwriting and painting.

展园面积：1049m²

该园以"绿杨春"为主题，通过书、画等形式的造园手法，展现扬州古典园林元素。

Introduction:

Focusing on the theme of "Green Spring of Poplar", the classical gardening elements of Yangzhou is displayed through the gardening skills of handwriting and painting. The entrance of the garden is a combination of traditional enclosure wall and half pavilion. Juxian Pavilion has fully utilized the terrain to form the layered fountain and flower stream. The Tu'er Pavilion on the north of the garden is cubic and grand. The overall style of the garden has carried on and develops the Chinese traditional gardening marrow – landscape theme; the garden has adopted the natural hill arrangement skill – "digging pond and piling hill" "extending hill and settling water". The unique charm of Yangzhou gardens are displayed by integrating the long historical culture and exerting the gardening advantages of Yangzhou.

Combining the landscape space and the height difference of the whole terrain, the density of the plants is high outside and low inside. Sheet planting, group planting and point planting are made use of. The trees are mainly local and meanwhile the Qiong Flower (municipal flower of Yangzhou), Chinese herbaceous peony, willow (municipal tree of Yangzhou), ginkgo and a large amount of aquatic

plants form the natural and ecological groups with regular order and proper density. The Tu'er Pavilion is exquisitely designed as a cross connected pavilion with woodblock. On the screen wall in the center of the pavilion, a window is wide open. The Yang style bonsai of five-leaved pine setting in the window has livened up the whole exhibition.

展园介绍：

扬州展园以"绿杨春"为主题，通过书、画等形式的造园手法，展现扬州古典园林元素。

园区入口以传统古典围墙与歇山半亭作为组合。"聚仙亭"充分利用地势形成"叠泉""花溪"，园北的兔耳亭，端正大方。园内整体造型，继承和发扬中国传统园林精髓——"山水"主题，展园采用自然式丘壑布局手法——"挖池堆山""延山理水"，融入扬州悠久历史文化，发挥扬州造园特长，展现了扬州园林的独特魅力。

扬州展园结合了景点的空间以及整个地形的高差，植物的栽植外部紧密而内部疏朗，片栽、组群、点植相结合，以地方树种为骨干，同时栽种扬州市花琼花、芍药，市树柳树、银杏，与大量的水生植物搭配成景，形成错落有致、疏密相间的自然生态小群落。展园内的兔耳亭造型别致，是一个交叉连接的厅式小亭，这里有雕版展示。亭中部的屏墙上，大大地开着个窗口，窗口中将放置扬派盆景五针松，起到了点睛作用，让整个展园活了起来。

Shandong Garden

山东园

Area: 1,150m^2

Shandong Garden has fully presented Shandong traditional culture and gardening feature with the main clue of Qi-Lu culture.

展园面积：1150m^2

该园以齐鲁文化为主线，充分体现山东传统文化、园林园艺特色。

Introduction:

Guided by the expo theme of "Eternal peace & harmony between nature & mankind, nurturing the future earth–a city for nature, co-existing in peace" and majored by "Harmonious Shandong", Shandong Garden has fully presented Shandong traditional culture and gardening feature with the main clue of Qi-Lu culture. Using the natural gardening, the mountain (Mount Tai), water (Baotu Spring), human (Confucius and Lunyu) and plants are organized to form the magnificent Qi-Lu Garden with rich culture, bright feature, unique means and exquisite sculpture.

展园介绍：

以本届世园会"天人长安·创意自然"的理念为指导，以"和谐山东"为主题，以齐鲁文化为主线，充分体现山东传统文化、园林园艺特色，运用自然式园林手法将"山"——泰山、"水"——天下第一泉(趵突泉)、"人"——孔子与《论语》、植物等有机结合，形成气势恢弘、文化丰富、特色鲜明、手法独特、精细雕刻于一体的齐鲁园。

觀瀾亭
趵突泉

Shenyang Garden

沈阳园

Area: 932 m^2

It tells the story of vicissitudes that the city has experienced through huge changes and reforms, highlighting the theme of "Shenyang in the flux of time".

展园面积：932 m^2

该园表现出沈阳脱蛹化蝶般的沧桑巨变，体现出"沈阳流年"这一设计主题。

Introduction:

The trilogies of "Shenyang History – Shenyang Flavor – Shenyang Impression" tell the story of vicissitudes that the city has experienced through huge changes and reforms, highlighting the theme of "Shenyang in the flux of time". Shenyang History: the history and long-existing culture of Shenyang is recorded by landscape arrangement featuring rich cultural content and splendid past of the city. Shenyang Flavor: The Hun River, the river flowing through the industrial city and nurturing the rich folk culture of the citizens, is under the spotlight here. It also contains special elements such as Fengtian Drum, popular rural folk dance and song-and-dance duet in Northeast China. Shenyang Impression: the Shenyang Impression gives full illustration of people's impression over Shenyang, the modern metropolis of China today.

展园介绍：

展园以“沈阳历史——沈阳味道——沈阳印象”为三部曲，表现出沈阳脱蛹化蝶般的沧桑巨变，体现出“沈阳流年”这一设计主题。以园林景观设计布置的手法记载沈阳这座城市古老的历史文明，苍劲的文化脉络，辉煌的城市印迹。重点表现穿越这座工业城市的母亲河——浑河孕育了沈阳儿女丰富多彩的民间文化：奉天大鼓、东北大秧歌、东北二人转等，充分描绘出沈阳今天这个现代化的大都市所给人们带来的城市印象。

Chongqing Garden

重庆园

Area: 1,150m^2

The overall planning of the garden features "Phoenix" and Chaotianmen Port, the landmark of Chongqing, showing the new image of the city.

展园面积：1 150m^2

该园以"火凤凰"和"朝天门码头"为整体构图形式，展现重庆新时代城市形象。

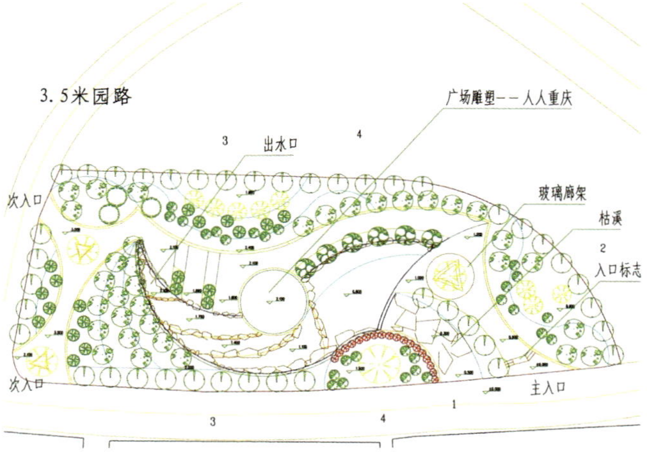

Introduction:

The garden is designed with the theme of "People's Chongqing". The overall planning of the garden features "Phoenix" and Chaotianmen Port, the landmark of Chongqing, showing the unique local culture and new image of the city. The harmonious atmosphere of the city is highlighted in the exhibition. Artistic creations such as waterfall —— running water, port square, spring —— jumping water, floating bridge, and flower mirror are presented to imitate the power of flowing rivers and to showcase Chongqing, the city sitting on the cross point of two major rivers, and the unique landscape of the mountain city. "Phoenix" is formed by the artistic curved lines.

展园介绍：

展园以"人人重庆"为设计主题，以"火凤凰"和重庆地标"朝天门码头"为整体构图形式，展现重庆独具特色的地域文化和重庆新时代城市形象，宣传重庆和谐共生的城市品味。展园内部主要结合跌水——奔流之水，码头广场，涌泉——跃动之水，浮桥，花镜等小品，表现江河奔流的宏大气势，展现重庆两江交汇及"山城"起伏跌宕的地域特色。而"火凤凰"则由园内艺术曲线勾勒出大形。

Guangzhou Garden

广州园

Area: 1,257 m²

With the theme of "silk road – integration and dialogue", it has presented the tourist routes of overland Silk Road and maritime Silk Road.

展园面积：1 257 m²

该园以"丝路－融合与对话"为主题，展现了陆上丝绸之路和海上丝绸之路的游线。

Introduction:

The theme of Guangzhou Garden is "silk road – integration and dialogue". The tourist routes are paved with a winding silk road, which indicates the overland Silk Road; and the pavement of Yunshan Mountain, winding stream, drop water of oyster shell wall, ecological pond and dry stream is also accomplished at a stretch. This presents a concentrated changing process of the water system from the beginning to the end, which symbolizes the maritime Silk Road. The Garden also includes such layout as "calm sea" memorial archway, sailing boat, wharf, solid green wall, steel rim of blue and white porcelain, etc. With the northern plants as its framework, the Garden is decorated with the characteristic plants of Lingnan. The plants in the Garden include pinus bungeana, frangipani, phoenix sylvestris, butia capitata, multi-shoot cycas recoluta, bamboo palm, alpinia sanderae hort, birds of paradise, spathiphyllum wallsii, Ixora chinensis, Boston fern, etc.

展园介绍：

展园以"丝路－融合与对话"为主题。以蜿蜒的丝路铺装为游线，寓意陆上丝绸之路，而云山一脉、蜿蜒小溪、蚝壳墙跌水、生态水池、旱溪铺装更是一气呵成，浓缩了从起点到终点的水系变化，象征着海上丝绸之路。全园包含"海不扬波"牌坊、启航帆船、埠头、立体绿化墙、青花瓷钢圈等情景空间。以北方植物为骨架，结合岭南特色植物的配置风格。白皮松、鸡蛋花、银海藻、布迪椰子、多头苏铁、棕竹、花叶良姜、天堂鸟、白掌、龙船花、波士顿厥等特色植物。

广东广州园
海不揚波
花城丝路

Shenzhen Garden

深圳园

Area: 1,516m^2

With the design theme of "innovative Shenzhen, gallery of paper folding art, Shenzhen story in image", Shenzhen city is compared to a piece of paper that is to be folded.

展园面积：1516m^2

该园设计主题是"创意深圳，折纸廊，情影深圳事"，通过一张小小的折纸来影现深圳的人和事。

Introduction:

The theme of the design: Innovative Shenzhen, gallery of paper folding art, Shenzhen story in image. Shenzhen city is compared to a piece of paper that is to be folded. Cities differ from each other as they have unique ways of folding themselves. Shenzhen, once a small fishing village, has become an international metropolis with ecological concern and strong design power in its 30 years' folding process.

Shenzhen Garden will tell the story of the city and its dwellers in the language of paper folding art. The top of the gallery of paper folding art is planted with azalea which is the symbol of Shenzhen city. The central waterscape area has shown the typical waterscapes and ecological landscape space of the seaside city – Shenzhen. Traditional architecture is presented in the simple design of our time, highlighting traditional colors, spatial arrangement, and materials. New technology and materials are also adopted to give expression to the local color and design strength of Shenzhen.

展园介绍：

展园设计主题是“创意深圳，折纸廊，情影深圳事”。城市意喻一张折纸，不同的城市折出的故事也是完全不同的效果。深圳从一个小渔村，通过 30 年时间的沉淀岁月的折叠，发展成为现在的国际化大都市、生态城市、设计之都。

深圳园将通过一张小小的折纸来影现，折纸廊空间通过动态转折处理的长廊空间叙述深圳这个城市的历史人物和事件。折廊的顶部景观化处理种植了代表性的深圳市花勒杜鹃。中央水景区展现的正是代表深圳这样一个滨海城市的特色水景与生态景观空间。展园以现代简约的设计手法表达传统民居建筑，提炼传统的色彩、空间布局、材料等元素，运用新技术新材料进行表达，在体现地方特色同时强调了深圳作为设计之都的创意力量。

Universiade SHENZHEN 2011

Chengdu Garden

成都园

Area: 893 m²

The Garden is named as "Tianfu Harmonious Garden". Various styles and elements in "Tianfu Harmonious Garden" are inspired by the style of the modern architecture in Chengdu to show the perfect combination of tradition and modern psychology.

展园面积：893 m²

该园取名"天府谐园"，其各种造型和元素取源于成都近代建筑造型，表现历史传统和现代心理浑然天成的创意。

Introduction:

Chengdu Garden is named as "Tianfu Harmonious Garden". "Tianfu" refers to Chengdu, Sichuan. "Harmonious Garden" refers to the harmonious and natural garden carrying on the culture of Chengdu and advocating creative development. In the Garden, there are several landmark buildings showing the flavor of dwellings in typical Sichuan lane.

Various styles and elements in "Tianfu Harmonious Garden" are inspired by the style of the modern architecture in Chengdu. Modern materials are used in the structure manufacturing to show the natural creativity of the historical tradition and modern psychology as well as to highlight the coexistence between ancient and modern cultures and the symbiosis between mankind and nature. The Garden has the attractions as follows: Typical Sichuan Lane, Tower Reflected in Stream, Landscape Wall Attraction, Deep Bamboo Forest, Sichuan Flower Gardening, etc. All parts of the Garden are both integrated as a whole and independent relatively, which emphasizes the innovation in the culture of Chengdu Garden. The Garden has such characteristic bamboo as Bambusa Beecheyana and plant sculpture "storytelling and ballad-singing figurine" (the origin of which is the style of "storytelling and ballad-singing figurine", the historical relic of Han Dynasty in Chengdu).

A 10m high bell tower stands in the center of the Garden. Along the central axis, there are two glass embossing walls inlaid with black bricks, featuring waterscapes inside the walls. There are fish tanks inside the walls with a dimension of 2m×0.8m×0.2m to imply "the water of Shujiang River has bred Sichuan and impregnated powerful vitality to it".

展园介绍:

成都园取名“天府谐园”，“天府”指四川成都，“谐园”则指成都文化传承、创新发展、和谐自然的展园。在整个园区内，几处标志性建筑再现了蜀巷民居韵味。

“天府谐园”中的各种造型和元素取源于成都近代建筑造型，采用现代材料结构制作，表现历史传统和现代心理浑然天成的创意，凸显古与今文化共存、人与物自然共生的意境。分为蜀巷情深、塔映花溪、景墙艺韵、竹篁森森、蜀花园艺等园艺景点。各部分即统一又相对独立，都突出成都展园文化的创意构想。园内种有吊丝球竹等特色竹子，植物雕塑《说唱俑》（取成都汉代文物“说唱佣”造型）也极具创意。

一栋 10m 高的钟塔矗立在展园中心位置。园区中轴线两侧还有两堵由青砖镶的玻璃浮雕墙，墙体内暗含水景。墙中镶嵌了一个高约 2m，宽 80cm、厚约 20cm 的“鱼缸”。寓意“蜀江之水孕育了天府之国，散发出勃勃生命力”。

天府谐园

Lanzhou Garden

兰州园

Area: 1,474 m^2

Landzhou Garden, also called "Lily (Bai He in Chinese) Garden" with the meaning that "everything goes well", fully displays the concept of "harmony".

展园面积：1 474 m^2

该园又名百合园：意喻"事事和好"，充分体现"和谐"的理念。

Introduction:

Landzhou Garden, also called "Lily (Bai He in Chinese) Garden" with the meaning that "everything goes well", fully displays the concept of "harmony". With the natural design layout, the Garden uses waterwheel that symbolizes Yellow River civilization and lily that represents the flavor of Lanzhou to showcase the beautiful scenery of Yellow River, the ecological environment construction of Southern and Northern mountains of Lanzhou City and the development of forestry science and technology. As a result, Lanzhou Garden turns into a landscape garden that unfolds a picture where human and nature live together in harmony, thus displaying the Yellow River civilization of Lanzhou City and improved living standard of local people.

The sculpture of lily is a horticultural relief wall which is 2.4m high and 12m long. The relief wall has introduced the greening achievements of the Southern and Northern mountains of Lanzhou City. In the landscape area of Yellow River civilization, two 4~5m high waterwheels are eye-catching. The color of timber combines with the sea of the flowers to make people close to nature.

展园介绍：

兰州园又名百合园，意喻"事事和好"，充分体现"和谐"的理念。本设计采用自然式布局，通过用代表黄河文化的水车和代表兰州特色植物百合的展示，反映兰州市黄河风情线和南北两山生态环境建设和林业科技发展，展示甘肃黄河文化，人民生活蒸蒸日上的人与自然和谐的景观园区。

园中的百合雕塑是一面高 2.4m、长 12m 的园艺浮雕墙，浮雕介绍了兰州市南北两山的绿化成果。兰州园里"黄河文化"景区中，两轮 4、5 米高的大水车显得格外惹眼，不加装饰的原木颜色与花海的融合，让人们感觉到了与自然最亲近的接触。

Harbin Garden

哈尔滨园

Area: 2,874 m^2

Themed by "sentiment for ice and snow", the Garden has displayed a landscape of ice and snow culture unique to the northern region.

展园面积：2 874 m^2

该园以“冰雪情怀”为主题，创造出独具北方特色的冰雪文化景观。

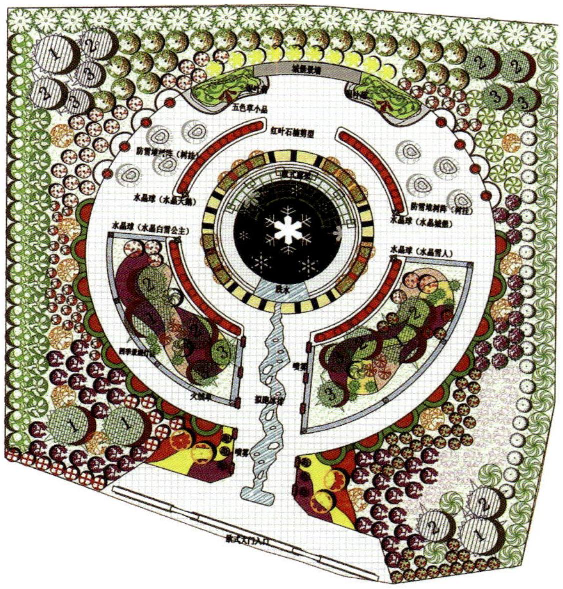

展园介绍：

展园以“冰雪情怀”为主题，整体布局为“一环、两轴、一中心”的形式，通过入口大门、跑冰排溪流、城堡浮雕景墙、欧式长廊、五色草地形小景、仿冰树挂、涌泉跌水、雪堆树池、四季景观灯柱、飘雪水晶球等景观小品，创造出独具北方特色的冰雪文化景观。

园内植物景观采用自然式配置，以雪松、栾树、红叶石楠等西安当地树种为基础；选择火绒草、银叶菊等白色观花观叶植物在显要位置进行点缀，突出冰雪主题。同时，利用哈尔滨园林标志性植物五色草进行造景。

Introduction:

Themed by "sentiment for ice and snow", Harbin Garden has its overall layout as "one circle, two axes and one center". The Garden displays a landscape of ice and snow culture unique to the northern region by showcasing such garden parergons as main entrance gate, stream with ice floe, landscape wall with castle anaglyph, European-style gallery, five-color grassplot scene, imitated rime, fountain and drop water, snow drift and tree grate, lamp pole for the landscape of four seasons, and crystal ball with snow effect inside.

The Garden chooses a natural layout for the plant landscape. Such local trees of Xi'an as cedrus, koelreuteria paniculata and photinia fraseri are used as the main plants; while such white ornamentals as edelweiss and senecio cineraria are put in the market place to highlight the theme of the Garden; also, coleus blumei, the symbolic plant in the Harbin Garden, is used for landscaping.

Tangshan Garden

唐山园

Area: 955 m^2

The theme of the Garden is "On the Wing", denoting Tangshan's miraculous rebirth as "the Nirvana of the Phoenix through Fire" and delivering the information that Tangshan will hold the International Horticultural Exposition of 2016.

展园面积：955 m^2

该园以“振翅”为主题，表达唐山的新生如“凤凰涅槃，浴火重生”般瞩目，也传达了2016年世园会在唐山举办的消息。

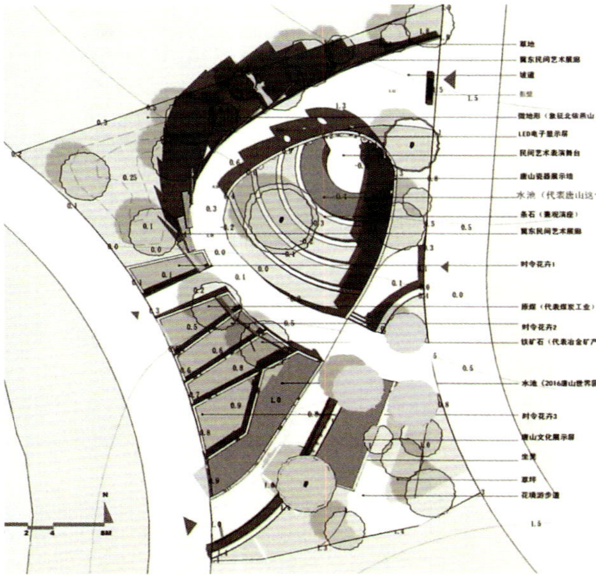

Introduction:

The theme of the Garden is "On the Wing", denoting Tangshan's miraculous rebirth as "the Nirvana of the Phoenix through Fire" and the promising future of the city development.

The Garden has delivered the information that Tangshan will hold the International Horticultural Exposition of 2016, and realized the ecological connection between Tangshan and Xi'an. With the local Eco Park of South Lake as the background, and the plane graph reported to BIPH as the basic, Tangshan Garden has displayed the achievements in ecological promotion.

The paths and outer line of the Garden form the shape of the Chinese character "凤" (meaning Phoenix), re-asserting the theme. The landscape, namely "Under the Yanshan Mountain" and "On the Beach of Bo Hai", feature the mountain on the north and sea on the south. They represent the history, culture and industrial prosperity of the city by putting together the solid mountains and flowing water. The major plantations are Chinese scholar trees, Chinese pines, lacebark pines, torch trees, China roses, weigela florida, winter creeper euonymus, and kudzu vines.

展园介绍：

该展园以“振翅”为主题，表达唐山的新生如“凤凰涅槃，浴火重生”般举世瞩目，城市发展前景如“振翅腾飞”般有力。

展园通过展示传达唐山作为2016年世园会举办地的信息，实现唐山与西安的生态链接。唐山展园以南湖（旧事）城市地方生态公园为布景，以向国际园艺出产者协会申报的规划方案平面图为根本，全面展现生态扶植的成绩。

整体布局上，由园路等硬质景观的线性与外轮廓线整体构成“凤”字造型，暗合主题。景观“燕山脚下”和“渤海之滨”，北山南海，刚柔并济，体现了唐山悠久的历史文化底蕴和现代工业的兴盛。植物选用能在西安良好生长的，具有唐山地方特色的乡土树种来烘托和展现唐山地域风情。主要有国槐、油松、白皮松、火炬树、月季、锦带花、扶芳藤、葛藤等。

曹妃甸
环城水系

Wenchuan Garden

汶川园

Area: 751m²

With the theme of "National Love, Thanksgiving and Advancing Bravely", Wenchuan Garden has showcased the achievements of post-disaster reconstruction and manifested the unique charm of national culture.

展园面积：751m²

该园是"民族大爱·感恩奋进"的主题立园，展示灾后重建成果，彰显独具魅力的民族特色文化。

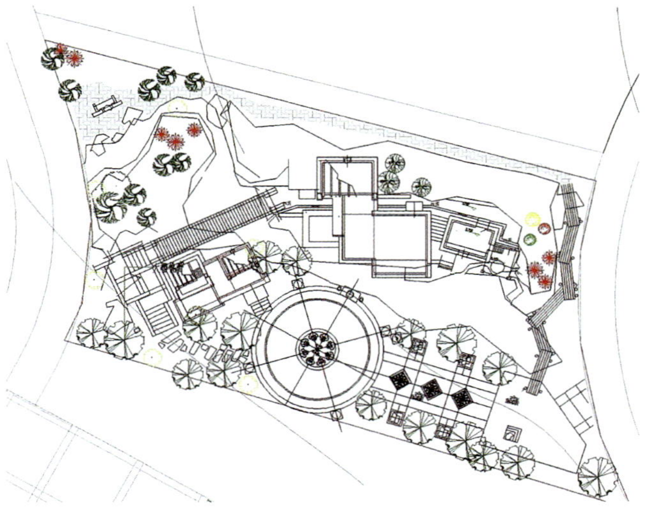

Introduction:

With the theme of "National Love, Thanksgiving and Advancing Bravely", Wenchuan Garden has showcased the achievements of post-disaster reconstruction and manifested the unique charm of national culture. The Garden has integrated natural and horticultural Wenchuan, thankful and striving Wenchuan, and harmonious Wenchuan in Xi' an International Horticultural Exposition. Visitors can experience the huge changes Wenchuan and its rebirth in the post-disaster reconstruction.

Wenchuan Garden has one theme, two structures and three sceneries. One-theme: Wenchuan on its way. Two-structure: exhibition hall under the slope and garden along the waterside supplement each other. Three-scenery: Qiang Family, Qiang Culture, and special local products. Qiang Family shows the unique buildings in Qiang Villages in Wenchuan; Qiang Culture includes the songs and dances of Qiang, Qiang embroidery shows, zaza wine, sheepskin drum, celebration of success and other lifestyle in Qiang villages; special local products are Chinese dove tree and sweet cherry trees.

展园介绍：

汶川展园是"民族大爱·感恩奋进"的主题立园，展示灾后重建成果，彰显独具魅力的民族特色文化。展园集中使自然园艺的汶川、感恩图强的汶川、天地和美的汶川在西安世园会绚丽绽放，让八方宾朋感受汶川发展的巨大变迁和灾后重建的涅槃重生。

展园分为一个主题、两大格局、三道风景。一个主题：汶川一路走来。两大格局：依坡造岩，修房筑寨，构成馆区；因势布园，理水植树，构成园区，前园后馆，相生相彰。三道风景：羌寨人家、羌寨风情、特色物产，羌寨人家是指汶川地区羌族特色建筑；羌寨风情是指羌歌羌舞、羌绣表演、喝咂咂酒、羊皮鼓舞，披红挂彩等羌寨生活风情；特色物产指栱桐、甜樱桃树等特色植物。

汶川园
承建单位
景观科技有限公司

Xining Garden

西宁园

Area: 1,515 m^2

The Garden features a Tang court – Tibet route, reflecting the beauty of Qinghai as the source of the Three Rivers, the origin of mountains, and the spirit of the nation, and highlighting the beautiful scenery and culture of the ancient city on Qinghai-Tibet Plateau.

展园面积：1 515 m^2

该园以“唐蕃古道”为主线，展现了“三江之源、万山之宗、民族之魂”的青海之美、高原古城秀丽的景色、多元的文化内涵。

Introduction:

The Garden features a Tang court – Tibet route from east to west. Along the route, there are theme spots including Beautiful Qinghai, Hehuang Culture and Colorful Xining. These designs aim at showing the importance of the route in Qinghai Province as well as the history of Hehuang Culture. The Garden reflects the beauty of Qinghai as the source of the Three Rivers, the origin of mountains, and the spirit of the nation, and highlights the beautiful scenery and culture of the ancient city on Qinghai-Tibet Plateau.

The landscapes such as “Babao Ruyi Tower”, “Sun & Moon Bronze Mirror”, and “Tea-Horse Trade Center” made of imitated copper have reproduced the prosperous image of Tang court – Tibet Trade. At the end of Tang court – Tibet route, a dwelling wall with “the six landscape spots of Big Nanshan Mountain Ecological Project” is built to reflect the big changes and beautiful landscapes of Xining. In the arrangement of plants, typical plants of Xining such as lilac, spruce, and China Savin are planted by large area to make people feel like staying in beautiful Xining.

展园介绍：

以“唐蕃古道”为主线，由东向西贯穿全园，沿青海境内的主线分别布置以“大美青海”、“河湟文化”、“锦绣西宁”为内涵的主题景点。以此反映河湟文化的悠久历史及唐蕃古道在青海的重要作用，反映“三江之源、万山之宗、民族之魂”的青海之美，以及反映高原古城秀丽的景色和多元的文化内涵。

其次展园用仿红铜材质的“八宝如意塔”、“日月铜镜”、“茶马互市”等景区再现唐蕃贸易往来的繁荣景象。在“唐蕃古道”的终点建设一堵以西宁民居墙形式的线刻“大南山生态工程六大景区景墙”，反映西宁今非昔比的变化和美景。在植物配置上，具有鲜明西宁特色的丁香花、云杉、圆柏等植物将被大面积栽植，让人仿佛置身于美丽西宁。

Tianjin Garden

天津园

Area: 2,231m²

The main sight of the Garden is a "Jin Zun" and other landscape creations, showing the characteristics of Tianjin.

展园面积：2231 m²

该园最显眼的"津沽金樽"和众多景观小品，体现了天津的特色。

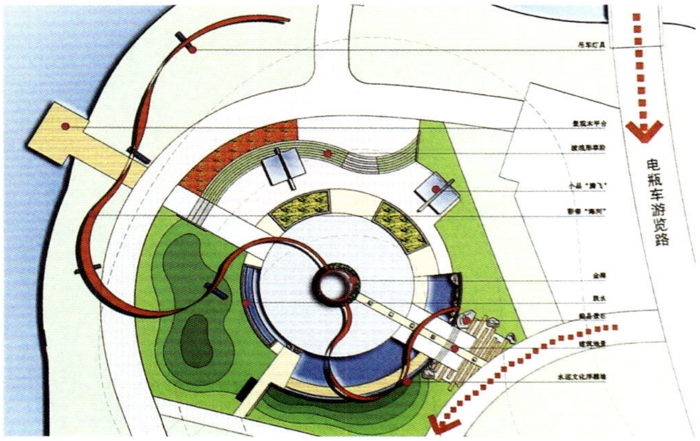

Introduction:

The design of "Tianjin Garden" is unique. Flowers are blooming in the Garden to provide beautiful scenery. The main sight of the Garden is a "Jin Zun" (golden wine jar) 12m in height (lower radius: 4m, upper radius: 5m) with hollow patterns bearing the characteristics of Tianjin City, including "Jingwei Fills up the Sea" and "the Spring of River and Sea". A red belt linking the west bank and the center of the Garden covers the nearby area of the "Jin Zun". It is cut with drop-point holes to create an overwhelming visual effect. The major artistic creation in the Garden is the relief wall illustrating the vicissitudes of the shipping and grain transport that have exerted great influence on the development of Tianjin. Other creations are four lamps and lanterns of wharf crane, six mini replications of International Architectures in Tianjin, shallow pond, mini waterfall wall, wooden plank roads, wooden platforms, and a "flying" structure.

展园介绍：

"天津园"在设计上非常独特，园内鲜花盛开，景色宜人。园内最显眼的是一个高达12m（下半径4m、上半径5m）的景观主体"金樽"，并在其上镌刻、透空出"精卫填海"、"海河之春"等体现天津城市特征的主题图案。"金樽"两侧是从西侧水面一直飘舞至场地之中的红色彩带"海河之韵"，彩带上透空出大小不一的水滴型圆孔，让人感觉到更震撼的景观效果。展园中其他小品主要有：体现影响城市发展的天津海运、漕运文化历史进程浮雕墙。4具码头吊车行驶的景观灯具，展园6块天津"万国建筑"地景。还设置了浅水池、跌水矮墙、木栈道、木平台、"腾飞"构架等景观小品。

Yunnan Garden

云南园

Area: 1,164 m²

Themed with "A Beautiful Place", Yunnan Garden has presented the architectural culture of the Dai nationality.

展园面积：1 164 m²

该园以"有一个美丽的地方"为设计理念，充分体现傣族建筑文化特色。

Introduction:

Themed with "A Beautiful Place", Yunnan Garden has presented the architectural culture of the Dai nationality. The design features the history, culture and nature of Yunnan Province, the place on the Peak of Rosy Clouds. The main buildings are miniatures of the key cultural heritage sites, the "White Tower" and "Jingzhen octagonal pavilion". They sit on the two entrances on the east and west sides of the Garden and are linked by the "Rainbow Path". The "Rainbow Path" is designed with concern of the view of Chang' an Tower. The river under the path added spatial richness of the landscape. The thick bamboo arrangement in the Garden and the traditional folk music such as "Fernleaf Hedge Bamboo under Moonlight" and "There is a Beautiful Place" played in the Garden bring visitors to the original Dai cultural atmosphere. Special Yunnan plants such as paeonia suffruticosa in Lijiang, talipot palm, bodhi tree, cattleya, panax notoginseng, and alpine rose will be introduced besides bamboo and palm, showing the diversity of plant species and the high-standard gardening skills of Yunnan.

展园介绍：

展园主题"有一个美丽的地方"，充分体现傣族建筑文化特色。设计上突出了彩云之巅的历史文化和自然生态特色。景观上选用了国家重点保护文物"白塔"、"景真八角亭"为主题建筑，分别位于东西两个入口，再用一条"彩虹道"相连接。"彩虹道"的设计巧妙地将"长安塔"借入园内，路下方是蜿蜒的溪水，增加了景观层次。园内种植大量竹林，播放"月光下的凤尾竹"、"有一个美丽的地方"等民族特色背景音乐体现傣家风情。除了种植竹子、棕榈，园内还会栽种丽江牡丹、贝叶棕、菩提树、洋兰、三七、高山杜鹃等具有云南特色的植物，充分展示云南植物种类多样性和云南园林园艺的高超水平。

Anhui Garden

安徽园

Area: 1,040 m²

With the theme of "new Anhui and new landscape", it indicates the core of layout- "inheriting the harmonious culture of Anhui and presenting the harmonious development in modern Anhui".

展园面积：1 040 m²

该园以"新安徽、新景观"为设计主体，表达了"传承安徽地域和谐文化之德，展示安徽现代和谐发展之音"的设计核心。

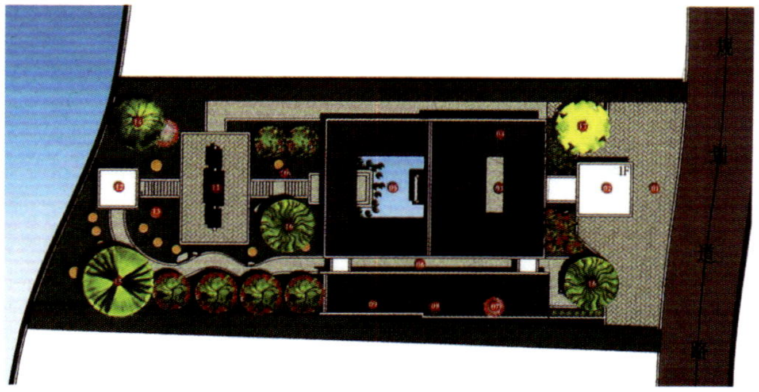

Introduction:

The layout of Anhui Garden is a modern expression of Hui culture. The Garden not only follows traditions, but also makes innovations in the layout theme of "new Anhui and new landscape". And it indicates the core of layout- "inheriting the harmonious culture of Anhui and presenting the harmonious development in modern Anhui". Such local landscapes as bonsai of Xin'an (Huizhou) and Lingbi stone, etc. have been introduced to the Garden. And the Garden chooses the plants near the water of the villages in Huizhou as its main green plants; the layout of its indoor space keeps a topological relation with that of the indoor space of the ancestral hall in Huizhou. Besides, the Garden also introduces the cultural symbols of the region, such as Matou Wall, patio, courtyard, six-foot wide lane and memorial archway.

展园介绍：

徽文化的现代表达，既有传统，又有创以"新安徽、新景观"为设计主体，表达"传承安徽地域和谐文化之德，展示安徽现代和谐发展之音"的设计核心。设计引入新安盆景、灵璧石等地方景观材料，以徽州村落水口植物为主要绿植选择；室内空间保持与徽州宗祠室内的拓扑关系，引入马头墙、天井空间、庭院空间、六尺巷空间、牌坊等地域文化符号。

Changchun Garden

长春园

Area: 843 m^2

This garden employs the modern and simple construction approach, and it is dominated by compositons of circles and circular curves representing "harmony".

展园面积：843 m^2

该园运用现代简约的造园手法，以“圆融和谐”为主要构图原则。

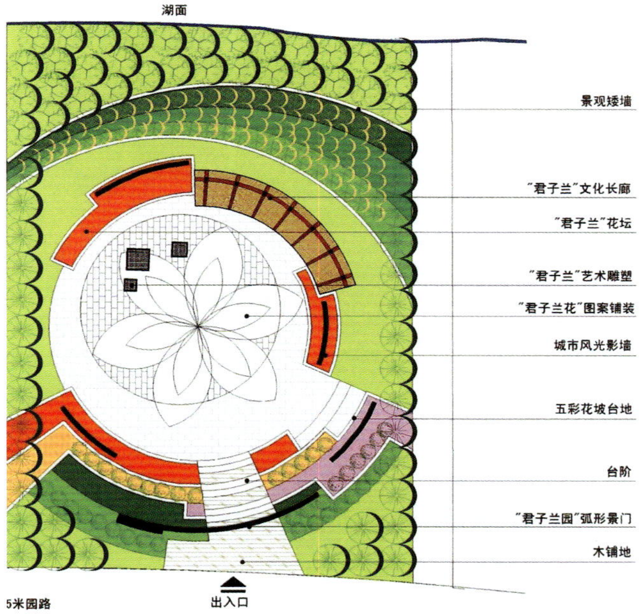

Introduction:

Employing the modern and simple construction approach, Changchun Garden is dominated by compositions of circles and circular curves representing "harmony". It uses kaffir lily, the city flower, and the urban view to present its characteristics. The Garden is spacious in the middle with tall trees around, which is created by the topography transformation, the construction paragons and the layout of plants. The overall appearance of the space is "transparent, grand and open", which showcases distinguishing features of northern gardens in Changchun.

展园介绍：

园区运用现代简约的造园手法，以“圆融和谐”的圆和圆弧线构图为主，以市花君子兰和城市风光作为表现特色，通过地形改造、建筑小品及植物配置，形成中间开敞，周边由高大树丛环抱。“通透大气，疏朗开放”整体空间外貌，展现了长春北方园林特色。

Liaoyuan Garden

辽源园

Area: 1,028m²

The design centers on the historical context of "Shengjing Hunting Park" and "Imperial Deer Garden" in Liaoyuan City to showcase the local historical culture featuring nature and simplicity as well as large territory and rich resources.

展园面积：1 028m²

该园以辽源市"盛京围场"、"皇家鹿苑"这一历史脉络为题材，表现出辽源自然古朴、地广物丰、的历史文化特色。

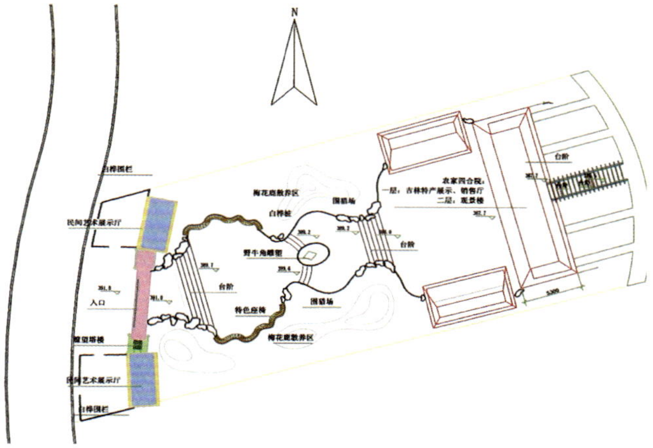

Introduction:

With the theme of "Eternal peace and harmony between nature and mankind, nurturing the feature earth", Liaoyuan Garden focuses on the thought of "people first, cultural heritage, innovation, development and harmonious nature". Following the principle of building gardens in line with local conditions, a landscape garden integrating horticultural characteristics and regional culture of Jilin Province comes into being. The design centers on the historical context of "Shengjing Hunting Park" and "Imperial Deer Garden" in Liaoyuan City. By using various design elements including the watchtower made from birch, buffalo horn and red lantern in northeast China, Liaoyuan Garden showcases the local historical culture featuring nature and simplicity as well as large territory and rich resources.

展园介绍：

辽源园围绕"天人长安，创意自然"的主题，强调以人为本、传承文化、创新发展、和谐自然的思想，遵循因地制宜建园的原则，塑造出具有吉林园艺特色和地域文化内涵的景观展园。展园的设计主要以辽源市"盛京围场"、"皇家鹿苑"这一历史脉络为题材，选用东北特有的白桦木瞭望塔、野牛角、红灯笼等设计元素，表现出辽源自然古朴、地广物丰的历史文化特色。

欢迎您再次光临

Nanchang Garden

南昌园

Area: 945 m^2

Aiming at "establishing a new city image by building an exquisite garden", Nanchang Garden brings the unique style of the traditional architecture into full play.

展园面积：945 m^2

该园以"打造园林精品工程，树立城市全新形象"为建设宗旨，将传统建筑的独特风格表达得淋漓尽致。

Introduction:

Nanchang Garden aims at "establishing a new city image by building an exquisite garden". The overall layout uses a "natural garden" mode, with the scenery of sunset and cloud, green and rippling water and Tengwang Tower as the main axis. Supplemented by natural plants and garden paths as the sub-axis, Nanchang Garden produces a magic visual effect that "every step affords different scenery". In terms of architectural style, the main building is designed in accordance with Tengwang Pavilion as the prototype, called "Tengwang Tower", which covers an area of 1,50 m^2. The architecture is mainly built with bricks and tiles as well as cast-in-place framework. By integrating the colorful paintings and brackets, Nanchang Garden brings the unique style of the traditional architecture into full play.

展园介绍：

南昌展园以"打造园林精品工程，树立城市全新形象"为建设宗旨。总体布局全园布局采用自然式园林手法的布局形式，以落霞云景、碧水微澜、腾王台为景观主轴，自然式的植物造景、园林小径等景点为辅轴，形成步移景异的观赏效果。展园主体建筑以腾王阁为设计原型，名曰"腾王台"，其建筑占地 150 m^2，建筑以砖瓦及现浇框架结构为主，绚丽多彩的彩绘及斗拱装饰融为一体，将传统建筑的独特风格表达得淋漓尽致。

绿色引领时尚
南昌园

Qingdao Garden

青岛园

Area: 1,089 m²

The Garden, with the emphasis on "one sea, one mountain and one city", perfectly interprets the charm and historical culture of Qingdao with the combination of local plants and landscaping approach.

展园面积：1089 m²

该园围绕"一海、一山、一座城"整体布局，结合具有青岛特色的植物、造景手法等更好的诠释青岛的人文风貌、历史文化。

Introduction:

Qingdao Garden exhibits the distinctive culture of Qingdao City. Tourists can feast their eyes on the imposing scenery of Qingdao, a good place featuring "nature's treasures and talented people". Moreover, they can appreciate the elegant charm of Qingdao, known as "City of Sails" where the blue sky is mirrored in the azure sea. In terms of the overall layout, the Garden emphasizes on "one sea, one mountain and one city" (one sea: the white sailing boat structure at the entrance and the sea-like lake in the west; one mountain: deep green Laoshan Mountain landscaping; one city: the miniature landscape buildings of European-style such as Qingdao Raiway Station).

It has perfectly interprets the charm and historical culture of Qingdao with the combination of local plants and landscaping approach. In Qingdao Garden, there is honeysuckle – the city flower of Qingdao, oriental cherry, wintersweet, prunus cerasifera, crape myrtle, Chinese flowering crabapple, etc. The miniature landscaping and the arrangement of the flowers have shown the elements and characteristics of Qingdao to depict the miniature of the green and ecological Qingdao.

展园介绍：

青岛展园展示青岛的文化情怀。观赏青岛园既可以饱览“物华天宝，人杰地灵”的青岛风光和领略拥抱“海天一色，帆船之都”的温雅风采。总体布局 展园围绕“一海、一山、一座城”整体布局，一海：门口白色的帆船造型加上园西似海的湖面；一山：墨绿色的崂山造景；一座城：青岛火车站等欧式建筑小城的微缩景观。

展园还结合具有青岛特色的植物、造景手法等更好的诠释青岛的人文风貌、历史文化——在整个“青岛园”，还有青岛市花耐冬，以及樱花、梅花、紫叶李、紫薇和海棠等。微缩造景与花卉植被的设计都显现青岛元素、突出青岛特色，共同描摹了绿色生态青岛的缩影。

Henan Garden

河南园

Area: 1,168 m^2

With the theme of "the Center of the Country, the Civilization of China", and in combination with the environment, the Garden has fully showcased the history of Central Plains and the culture of Luoyang.

展园面积：1 168 m^2

该园以“天下之中，华夏文明”为主题，结合园区环境做为承载充分展示中原历史，河洛文化。

Introduction:

With the theme of "the Center of the Country, the Civilization of China", Henan Garden includes the peonies of Luoyang, the sculptures of "The Empress Wu appreciating the flowers", landscape stones, festoon gate, waterscapes, etc. In combination with the environment, it has fully showcased the history of Central Plains and the culture of Luoyang.

Henan Garden has been designed with three attractions, the "spring scenery of Yu Garden", the "center of the world" and the "national grace". In the "spring scenery of Yu Garden", the abstract style of peony as the subject and shrub plants which mainly bloom in spring around the Garden highlight the effect of brilliant flowers in full bloom in spring. In the "center of the world", there is a style of the Chinese character "中" through the square layout. The solid wood stools are arranged are arranged in each corner of the square, and a fountain is designed in the center of the square to increase flexibility and enjoyment of the square. In the "national grace", a style of peony has been laid out by the path inside the Garden. Different colors of peony flowers are fully planted in the Garden. The sculpture of a five-meter high peony fairy is placed in the central stamen. Around the sculpture, there are flowering shrubs, such as Chinese herbaceous peony, Chinese rose and crape myrtle, adorned with stones in order to enhance the effect of the scenery.

展园介绍：

河南展园以“天下之中，华夏文明”为主题，以洛阳牡丹、武后赏花组雕、景观石、垂花门、水景等景观，结合园区环境做为承载充分展示中原历史，河洛文化。

园中有设计豫园春色、天下之中、国色天香 3 个景点。豫园春色以抽象的牡丹造型为主体，周边种植以春季开花为主的乔灌植物，突出春花烂漫、春意盎然的景观效果。天下之中以广场铺装的形式做出“中”字造型，广场四角布置实木坐凳，中部设计水景喷泉，以增加广场的灵动性、趣味性。国色天香以园路铺设出牡丹花的造型，满栽各色牡丹，中心花蕊部位放置 5 米高牡丹仙子雕塑，周边栽植芍药、月季、紫薇等花灌木，其间点缀放置景石，以烘托景观效果。

豫园春色
盖章
盖章

Kaifeng Garden

开封园

Area: 612 m²

With the theme of "Flavor of Song Dynasty and Fragrance of Chrysanthemum", Kaifeng Garden puts different elements together such as Song Prose, landscape, ancient city wall, Bao Zheng (a famous officer in Song Dynasty) and chrysanthemum, to fully display the profound culture of Song dynasty in Kaifeng City and the unique charm of new look of the ancient city.

展园面积：612 m²

该园以"宋韵·菊香"为主题，通过对宋词、景区风貌、古城墙、包公、菊花等要素的有机组合，展示了开封大宋文化深厚底蕴和古城新貌独特魅力。

Introduction:

With the theme of "Flavor of Song Dynasty and Fragrance of Chrysanthemum", Kaifeng Garden uses the traditional gardening method to put different elements together such as Song Prose, landscape, ancient city wall, Bao Zheng (a famous officer in Song Dynasty) and chrysanthemum, so as to fully display the profound culture of Song dynasty in Kaifeng City and the unique charm of new look of the ancient city.

All the buildings of Kaifeng Garden are imitating the dwellings in the painting of "Riverside Scene at Qingming Festival". The Garden is divided into four parts: the atrium, Song Prose Garden, Chrysanthemum Garden, and the Facial Makeup area of Bao Zheng. The atrium is the main landscape area of the Garden enclosed by the gate, hall and landscape walls of Song style. Song Prose Garden is located on the east of the Garden with a landscape gate and a screen wall facing the gate to be the front yard of another courtyard. Chrysanthemum Garden is the main exhibition area for chrysanthemum – the city flower of Kaifeng. The 3.5m high Facial Makeup area of Bao Zheng features the highest land of the whole garden to be the symbol of Kaifeng Garden.

展园介绍：

开封园以"宋韵·菊香"为主题，运用传统造园手段，通过对宋词、景区风貌、古城墙、包公、菊花等要素的有机组合，形成具有浓郁宋代风情的精致园林庭院，充分展示了开封大宋文化深厚底蕴和古城新貌独特魅力。

开封园中的所有建筑都将以《清明上河图》中的民居为蓝本进行建设，整个园区被分为4个功能区，分别是中庭、宋词园、菊圃、包公脸谱区。中庭是开封园的主景区，由宋式风格的门、厅、景墙等围合而成。宋词园将位于全园的东侧，通过景门、照壁的设置，形成另一个庭院的前庭。菊圃则是开封市花——菊花的主要展示区。包公脸谱区是全园地势最高的区域，3.5m高的包公脸谱是开封园的标志。

祥
CAUTION
WET FLOOR
小心地滑

吉
祥

Shanxi Yangling Garden

陕西杨凌园

Area: 831 m²

The theme of the exhibition is Technological Yangling, Open Yangling, Ecological Yangling.

展园面积：831 m²

该园的设计主题为科技杨凌、开放杨凌、生态杨凌。

Introduction:

The theme of the exhibition is Technological Yangling, Open Yangling, Ecological Yangling. Through open touring routes, the exhibition park is divided into entrance square, vines displaying zone (flower shelves), resting zone, central flowers zone, aquatic plants zone and moving forest zone. The special plants include Japanese peach, potted strawberry, dahlia, early chrysanthemum and some crops. There are sights like wheat straw wall, DNA duplex shelf, wooden platform, sculpture group, central flower zone, water flower zone and moving forest zone.

展园介绍：

展区以科技杨凌、开放杨凌、生态杨凌为主题。展园通过开放式空间布局组织全园游览路线，同时将展园划分成为入口广场区，藤本展示区（花架），休息活动区，中心花卉区，水生植物区，移动森林区。其特色植物有：日本丽桃、盆栽草莓、大丽花、早菊以及部分农作物。园内景点主要有麦秸秆景墙 、DNA 双螺旋花架 、木铺装平台 、雕塑群 、中心花卉区 、水生花卉区 、移动森林区 。

Shanxi Shangluo Garden
陕西商洛园

Area: 1,026m^2

Focusing on the theme "Landscape in Qinling, Humanity in Shangluo", the garden gives the different sights of southern Shanxi from the other areas.

展园面积：1026m^2

该园以"山水秦岭·人文商洛"为主题，使人更深刻地感受了其他地区截然不同的景致。

Introduction:

Focusing on the theme "Landscape in Qinling, Humanity in Shangluo", the exhibition park is divided into four zones: culture zone, waterscape zone, special plants zone and resting zone. Most of the plants are local trees arranged with proper density, orderly layers and rich color. The flowers and bushes of four seasons are collocated to keep the exhibition park green all year round with flowers in three seasons. The garden ornaments are appropriately arranged to present the creativity of the design. The walnuts, persimmons, chestnuts and flower drums are displayed as sculptures, corresponding to the constructions. The green land around not only matches the ornaments but also becomes the landscape independently, giving the different sights of southern Shanxi from the other areas.

展园介绍：

以"山水秦岭·人文商洛"为主题，展园分为文化景观区、跌水景观区、特色植物景观区、观景休憩区四个区域。植物配置以乡土树种为主，疏密适当，高低错落，形成一定的层次感，色彩丰富。四季不同花色的花灌木进行搭配，使展园达到四季常绿，三季有花。适当安置园林小品，力求小品设计在造型、颜色、做法上有新意，核桃、柿子、板栗及花鼓等都是以雕塑形式出现，使之与建筑相适应。周围的绿地不仅可以对小品起到延伸和衬托，又独立成景，使人更深刻地感受了陕南与其他地区截然不同的景致。

Shanxi Ankang Garden

陕西安康园

Area: 1,013 m^2

Exhibition area is themed as "Enjoyable Mountains and Waters of Qinling-Bashan Region". It has a ringlike layout surrounded the lake as a whole.

展园面积：1 013 m^2

该园以写意秦巴山水为主题，整体绕湖环形来布局。

Introduction:

Exhibition area is themed as "Enjoyable Mountains and Waters of Qinling-Bashan Region". The Ankang Garden has a ringlike layout surrounded the lake as a whole. The building and the overall layout provide Ankang Garden with a "high-low-high" terrain, which is composed of numerous attractions, including memorial archway Xiangxi, bonsai of Nangong Mountain, plank road of Xiangxi, memorial archway of tea garden, farmland of Hanyin and dragon boat of Han River. Ankang Garden is a construction with geographical features, which reflect the characteristics of rural tile-roofed houses in Ankang. The characteristic plants in the Garden are cinnamomum camphora, eucommia ulmoides, magnolia officinalis, taxus mairei, liriodendron chinense, physochlaina macrophylla bonati of Ziyang, rhododendron mucronulatum, camellia japonica, and fragrant flower. The scenic spots are mainly about memorial archway of Xiangxi Cave, terrace of Nangong Mountain, a place for singing in the tea garden, waterwheel, farmland of Hanyin, dragon boat, Yinghu Lake, etc.

展园介绍：

展区以写意秦巴山水为主题，整体为绕湖环形布局。从建构筑和整体布局上来看，安康园由香溪牌坊，南宫山盆景，香溪栈道，茶园牌坊组成，而汉阴农田——汉水龙舟是一个由"高"——"低"——"高"的过程。其建筑风格反映了安康农家瓦房特征的地域特色建筑。展园的特色植物有：樟树、杜仲、厚朴、红豆杉、鹅掌楸、紫阳大叶泡、迎红杜鹃、山茶、香团刺等。园内景点介绍主要有香溪洞牌坊、南宫山平台、茶园放歌台、水车、汉阴农田、龙舟、瀛湖等。

水車園

水車清音

水車園

EXPO2011XI'AN
安康园
Ankang Garden

绿色创意自然
请您爱护花木
漢陰農園

Shanxi Baoji Garden

陕西宝鸡园

Area: 1,350 m^2

Exhibition area is themed as "Green Baoji, city of forest" to demonstrate the ecological civilization of Baoji.

展园面积：1 350 m^2

该园以"绿色宝鸡·森林城市"为主题，展示宝鸡生态文明。

Introduction:

Exhibition area is themed as "Green Baoji, city of forest". In terms of technique of expression, it takes the ancient famous trees, city flowers, forestry and urban landscape as main subjects; in terms of historical culture deposits, it highlights the historical marks of Emperor Yan's Hometown and Zhou and Qin civilization, cultural mark of World Buddha Famen Temple, and geographical indication of Chinese stone drum; in expressing achievement, it reflects the great accomplishment of Baoji garden urban construction; in manifesting ecological landscape, it reflects the substantial results in creating green Baoji and in urban and rural integration, and the natural scenery, like Taibai Mountain; in displaying modern elements, it highlights lighting design, reflects the ideas of city construction, humanistic naturalness and harmony. By doing so, it is about to achieve the goal to demonstrate the ecological civilization of Baoji, to lead green culture and to pass green visiting card.

展园介绍：

展园以“绿色宝鸡·森林城市”为主题。在表现手法方面，以宝鸡古树名木、市树市花、林业产业、城市景观为主体；在表现历史文化积淀方面，突出展示炎帝故里和周秦文化的历史标志、世界佛都法门寺的文化标志及中华石鼓园的地理标志；在表现发展成果方面突出展示宝鸡园林城市建设取得的伟大成就；在表现生态景观方面，突出展示创建绿色宝鸡、森林城市、城乡一体化等方面取得的丰硕成果以及太白山等自然风光景观特点；在表现现代元素上，突出灯光设计，体现宝鸡城市建设、人文自然及和谐包容的意念。通过以上内容，达到展示宝鸡生态文明、引领绿色文化、彰显绿色名片的目的。

Shanxi Hanzhong Garden

陕西汉中园

Area: 975 m^2

By means of getting to know more about history, culture and natural scenery of Hanzhong, the design of Hanzhong exhibition area has formed a high-quality modern garden.

展园面积：975 m^2

该园通过对汉中历史文化、自然风光的深度解读形成融古典造园艺术。

Introduction:

Hanzhong is known as the birthplace of Chinese and the treasure of China. By means of getting to know more about history, culture and natural scenery of Hanzhong, the design of Hanzhong exhibition area has formed a high-quality modern garden to reflect the theme "harmonious coexistence of city and nature" of Xi' an World Horticultural Exposition. In terms of constitute of scheme, it mainly exerts the conflict in form between Han-imitating ancient buildings and modern city; in artistic conception of garden, it reveals the beautiful natural scenery of Hanzhong, the dwellings near the mountain and by the river, and the inheritance, unification of profound historical culture deposits and urban development.

展园介绍：

汉中，素有"汉家发祥地，中华聚宝盆"的美誉。本次汉中展园的设计通过对汉中历史文化、自然风光的深度解读形成融古典造园艺术、现代生态理念和先进科技手段于一体的现代园林精品，体现西安世园会"天人长安、创意自然——城市与自然和谐共生"的主题。在方案构成上主要表现出汉江及秦岭南麓依山傍水的仿汉古建筑与现代城市形式上的冲突，在园林意境中展现汉中秀美的自然风光、依山傍水的民居、深厚的历史文化底蕴与城市发展的传承、和谐、统一。

绿色引领时尚

Shanxi Tongchuan Garden

陕西铜川园

Area: 1,015m²

The architectural feature of furnace in Tongchuan Chenlu is adopted to reflect the kiln culture of Tongchuan Yaozhou.

展园面积：1015m²

该园突出了铜川陈炉“窑炉”的建筑特点，展现了铜川耀州的窑文化。

Introduction:

The exhibition park is displayed like a story. On the aspect of constructions and terrain, the overall layout descends from north to south, presenting the rich terrain of Tongchuan: the majestic mountain – the humble layers of loess tableland – the vast plain – the tranquil valley. The architectural feature of furnace in Tongchuan Chenlu is adopted to reflect the kiln culture of Tongchuan Yaozhou. There are special plants like silk tree, horse chestnut, silver birch, Chinese pine, Spiraea pubescens, Cyperus fuscus, etc. The sights include Chenlu Terrace, West Palace Wall, Pouring Pot, Lanzhi Valley Stream, Medicine Pool, etc.

展园介绍：

展区以铜川的故事展开。总体布局从建构筑物和地形上来看，由北向南是一个由高到低的过程，体现了铜川“高耸巍峨的山地——层层叠叠朴厚的黄土台地——一望无际的辽阔平原——幽静私密的溪涧山谷”的丰富的地势地貌。其提取了铜川陈炉“窑炉”的建筑特点来展现铜川耀州窑文化。特色植物有合欢、七叶树、白桦、油松、柔毛绣线菊、褐穗莎草等。园内景点有陈炉台地、西宫墙、倒注壶、兰芝谷溪、洗药池等。

铜川园

Shanxi Weinan Garden

陕西渭南园

Area: 1,040 m^2

Exhibition is themed as "Culture and landscape of Weinan" to highlight its culture and design.

展园面积：1040 m^2

该园以人文渭南、山水渭南为主题，突出其文化和设计。

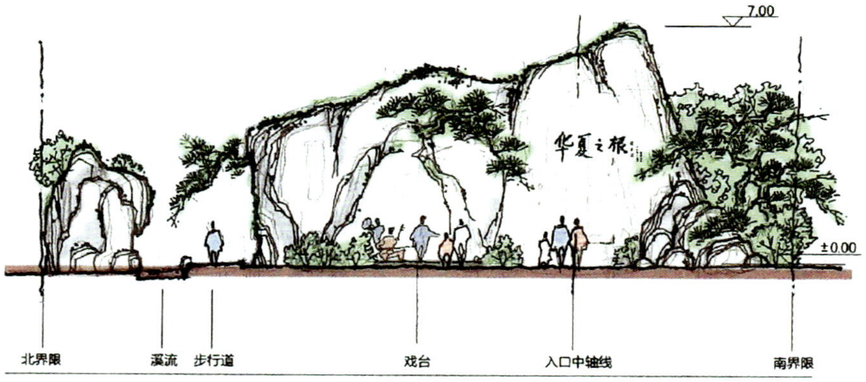

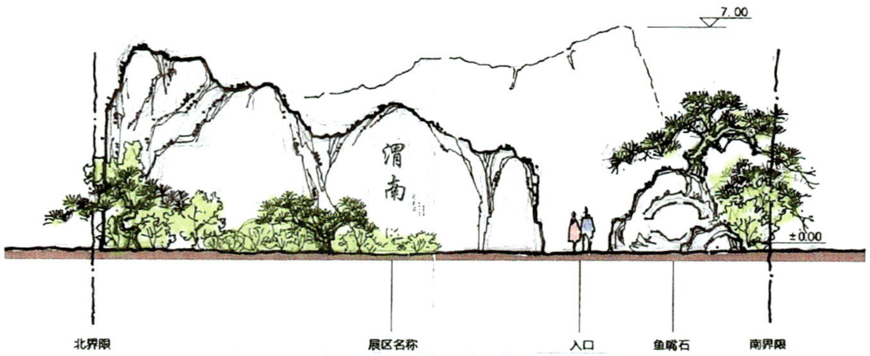

Introduction:

Exhibition is themed as "Culture and landscape of Weinan". Weinan Garden has four theme sections, namely, entrance section, inner-square exhibition section, valley exhibition section, and landscape exhibition section. The unique plantations are China Armand pine, apple tree, and persimmon tree. Sight spots are simulative landscape of Hua Mountain, such as North Peak, Baichi Gorge, Fish-mouth stone, and Jingxin Stone; waterfall landscape, free overfall, and Yellow River wetland; statues of the three Saints: Sima Qian, author of Records of the Historian, Du Kang, the wine-maker, and Cang Jie, the inventor of Chinese characters in legend; big LED screen; Silhouette figure cut and hitching posts.

展园介绍：

展区以人文渭南、山水渭南为主题。渭南园展区划分四个主题区：入口区、内广场展示区、峡谷展示区、山水写意展示区。其特有植物有华山松、苹果、柿子等。园内景点主要有模拟华山景观、北峰、百尺峡、鱼嘴石、警心石等景观，水系景观有瀑布、跌水、黄河湿地等；此外还有"司马著记"、"杜康酿酒"、"仓颉造字"三圣人雕塑，ED 大屏幕、皮影雕塑、拴马桩等。

Shanxi Xianyang Garden
陕西咸阳园

Area: 1,295 m²

The theme of the exhibition park is Qin People, Qin Culture and Xianyang Ecology, outlining the Shanxi culture.

展园面积：1295 m²

该园以秦人、秦风、生态咸阳为设计主题，突出陕西文化。

Introduction:

The theme of the exhibition park is Qin People, Qin Culture and Xianyang Ecology. The exhibition area is divided into three zones: entrance square, culture exhibition zone and gardening exhibition zone. The architectural style is represented by the pavilion, the great wall and the beacon tower with features of Qin Dynasty. There are special plants like sophora japonica, crape myrtle and ginkgo. The main sights include wrought copper, glass slips that is carved with poems on the back, Qin Straight, beacon tower on the great wall, bronze statue of Qin Shi Huang, the green sculptures, etc.

展园介绍：

展园以秦人、秦风、生态咸阳为主题。园区分为三部分：入口广场、文化展示区、园艺展示区。建筑风格为大秦风格的亭子、具有秦韵的长城与烽火台。特色植物有：国槐、紫薇、银杏。园内景点主要有锻铜、玻璃竹简其阴刻诗文、秦直道、长城烽火台、秦始皇铸铜雕像、绿雕等。

咸陽

Shanxi Yulin Garden

陕西榆林园

Area: 992 m^2

In this exhibition park, the miniature of Yulin Zhenbei Terrace is the landmark assisting by the desert scenery.

展园面积：992 m^2

该园以榆林镇北台缩影作为标志性建筑物，辅以特有的大漠风光来设计。

Introduction:

Yulin is a red land with long history, a golden land with rich resources, a green land with efficient ecological construction. In this exhibition park, the miniature of Yulin Zhenbei Terrace is the landmark assisting by the desert scenery. The gardening method integrates ancient and modern times, combines the dynamic and the static and uses the macro and the micro. The four zones and fourteen sights fully present the historical and cultural feature, the abundance of resources, the accomplishment of desertification and the spirit of braveness of Yulin, expressing the harmonious, ecological and developing New Yulin.

展园介绍：

榆林历史悠久，是一块红色的土地；榆林资源富集，是一块金色的土地；榆林生态建设成效显著，是一块绿色的土地。本园以榆林镇北台缩影作为标志性建筑物，辅以特有的大漠风光，采用了融通古今，动静结合，宏微并用的造园手法。四大区域，十四处景点充分展示了榆林历史文化之特色，能源储备之丰富，防沙治沙之成就，不畏艰难之精神，体现了一个和谐、生态、跨越发展的新榆林。

Shanxi Yan'an Garden
陕西延安园

Area: 1,015 m²

The theme of the garden is "Sacred Place of Revolution and Harmonious Yan' an", striking the odes of splendid and harmonious Yan' an.

展园面积：1015 m²

该园以"革命圣地，和谐延安"为主题，奏响辉煌的和谐延安歌曲。

Introduction:

Setting the theme "Sacred Place of Revolution and Harmonious Yan' an", the exhibition park consists three parts: Yan' an Impression, A Beautiful Place with Outstanding People, Harmony of Human and Nature. The natural landscape layout of "three mounts erect and two rivers converge" and the ecological feature of the Integration of Mountain, Water, Town, Garden in Yan' an are displayed by the Chinese traditional gardening means. The whole exhibition park put together the mount, the tower, the river and the kiln to present the featured style of Yan' an, recalling the poem that praises the Chinese revolution.

展园介绍：

延安园以"革命圣地，和谐延安"为主题，由"延安印象"、"人杰地灵"、"天地人和"三个部分组成，表现延安"三山雄立、两水汇流"的自然山水格局，以中国传统园林叠山理水的艺术手法展现延安"山、水、城、园"一体的生态特质。整个展园以"一山"、"一塔"、"一河"、"一窑"将延安"两黄两圣"的风貌融于方寸之间，引发人们联想起"中国革命峰回路转、两黄两圣源远流长、红色记忆奏响辉煌、革命圣地和谐延安"的诗句。

Corporate Park

企业园

Corporate Parks are comprised of nine parks such as Flawless Building, Energy-saving & Environmental Protection Hall, Magnificent & Creative Garden, China Telecom Science and Technology Hall, Northwest Agriculture & Forestry University Ecological Science and Technology Hall, Taxus Exhibition Park, Greenbelt Ecological Science and Technology Hall.

Flawless Building covers an area of 782 m^2, main building is dominated by mirror stainless steel, overall design utilizes the element of water, interprets the harmonious relationship between human and ecological environment in the future; Beautiful Landscape Creative Park covers an area of 802 m^2, regards nature, green, ecology, energy-saving as the design concept, creates a low-carbon landscape exhibition garden by combing manual work and nature; China Telecom Science and Technology Hall covers an area of 1,816, it follows the energy-saving design concept, adopts modern science and communication technology to reflect the relationship between technology and future life; Energy-saving Environmental Protection Hall covers an area of 1,976 m^2, it connects traditional culture with modern technology closely through the energy-saving design concept, thus realized the coexistence of human and nature… Corporate Park is interpreting the theme, Eternal peace & harmony between nature & mankind, of the 2011 Xi' an World Horticultural Exposition with ultramodern technology.

企业园共有无痕的建筑、节能环保馆、豪情创意园、中国电信科技馆、西农生态科技馆、红豆杉展园、绿地生态科技馆、天鹅湖展园和昆虫馆 9 个。

无痕的建筑展园面积 782m^2，建筑主体以镜面不锈钢、玻璃为主，整体设计以水为元素，很好地诠释了未来人与生态环境的和谐关系；豪景创意园面积 802m^2，以和谐、自然、绿色、生态、节能为设计理念，采用人工与自然相结合的手法打造了一个低碳环保的景观展园；中国电信科技馆面积 1816m^2，遵循节能环保的设计理念，运用现代科技和通信技术，体现了科技与未来生活的关系；节能环保馆面积 1976m^2，通过节能环保的设计，将传统文化与现代技术紧密结合，从而实现了天人合一，城市与自然和谐共生……企业园正用超现代的设计手法诠释了 2011 西安世园会"天人长安·创意自然——城市与自然和谐共生"的主题。

Greenland Ecological Sci-tech Garden

绿地生态科技馆

Area: 1,029 m^2

The garden is designed as a tree stump left over after human cutting the tree, responding to the theme of horticultural exposition with superior modern expressing skill.

展馆面积：1 029 m^2

该馆将建筑设计成一颗像被人类砍伐而残留的树根，用超现代的设计表达手法契合世园会的主题。

Introduction:

There are two storeys over land for the garden, containing greening ecological technology exhibition, future home exhibition, sunshine room, assisting room, main entrance, greenland corporation exhibition and VIP room. The underground floor is for the mechanical and electrical equipments.

The garden is designed as a tree stump left over after human cutting the tree, responding to the theme of horticultural exposition with superior modern expressing skill. The construction emphasizes creativity and beauty with the shape of "0" to imply the smoothest conversation between construction and environment. Through the future sustainable developing technology, the concept of Zero-energy, Zero-emission and Zero-interference to the environment is spread. Inside the exhibition, a great deal of displaying skills like sound, light and electricity are used to present the self advantages of ecological constructions, with the focal point of the nine systems of Zero-energy. Forty technologies could be used to make the energy of the construction approach to zero. The new decorating materials have shown the feature of low-carbon.

展馆介绍：

绿地生态科技馆地上有两层，分别为绿色生态技术展示区、未来之家展区、阳光房、辅助用房、主入口、绿地企业展示区、VIP 室等；地下一层，为各机电设备用房。

该馆具象地将建筑设计成一颗像被人类砍伐而残留的树根，用超现代的设计表达手法契合世园会的主题。建筑强调创新与美感，形态似“0”，以体现建筑与环境最流畅的对话。通过展示未来可持续发展技术，传递“0”能耗、“0”排放、对环境“0”介入的理念。展馆内大量应用了声、光、电等多样的展示手法来展示生态建筑的自身优势，重点展示了“零”能耗的九大系统，四十项技术可以使建筑能耗接近零，并采用新型装饰材料体现低碳经济特性。

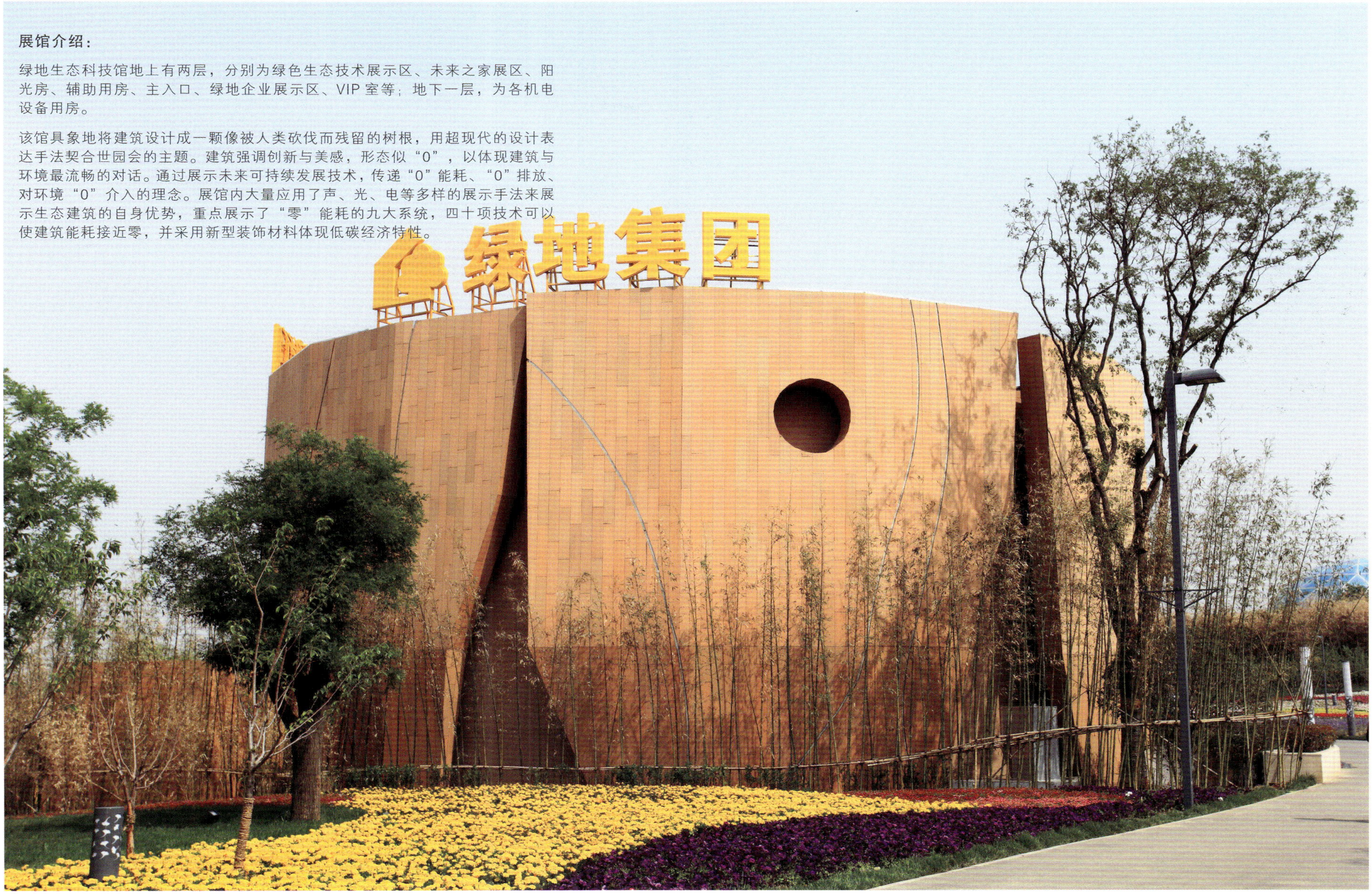

Magnificent & Creative Garden

豪景创意园

Area: 802 m^2

This exhibition garden centered as ecological city energy-saving, protecting nature.

展园面积：802 m^2

该馆的设计主题是以营造生态城市节能为中心，保护地球自然生态。

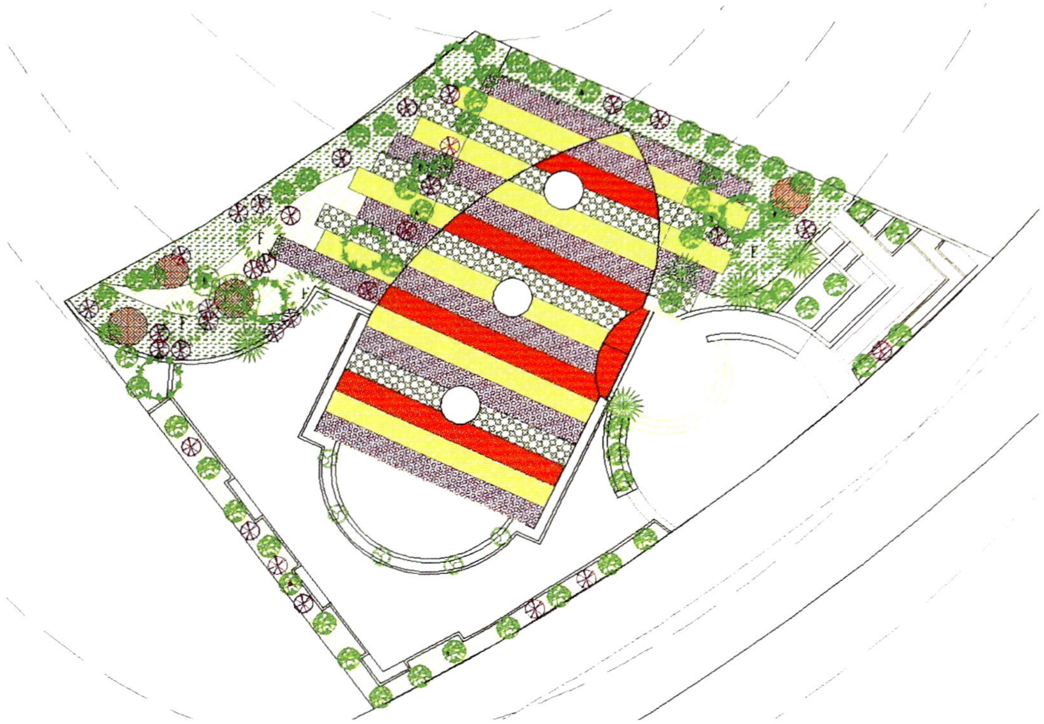

Introduction:

This exhibition garden centered as ecological city energy-saving, themed as protecting nature, it adopts the combination of artificial and natural style. In the concept of harmony, nature, green, ecology and energy-saving, it covers vertical greening on the artificial rocks on the outer walls of the building, presents a fragment of nature to the people living in the city. The advantage is selecting the combination of large potted plants and artificial rocks, low-carbon green, gives you a sense of returning to nature.

展园介绍：

展园以营造生态城市节能为中心，保护地球自然生态为主题，采用人造与自然相结合的风格，以和谐、自然、绿色、生态、节能为理念，在建筑物外墙面人工山石上进行垂直绿化，把大自然的片段呈现给久居都市中的人们。其优点是选用造型较大的盆栽植物与人工山石相结合，低碳环保，并给人们一种回归自然的感受。

西安2011世界园艺博览会
厂家直销 特价优惠
铁观音

西安2011世界园艺博览会
厂家直销 特价优惠
铁观音
普洱茶
创意园
云南普洱茶
入馆品茶

Traceless Building

无痕的建筑

Area: 782 m^2

Through the transformation of time and space among the circumstance, construction and human, the human living environment and harmonious relation in the future are perfectly explained.

展园面积：782 m^2

该园通过周边环境、建筑、人之间时间与空间概念的转化，完美地诠释出未来人类生存环境的和谐关系。

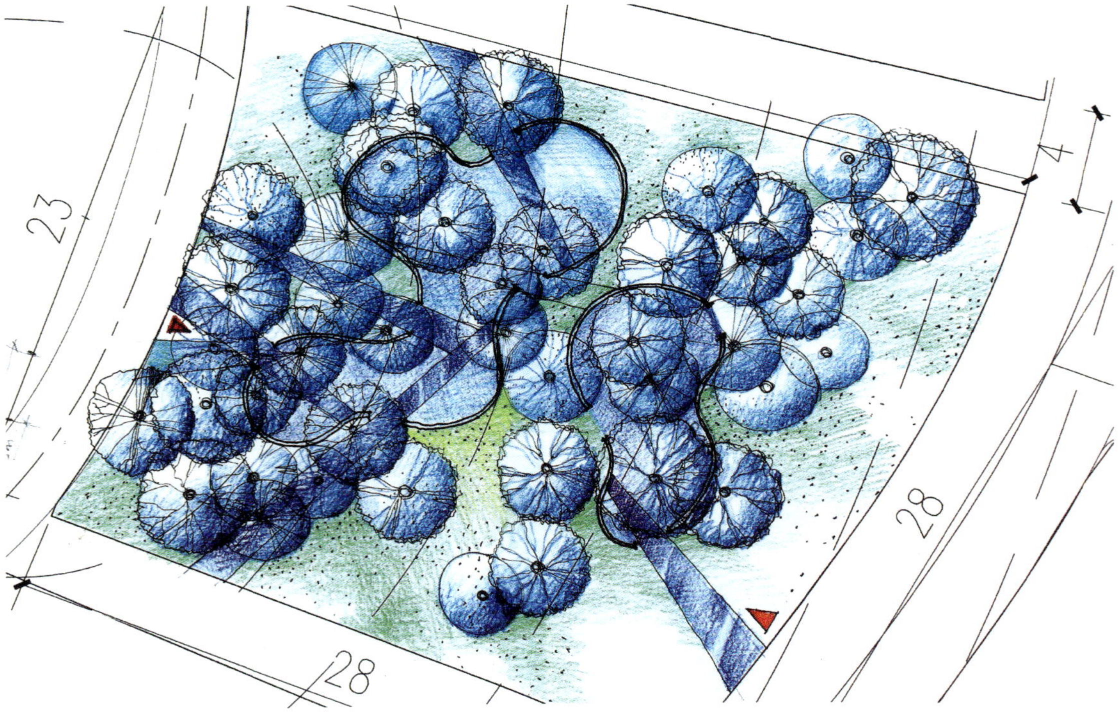

Introduction:

Traceless Building theme landscape part is made up of unconnected steel plates decorated by glass mirrors without roofs. The landscape introduction is totally inserted inside the glass mirrors.

Compared with other exhibitions of the horticultural exposition, Traceless Building is a small one with only about 700 m^2. However, the appearance is unique without roofs or other materials. Even the portal could not be told. The exhibition is divided into two enclosure areas. Outside the main construction, six natural passages connect the whole construction as one. .

This exhibition is a conceptual space artistic work, aiming at producing the complex and subtle circumstance consciousness through the conversion of the essential relationship among environment, construction and human during the visual process. The main materials are mirror stainless steel and glass. Water is the creative element of the overall design. The architectural forms and materials are in the harmoniously changing relation. Through the transformation of time and space among the circumstance, construction and human, the human living environment and harmonious relation in the future are perfectly explained.

展园介绍：

“无痕的建筑”主题景观园是由彼此并不连接、用玻璃镜面装饰的钢板构成，完全没有顶盖，景观内容介绍全面嵌入玻璃镜面内。

和世园会其他展馆相比，“无痕的建筑”个头很小，只占地七百多平方米，但是样貌却很独特，没有顶盖，没有出现更多的建筑材料，甚至也看不出建筑的正门，展馆主要分为两个围合区域，主体建筑之外有6条自然通道将整个建筑错落串联在一体。

此展区是一个概念型的空间艺术作品，意图通过环境、建筑、人之间的本质关系在视觉过程中的相互换位，自然的产生出错综复杂的、微妙的情境环境意识。建筑主体以镜面不锈钢、玻璃为主，整体设计以水为创意元素，建筑形态与建筑材质形成统一协调的变化关系。并通过周边环境、建筑、人之间时间与空间概念的转化，完美的诠释出未来人类生存环境的和谐关系。

Energy-saving & Environmental Protection Pavilion

节能环保馆

Area: 1,976 m^2

The Energy-saving & Environmental Protection Pavilion is the window propagating the public welfare concept of "Harmony of Human and Nature". It is the only exhibition focusing on Energy-saving & Environmental Protection in the horticultural exposition.

展馆面积：1 976 m^2

该馆是传播"人与自然和谐共生"公益理念的窗口，是世园会内唯一以节能环保为主题的展馆。

Introduction:

The Energy-saving & Environmental Protection Pavilion is the window propagating the public welfare concept of "Harmony of Human and Nature". The world advanced energy-saving and environmental protection technology, products and service are displayed here. It is the only exhibition focusing on Energy-saving & Environmental Protection in the horticultural exposition. The main construction closes up from the bottom to the top and the section is a isosceles trapezoid. Adopting the hollow glass and the adjustable Venetian sampan, the whole construction absorbs the energy mostly from the sampan, which is a complete self-sufficient greening circulating pattern. Meanwhile, the materials for heat preservation, fireproofing, quakeproofing and underfloor heating facilities are used to save energy; the LED lighting is used to save power. This pavilion is an intimate combination of traditional culture and modern technology, realizing the harmony of human, city and nature.

展馆介绍：

节能环保馆是传播“人与自然和谐共生”公益理念的窗口，其展示世界先进节能环保技术、节能环保产品、节能环保服务，是世园会内唯一以节能环保为主题的展馆。建筑主体由最低层向上逐渐收拢，其剖面形式呈等腰梯形。采用中空玻璃，可调控百叶舢板，整个建筑每日所消耗的能量大部分来自于舢板所吸收的能量，是一种完全的自给自足的绿色循环模式。同时馆内利用保温、防火、防震材料以及地暖设施节能；采用 LED 照明充分节电及其它节能新材料、新技术。节能环保馆的设计是传统文化与现代技术的紧密结合，从而达到天人合一，城市与自然和谐共生。

Taxus Chinensis Garden

红豆杉园

Area: 1,075 m^2

The Garden has further introduced the endangered plant – Taxus Chinensis comprehensively.

展园面积：1075 m^2

该园很好地展示了红豆杉这一珍稀物种，并且进行了全面的深入介绍。

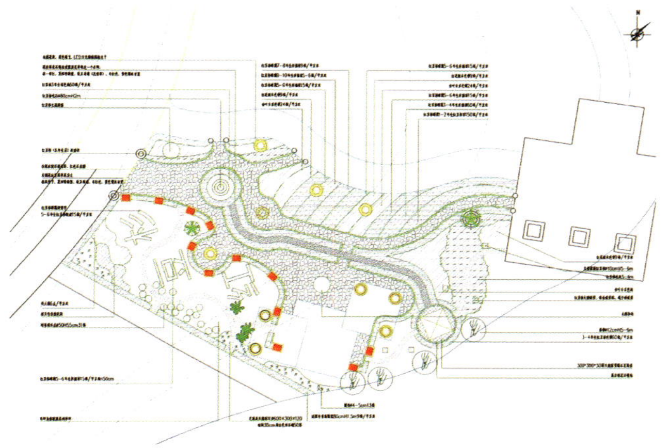

Introduction:

Taxus Chinensis is a rare and endangered species handed down from Quaternary Ice Age 2,500,000 years ago. Forty-two countries featuring Taxus Chinensis all call it as "the national treasure". It is worthy of the name of "panda in plants".

The Garden is divided into the indoor part and outdoor part. The outdoor part has shown Taxus Chinensis in different forms, such as the green tower in the shape of Taxus Chinensis and the shrub colored belt of Taxus Chinensis. There is glass greenhouse inside the Garden to show the cultivation of seeds and the process of growth in detail. On the other side, it has introduced the medicinal value of Taxus Chinensis and the technological process of extracting "paclitaxel". On these two aspects, it has further introduced the endangered plant – Taxus Chinensis.

展园介绍：

红豆杉是250万年前第四纪冰川时期遗留下来的珍稀濒危物种，全世界42个有红豆杉的国家均称其为"国宝"，是名副其实的"植物大熊猫"。

红豆杉园内展示区分为室内、外两部分。室外部分充分的将红豆杉以各种形式穿插其中，包括红豆杉造型绿塔、红豆杉灌木色带等。园区内建有玻璃温室。详细介绍红豆杉的育苗、培育、生长的过程。另一方面介绍红豆杉的药用价值及其提取"紫杉醇"的工艺流程。通过两个方面全面深入地介绍红豆杉这一濒危珍贵植物。

Taxus theme park
红豆杉
主题园

Swan Lake Garden

天鹅湖园

Area: 1,054 m^2

With the theme of swan, it has reflected the natural architectural concept aiming at creating the harmony between human and nature.

展园面积：1 054 m^2

该园以天鹅为主题，完美体现了人与自然和谐相处的自然建筑艺术理念。

Introduction:

Swan Lake Garden takes advantage of the 333,000m^2 wetland around to create a distinctive tourist destination integrating flying swans, rare species on the water surface and shellfish on the bottom of the lake.

The Garden looks like a DV camera which records the dynamic beauty of the birds in Swan Lake. The buildings are designed with steel structure to be a healthy exhibition hall integrating environmental protection, health, comfort and unique style together. In addition, there is reasonable ring road, waterfront platform and rest platform for the visitors. It has reflected the natural architectural concept aiming at creating the harmony between human and nature.

展园介绍：

在园区的天鹅湖展馆，巧妙的利用周边 333 000 m^2 的水域湿地，打造了一个集天鹅展翅飞翔、水上珍奇戏水、水底贝类蜇伏为一体，特色鲜明的观光圣地。

天鹅湖展馆主题外形酷似一台 DV 摄像机，将整个天鹅湖的百鸟飞翔的动态美尽录其中。天鹅湖展园建筑用钢结构框架，集绿色环保、健康、舒适、设计风格独具个性等诸多优势于一身的健康型展馆，同时辅以合理的人流环线，亲水平台，休息平台等人性化设计。完美体现人与自然和谐相处的自然建筑艺术理念。

China Telecom Sci-Tech Garden

电信科技园

Area: 1,816 m²

The exhibition garden abides by the energy-saving principle, it symbolizes China Telecom' s concept and pursuit that keep exceeding.

展园面积：1 816 m²

该馆遵循节能环保的建园原则，象征了中国电信不断超越的理念和追求。

Introduction:

The exhibition garden abides by the energy-saving principle, it adopts steel structure, glass curtain wall, modern and transparent in model, appearance material is novel, energy-saving environmental protection and can be reused. The hall is divided into two-storey, the second floor is semi-open, you can take a rest on the platform to overlook the distant landscape, with night light design and the water refection, let you filled with infinite fancy. Membrane shape on the roof pulling off like open wings, symbolizes China Telecom' s concept and pursuit that keep exceeding, entryway adopts the modern landscape design, coordinates with the main part of the exposition. The latest display of modern science and communication technology reveals the close relationship of technology with future life.

展园介绍：

展园遵循节能环保的建园原则，场馆采用钢结构、玻璃幕墙，造型现代、通透，外观材料新颖、节能、环保，且可重复使用。场馆分为上下二层结构，二层为半遮露天形式，客人可以坐在二楼平台上休息，眺望远处景色，加上夜景灯光的设计，配合水边倒影，让人充满无限遐想。屋顶采用帐拉膜造型，像张开的翅膀，展翅遨游，象征了中国电信不断超越的理念和追求，入口通道使用现代的园林设计风格，与世园会主体交相辉映，场馆内现代科技和通讯技术的最新展示，揭示科技与未来生活的紧密联系。

中国电信
CHINA TELECOM
天翼
中国电信 信息馆
我的e家

Northwest Agriculture & Forestry University Sci-tech Ecological Garden

西农生态科技馆

Area: 817 m^2

The pavilion presents the feature of “farming, forest, water” as the external display window of Northwest A&F University.

展园面积：817 m^2

该园体现了“农、林、水”的特色，是西北农林科技大学对外展示的窗口。

Introduction:

The pavilion is the external display window of Northwest A&F University, displaying the agricultural products of the university. The main construction is the greenhouse of 258 m^2 and several landscape columns. Outside the greenhouse room is the artistic spay printing subjecting “Burning Green Fire”; while inside the energy-saving room the plants cultivated by means of agricultural science and technology are displayed.

The pavilion presents the feature of “farming, forest, water” through the garden construct, water and the arrangement of plants. There are three landscape sections: the landscape at the entrance is design in accordance with the theme “farming”; the featured landscape columns are refined based on the image of ear of wheat; the central landscapes are made up of waterscape, garden landscape and flexible plant landscape.

展园介绍：

本展园是西北农林科技大学对外展示的窗口，主要展示学校的农业产品，主体建筑为258m^2的温室和若干个景观柱。温室景观区室外是以“燃烧绿色的火焰”为主题的艺术喷绘，节能温室内展示区主要展示一些利用农业科技手段培育的植物。

展园在园林建筑、水体和植物的配置上体现“农、林、水”的特色。分为3个景观区：入口处的景观设计围绕着农的主题进行设置，特色的景观柱以麦穗的形象为原型进行艺术的提炼；中心景观区则由水系景观、园林小品景观和丰富变化的植物景观构成。

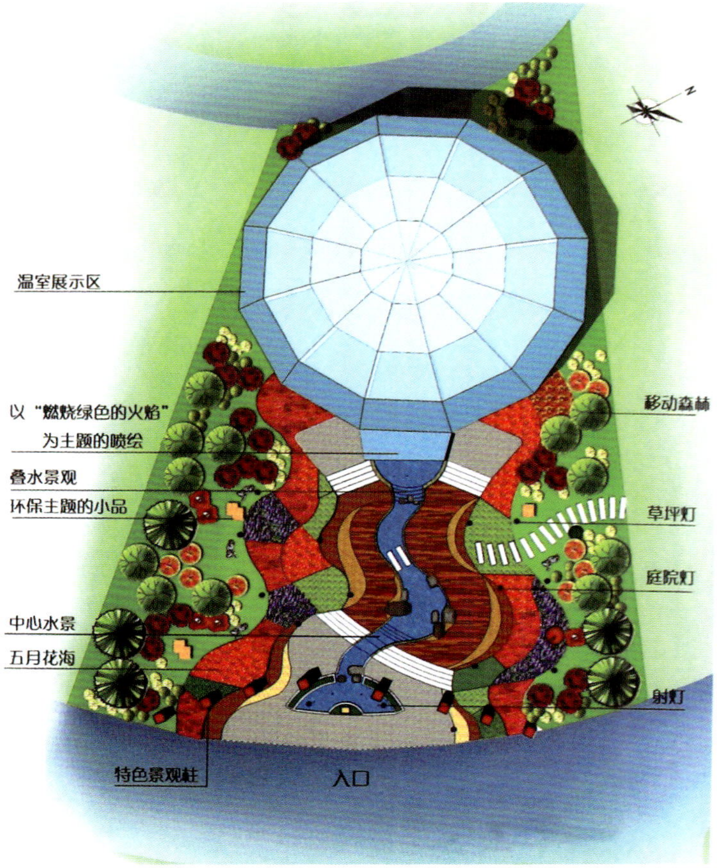
温室展示区
以"燃烧绿色的火焰"
为主题的喷绘
叠水景观
环保主题的小品
中心水景
五月花海
特色景观柱
入口
移动森林
草坪灯
庭院灯
射灯

出口
入口

Feature Garden

专类园

Feature Gardens of 2011 Xi' an World Horticultural Exposition consist of Chang' an Garden, Bonsai Garden, Qinling Garden and Poetic Landscape in Chang' an Garden.

Chang' an Garden is located on the Small Zhongnan Mountain, covers an area of 56,000 m^2, close to Chang' an Tower and lake, it takes bamboo as landscape element, creates a poetic and elegant landscape exhibition area, expresses people' s pursuit of comfort life; Bonsai Garden is located on the Small Zhongnan Mountain, too, adjacent to Qinling Garden and Chang' an Tower, total area is 17,000 m^2, in the world arts, bonsai art is one of the art treasures in the Orient with natural implications, and one of Chinese characteristic traditional landscape art···

2011 西安世园会专类园包括长安园、盆景园、秦岭园、人文山水·诗意长安园等。

长安园位于小终南山上，面积约 5.6 万 m^2，傍长安塔、依湖水而建。长安园以竹为景观元素，打造了一个诗意典雅的景观展园，表达了人们对舒适生活的追求；盆景园位于小终南山上，紧邻秦岭园和长安塔，总面积约 1.7 万 m^2，在世界艺术之林中，盆景艺术是富有自然情趣的东方艺术精品之一，也是中国独特的传统园林艺术之一……

Chang'an Garden

长安园

With the design theme of "interpretation of new concepts of green, ecology and Chinese style", the Garden has applied many landscape elements with rich Chinese characteristics.

该园是以绿色、生态、中国风的"新概念诠释"为主题，采用了极具中国韵味的景观元素进行设计。

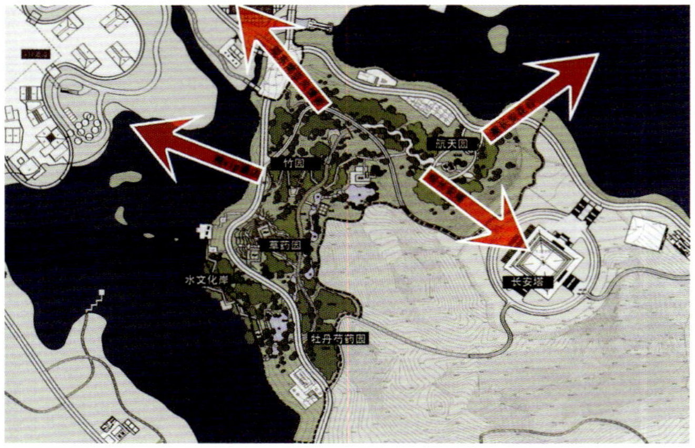

Introduction:

Chang' an Park has a design theme as "interpretation of new concepts of green, ecology and Chinese style". It has Space Plants & Space Science Education Exhibition Zone, Bamboo Culture Exhibition Zone, Herb Culture Exhibition Zone, Cultural Water Front Exhibition Zone, and Peonies Exhibition Zone.

Bamboo Culture Exhibition Zone, which is also called "Bamboo Garden", is the interpretation of new concept of Chinese Style. The entrance, sculpture, bamboo corridor, bamboo plank road, bamboo house, teahouse, and other landscape elements with rich Chinese characteristics are made by weaving bamboos slices in the traditional way. Herb Culture Exhibition Zone is shaped by layers of terraces, fresh and natural. When seeing people farming here, you will think of "While picking asters beneath the Eastern fence, My gaze upon the Southern mountain rests" – the verse of Tao Yuanming, a poet of the Jin Dynasty. Reed, lotus pond, trestle bridge, spring, stream, painting cottage, and a green island together constitute a rich poetic ink-and-wash landscape painting in the Cultural Water Front Exhibition Zone. There is painting in poetry and poetry in painting. Peonies Exhibition Zone is full of beautiful tree peonies and herbaceous peonies. Xu Ning, a poet in the Tang Dynasty, wrote in a poem, "Who do not love the peony? Its beauty surpasses that of all other flowers in the city."

展园介绍：

长安园以绿色、生态、中国风的“新概念诠释”为主题，包括航天植物及航天科普教育展示区、竹文化展示区、药草文化展示区、文化水岸展示区、牡丹芍药展示区。

竹文化展示区又被称为“竹园”，其诠释中国风的新概念，运用中国传统的竹编织出景区大门、雕塑、竹廊、竹栈道、竹舍、茶室等极具中国韵味的景观元素。药草文化展示区则以层层的台地塑造清新自然的药草园，人于其间躬耕，不由想起晋代诗人陶渊明的“采菊东篱下，悠然见南山”。文化水岸展示区以芦苇、荷塘、栈桥、泉水、溪流、画庐、绿岛共同绘出一幅诗意盎然的写意水墨风景画，可谓画中有诗，诗中有画，意境无穷。在牡丹芍药展示区是满园娇妍的牡丹、芍药。唐代徐凝诗云：何人不爱牡丹花，占断城中好物华。

Poetic Landscape in Chang'an Park

人文山水·诗意长安园

With the theme of "recreating poetic glamour in landscape and art; and reproducing the beauty of the plants described in The Book of Songs". the Park has created the best ecological environment.

该园以"重塑山水诗画意境·再现诗经植物风华"为主题，挑战最佳生态环境的创造。

Introduction:

The design theme of the Park is "recreating poetic glamour in landscape and art; and reproducing the beauty of the plants described in The Book of Songs". The design content includes cultural landscape, botanical garden of The Book of Songs, and botanical garden of wetland.

The Park displays the natural sceneries combining traditional Chinese poetry and paintings and classic gardens according to The Book of Songs, Tang poetry, and the Wang Chuan Ji written by Wang Wei (a famous poet in Chinese history). Taking the essence from the description of Shaanxi landscape in Guo Feng – The Book of Songs and the Tong poem Wang Chuan Bie Ye written by Wang Wei, the Park is designed as a traditional Chinese garden with themes of "Love" and "Harmony". Aiming at showing the oriental characteristics, it puts stress on addressing the ecological crisis and cultural challenge in the 21th century.

The sceneries are designed with cultural deposition. It is a mirror of the rich culture and long history of the oriental world. By showing the vicissitude of Shaanxi and addressing the ecological crisis of our time, the Park introduces real landscape space and makes the best manifesto of ecological environment.

展园介绍：

展园设计主题为“重塑山水诗画意境·再现诗经植物风华”，设计内容主要包括人文山水、诗经植物园、湿地植物园。

展园集中体现了诗经、唐诗以及王维《辋川集》中反映中国传统诗画意境与古典园林结合的生态自然景观。以《诗经·国风》中秦地风光诗句和唐代王维《辋川别业》诗句意境为境界，用中国传统园林构景手法，以塑造时代文化主旋律——“爱”与“和谐”造林思想为纲要，规划一个突显东方文化特色，应对时代“生态危机”、“文化挑战”的 21 世纪时代新园林。

景观设计中运用适当的笔墨塑造富有人文气息的园林景观。利用园林景观辐射悠久历史的东方文化，展示秦地的沧桑历史，应对时代的“生态危机”挑战，以实地山水园林空间发布创造最佳生态环境的宣言。

Aerospace Plant Garden
航天植物园

The Garden has reproduced the process of cognizing and exploring the aerospace, and walking into the universe, showing the new development of modern agricultural technology.

该园再现人类认识、探索太空，走向宇宙的历程，展示了现代农业技术新水平。

Introduction:

The Garden is built back to the mountain with the atmosphere of modern science and technology. Arc track and spacecraft model have shown the process of cognizing and exploring aerospace, and walking into the universe. There are new species of poinsettia, spiderflower, giant pumpkin, ornamental pumpkin, tomato and eggplant that are bred in aerospace, showing the new development of modern agricultural technology. It is an exhibition park full of novelties and scientific education.

展园介绍：

展园依山而建，整体造型富有时代科技气息，通过弧形轨道、航天器模型等，再现人类认识、探索太空，走向宇宙的历程。园中种植航天育种的一品红、醉蝶花、巨人南瓜、观赏南瓜、番茄、茄子等花卉蔬果新品种，展示了现代农业技术新水平，是一个极具新奇性和科普教育意义的展园。

CZ-3C
中国航天
航天科技

天王星
幼苗期

Qinling Park

秦岭园

With the theme of "presenting natural scenery, rare animals and plants which are characteristic of Qinling Mountain", all sceneries are built basing on the real landscape of Qinling Mountain to enable people to experience the beautiful landscape and profound culture of Qinliing.

该园以"展示秦岭自然山水、珍稀动物、特色植物"为主题，力求通过秦岭片段的移植展示，让人们观感秦岭山水之奇峻壮美，文化之深厚悠远。

Introduction:

Located on the Little Zhongnanshan Mountain, the park is designed with the theme of "presenting natural scenery, rare animals and plants which are characteristic of Qinling Mountain". The Park has four gardens and eight sceneries. Rockwork and water management are arranged in the Park according to the actual geography. All sceneries are built basing on the real landscape of Qinling Mountain to be open and deep. On the Little Zhongnanshan Mountain, there are many stone tablets and cliffs inscribed with poems describing the landscape of Qinling written by people from ancient times. These poems create profound landscape culture of Qinling Mountain.

Aiming at displaying the bio-diversity of Qinling Mountain and appealing to the public to love nature and protect animals, the Xi' an Horticultural Expo introduces to the exhibition the "Four Treasures of Qinling", the unique Shaanxi animals including panda, takin, crested ibis, and golden monkey, for the first time. In order to show the harmony between human and nature, the design of animal cages and exercise yards in the Qinling Park are skillfully done according to the actual land shape, following tightly the principle of merging into the environment.

To show the rich varieties of plants in Qinling, the Qinling Park will plans over 30 threatened plants such as Kingdonia uniflora, Paeonia rockii, and Cercidiphyllum japonicum, and over 40 conservative plants of Shaanxi Province such as Saruma henryi, Callianthemum taipaicum, and Aristolochia manshuriensis.

展园介绍：

秦岭园是位于园区小终南山上，以“展示秦岭自然山水、珍稀动物、特色植物”为主题。园区共有“四园八景”。园区结合周边地形地貌堆山叠石、理水种植，一切景点均截取秦岭山水片段建造，开阔且幽深。在小终南山上，随处可见镌刻有古人描写秦岭风光诗词的石碑、石崖，秦岭的山水文化令人赞叹回味。

西安世园会首次将极具陕西特色的“秦岭四宝”——大熊猫、羚牛、朱鹮、金丝猴引入展园，目的在于展示秦岭生物多样化，呼吁人们爱护自然，爱护动物。为了彰显人与自然的和谐相处，秦岭园中的动物笼舍、活动场地力求与山体和周边自然地形巧妙结合，尽量做到藏而不露，露而不显。

为展示秦岭丰富的植物景观特色，展园结合山水地形，种植独叶草、紫斑牡丹、连香树等国家濒危植物30余种以及马蹄香、太白美花草、木桶马兜铃等陕西省保护植物40余种。

Bonsai Garden

盆景园

Bonsai is one of the art treasures in the Orient with natural implications. It has reserved the most natural and creative interest of the Bonsai.

该园是富有自然情趣的东方艺术精品之一，保留了盆景最天然、最具创意的情趣。

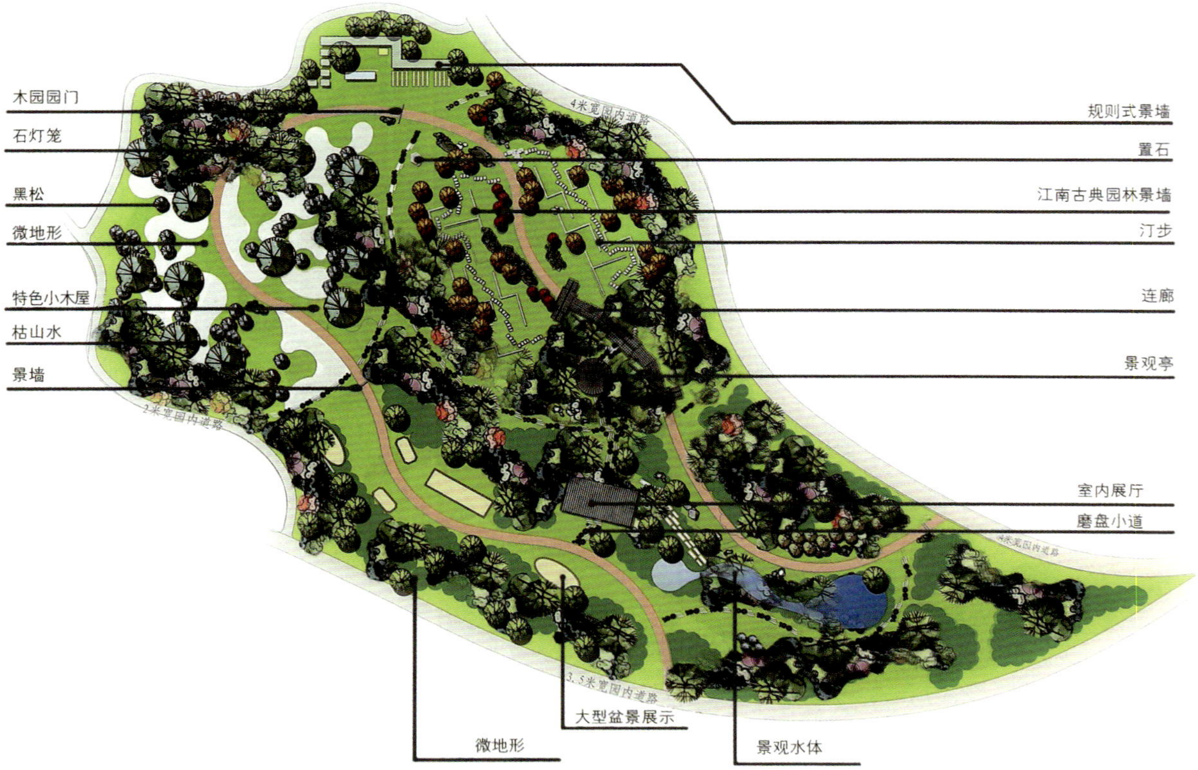

Introduction:

Bonsai is one of the art treasures in the Orient with natural implications. Passing through the main entrance—Chunhua gate that stands in front of Bonsai Garden. Ascend the stairs along the bai shi zi road and mo pan path, you will see, the entire Bonsai Garden is an open exhibition area. It constructs according to the terrain with regional featured stone piers, manger and hitching post around. Four characteristic exhibition areas in Bonsai Garden are: Guangzhong Garden, Five Genre Garden, Wood Garden and Tianqu Garden. Guangzhong Garden is full of Shaanxi folk custom; in the Five Genre Garden, the scattered landscape stone and the verses carved in intaglio on it, make you feel just like cross back to ancient times; Wood Garden derives from the Bonsai of Edo Era in Japan, Japanese black pine' s stems spreading out, the crown is like a canopy, the needle-point leaf is deep green all the year round, the tree performance is elegant, landscape ornaments in Japanese style such as black stone, stone lantern express cultural meaning; Tianqu Garden reserves the most natural and creative interest of the Bonsai

The plants in the Park are mainly from Xi' an, including ligustrum lucidum ait, sakura trees, pomegranates, red-leaf flowering peaches, photinia fraseri, and lodge pole pines.The outskirt of the Park is a mixed forest of ligustrum lucidum ait and willows. It is arranged with consideration of the color and height of the trees as their crown diameter varies with seasons. Large flowerbeds and local trees around the entrance are to give pleasant sensual experience in terms of vision, touch and smell.

展园介绍：

盆景艺术是富有自然情趣的东方艺术精品之一，穿过屹立在盆景园前的主入口春花门。沿着白石子路和磨盘小道拾级而上可以看到整个盆景园是一个开放式的展园，在小山包上依势而建，极具关中地域特色的石墩、饮马槽、拴马桩遍布四周。盆景园中的四大特色展园分别是关中园、五大流派园、木园和天趣园。关中园极具有陕西民俗风；五大流派园中散置的风景石、上面凿平阴刻诗句，使人体会到与古代交错的微妙感觉；木园取自日本江户时代的盆景，源于日本的黑松枝干横展，树冠如伞盖，针叶浓绿，四季常青，树姿古雅，日式山水、黑石、石灯笼等特色日式小品，为“文人植木”之意；天趣园则保留了盆景最天然、最具创意的情趣。

在植物的运用选择上大多选用西安当地植物，如大叶女贞、樱花、石榴、红叶碧桃、红叶石楠、黑松等。展园的外围是以混交林如大叶女贞、柳树等作为背景，并按照季象及冠幅的变化考虑色彩高度上的搭配。入口选用大片花境搭配乡土树种在视觉、触觉、嗅觉上达到赏心悦目的感官效果。

Characteristic Creations

特色创意

The waters, bridges, small ornaments, sculptures and the surrounding green landscapes in the gardens which represent the creative ideas of the International Horticultural Exposition 2011 Xi' an, feature distinct characteristics of the times, profound local features and creativity, and magnificent masters' style, to be another brightspot of the International Horticultural Exposition 2011 Xi' an.

With the delicate details, the international, marketing, cultural, ecological and harmonious Xi' an has shown people the excellent horticultural skills and the pursuit for perfectness.

代表 2011 西安世园会创意思想的园区的水、桥、小品、雕塑及周边的绿化等都紧扣世园会的主题，表现出鲜明的时代特色、深刻的本土特色和创意、大气的大师风范，成为 2011 年西安世园会的又一亮点。

正是从世园会的细微之处可以看出，一个国际化、市场化、人文化、生态化的和谐西安正向世人展示其园艺技术高超、追求完美的一面。

Horticultural Plaza

世园广场

The environmental design for the landscape plaza is developed in accordance with the local condition. The style and materials are flexible and the colors are simple, integrating with the landscape around.

景观广场环境设计因地制宜，设计手法、材质元素多变，用色朴素，与周边景观浑然一体。

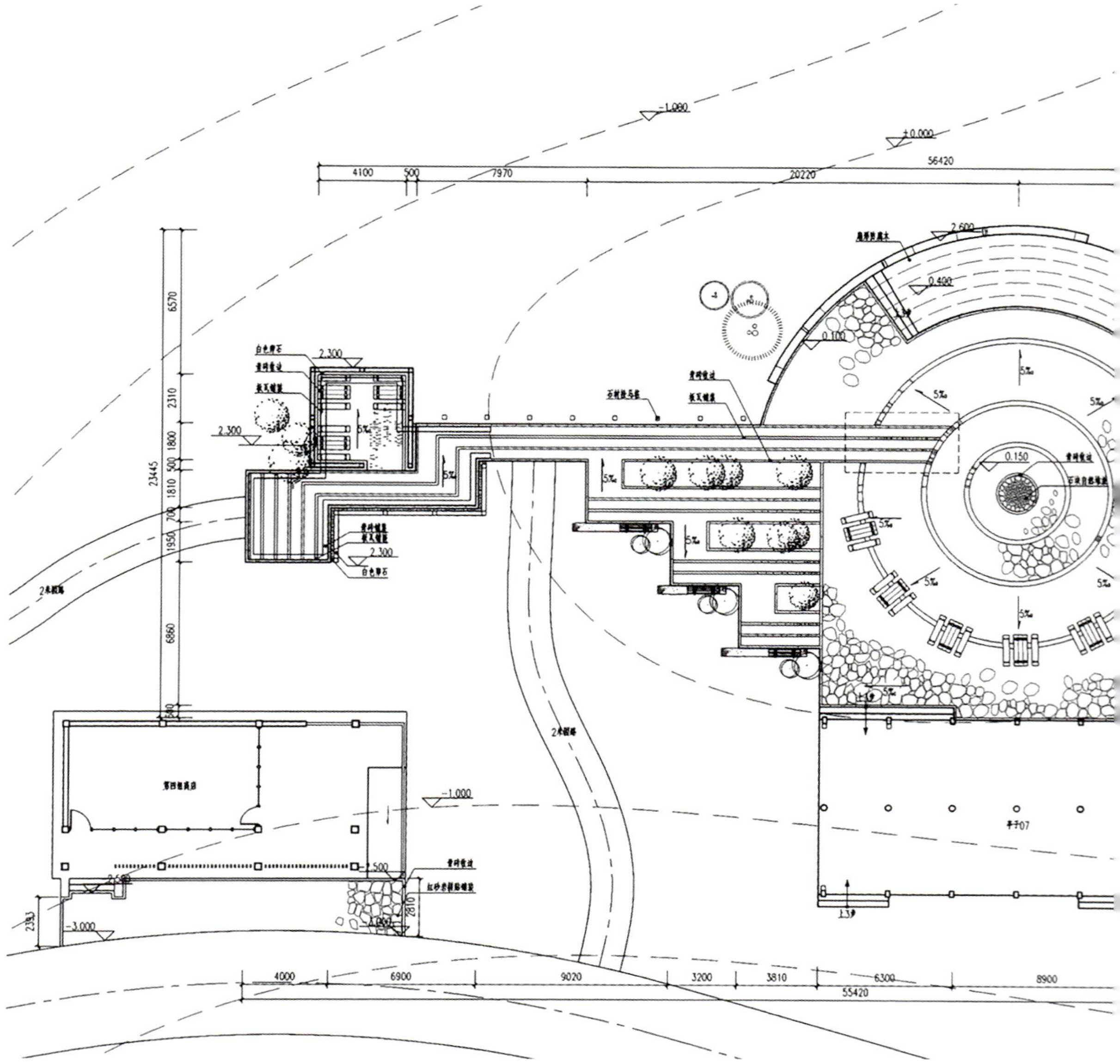

Introduction:

There are 33 landscape plazas in Horticultural Exposition, containing landscape plazas and service plazas. The environmental design for the landscape plaza is developed in accordance with the local condition. The style and materials are flexible and the colors are simple, integrating with the landscape around and better making each theme park stand out. The designs of Plaza No. 2, No. 9, No. 10 and No. 15 are especially remarkable.

Plaza No. 2 stands at the cross of International Garden and Chang' an Garden with area of 3, 255 m^2 in design of service function. The north side is resting space with the main pattern of chairs beside trees and ponds, combining stones and terracotta; the south side is traffic space decorated with paint and concrete. Located on the north of Romance by the Ba River, Plaza No. 9 is strip shaped stretching to the dock by the lake with area of 2, 662 m^2. The modern broken line design matches with the style of Romance by the Ba River and form the extension of space. Plaza No. 10 is a large complex service and landscape plaza on the west side of World Garden with area of 4, 180 m^2. Multi-layer spaces are created according to the terrain. Plaza No. 15 is located in the middle of Master Garden with area of 2, 897 m^2. The arbors in this area are separated by the modern broken lines to present dynamic feeling and harmonious relation for the area.

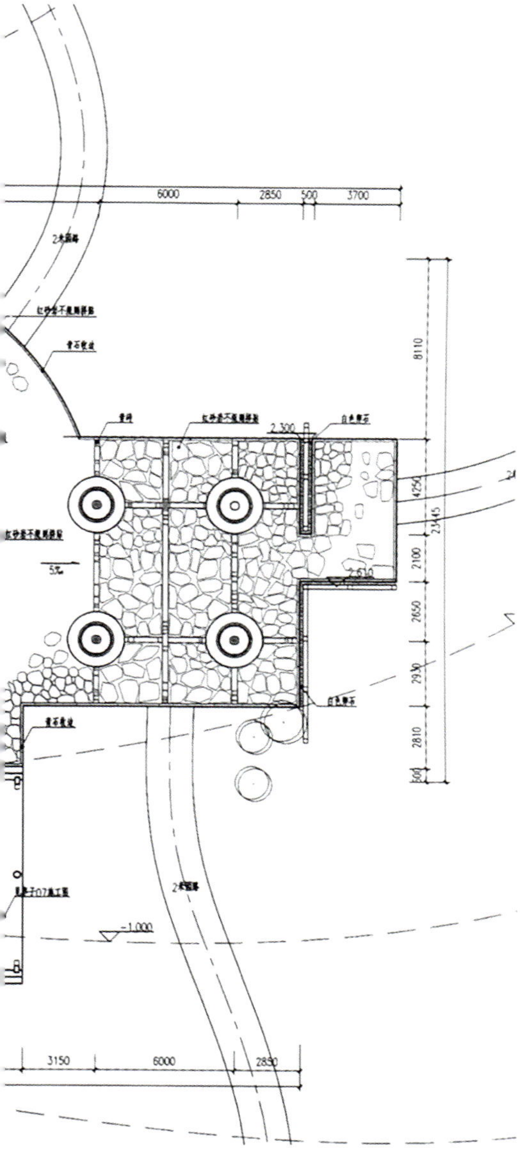
6000
2850
500
3700
8110
2.300
4250
23445
2100
2650
2930
2810
-1.000
3150
6000
2850

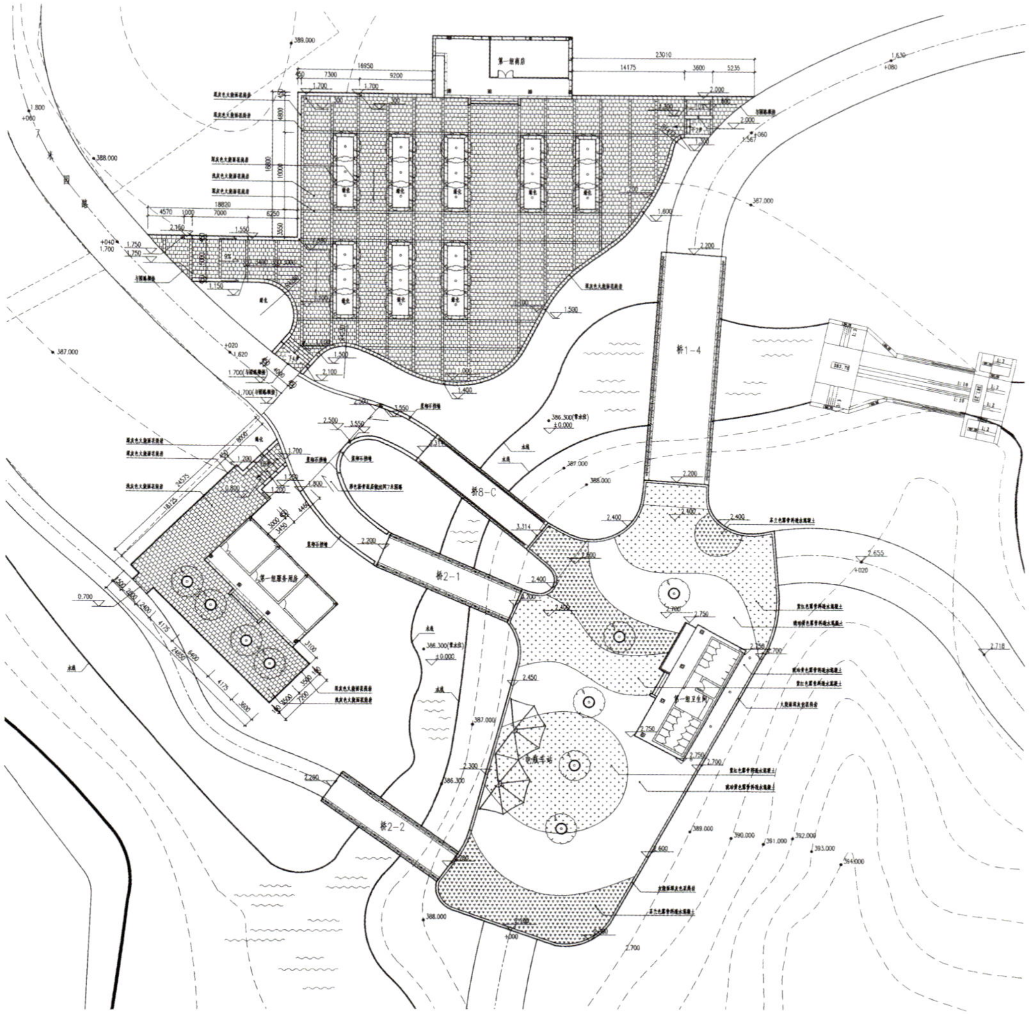

展区介绍：

西安世园会景观广场有33个，分为景观广场和服务区广场。景观广场环境设计因地制宜，设计手法、材质元素多变，用色朴素，与周边景观浑然一体，更好地衬托出各个主题展园。其中以2号、9号、10号、15号广场的设计尤为显著。

2号广场位于国际园与长安园交界处，面积3 255 m²，设计以服务功能为主，北侧为停留空间，运用石材与陶砖结合，树池座凳形式为主，南侧为多彩透水，混凝土铺装的交通空间；9号广场位于灞上人家北侧，为条形广场，延伸至湖畔码头，面积2 662 m²，设计以现代折线感为主，与灞上人家设计风格统一，形成空间感觉上的延续；10号景观广场位于世界庭院西侧，为大型综合性服务景观广场，面积4 180 m²，依地形而变化，创作出多层次的丰富空间；15号景观广场位于大师园中部，面积2 897 m²，广场设计通过极具现代感的折线对地形内保留大型乔木进行划分，使场地具有流动感，并体现出人工与自然的和谐共生。

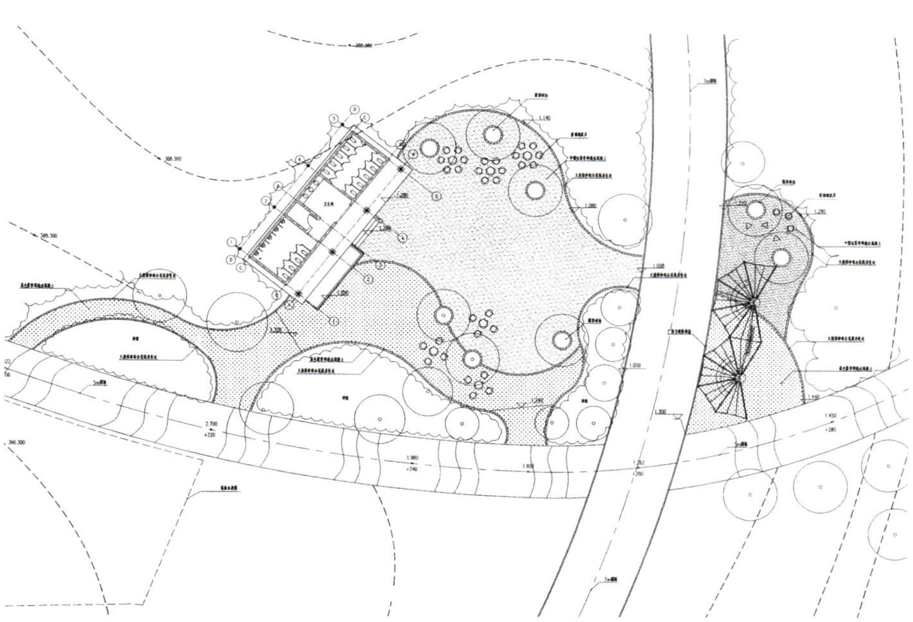

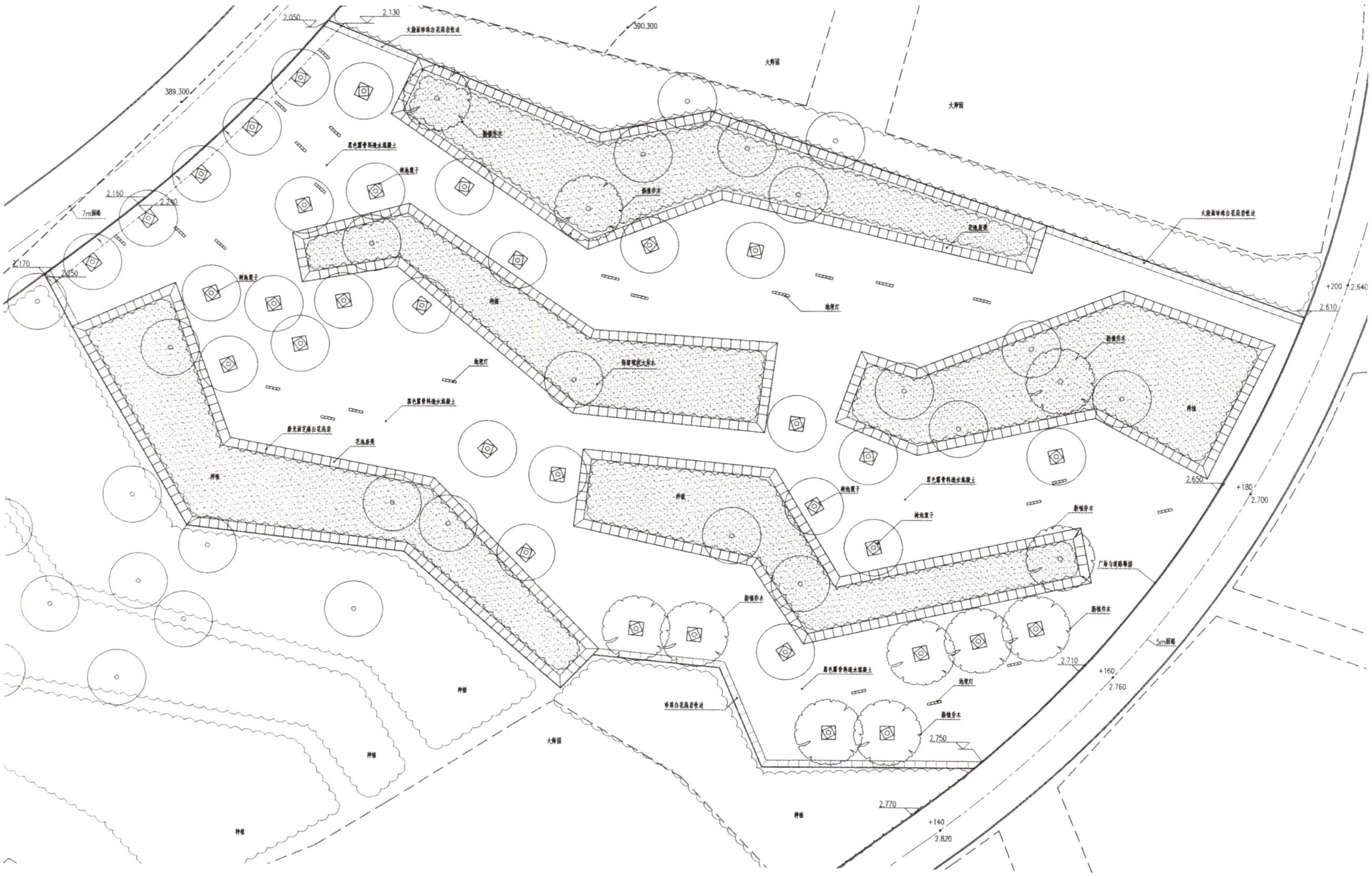

东沃创意广场

DOW Creative Plaza

Horticultural Expo Greening

世园绿化

The bright point of the Expo is green which highlights the trend of fashion.

世园会的亮点是绿色，让绿色引领时尚成为亮点。

Introduction:

The bright point of the Expo is green. The green trees and flowers are exquisitely selected and planted to enhance greening and landscaping. There are large-scale green plants, green sculptures and flowers on both sides of the road and in the main gateway to improve the level of greening and landscaping.

The greening level of the ecological area is the key to displaying the "Green to Fashion" theme of Xi' an Horticultural Expo. The greening would be enlarged in Chan-Ba ecological area in accordance with the standards of green planting: 5000 trees with semi tree crown and 20cm long radius would be planted. Additionally, the species and sizes of trees, the standards of frames, the height of the branches, the depth for exchanging the earth and transplanting are unified. Moreover, the greening density would be raised; the arbors, bushes, flowers and grass would be added to form the tridimensional greening; the flower baskets would be hung on the road lamps; the flower boxes would be set on the railings and sidewalks - the area of greening would be enlarged by adding not only standing boxes but also tridimensional frames.

展园介绍：

世园会的亮点是绿色。世园会精心挑选绿化苗木、花卉种类选择与种植，大力实施绿化提升和景观美化。道路两侧及区域主要出入口有大型绿植和绿雕、花雕、花卉，提升绿化和景观层次。

要展示西安世园会“绿色引领时尚”的主题，生态区的绿化水平亦是关键。浐灞生态区将按照绿化栽植的具体标准加大全区绿化工作：今年种植5000棵大树，而且争取是半冠以上、胸径20cm以上的大树。此外，对树的品种、规格、支架的标准、树木开叉高度、换土深度、营养土换植等要求予以统一。此外，还将加大绿植密度，增加乔木、灌木、花草，形成立体绿化；将在区内路灯上挂上花篮，栏杆、人行道上要增加花箱、花拱，不仅仅是竖箱，还要增加立体盆架，大面积增加立体绿化。

Waterscape

世园水景

Xi' an Horticultural Expo in not only a pageant of plants, but also a feast of waterscape displaying the bright city of water.

西安世园会不单是一场植物花草的盛会，更是一次“水秀古城，灵动西安”的“水景盛宴”。

Introduction:

The area of garden section is 4, 180, 000m2 and the area of water section is 1, 880, 000m2, both of which are the largest among all the horticultural expos.

Water is the distinguished element of the expo this year which is introduced to show the grand beauty of Xi' an Chan-Ba.

There is a vivid name about the water of the expo – Fairview Lake. The greatest feature of the lake which is not very big is that the running water of the lake comes from Ba River and goes back to Ba River.

Guangyun Lake ecological landscape area, covering about 11.67km2 with about 4, 000, 000m2 of water area added, is located in the middle and lower reaches of Ba River. It used to be one of the main ports in the climax period of Tang Dynasty and called "Bashang" area with long history and rich culture heritage.

Through the reasonable utilization of the natural water system and wetland of Ba River, the waterscapes such as boat street, regatta and water stages are created, interweaving the themes like urban river management, water and city life, future water technology, to realize the wonderful combination of "Flower • Greening • Water" and to paint the poetic gardening drawing.

Additionally, there is a flying silver dragon on the water in the Horticultural Expo designed by the famous sculptor Ren Jun. It is 36m high with concrete structure and stainless steel mirror and named "Flying Dream". The water dragon sculpture is of abbreviated style, sparkling in the sun and blurring in the moonlight.

展园介绍：

西安世园会418公顷园区规划面积，创历届世园会之最；188公顷水域面积，创历届世园会之最。

水，正是西安世园会与历届世园会的最大不同之处。通过引入水元素，西安世园会做足水文章，尽显西安浐灞的大水之美。

提起世园会的水域，还有一个很传神的名字——锦绣湖。锦绣湖不单面积大，还有一个最大的特点是这里的水都是从灞河而来，最终又回到灞河里去的活水。

说起广运潭，广运潭的生态景区面积约11.67平方公里，其中新增水面就有6500亩。广运潭位于灞河中下游，是盛唐时期主要的港口之一，此地史称"灞上"地区，历史悠久、文化积淀深厚。

世园会通过合理利用浐灞河自然水系和湿地，营造船街、舟会、水上舞台等特色水景观，将城市河流综合治理、水与城市生活、未来水科技等主题穿插其中，实现"花·绿·水"的精彩结合，营造诗意的园林画卷。

另外，世园会的水里有一条身贴水面、昂首欲飞冲云霄的"银色巨龙"。这是由著名雕塑家任军设计的作品，高36米，材质为钢结构，不锈钢镜面，名为"让梦想飞"。水龙雕塑为写意龙，势如石溅水起，水花形成一条腾跃于锦绣湖面的矫健巨龙，阳光下粼粼生辉，月色里幽幽成趣。

Bridge

桥

Across the ages of Chinese and western, bridge art expo is a kind of icing on the cake of the exposition.

跨越了中西各个时代，可谓锦上添花的“桥”艺博览。

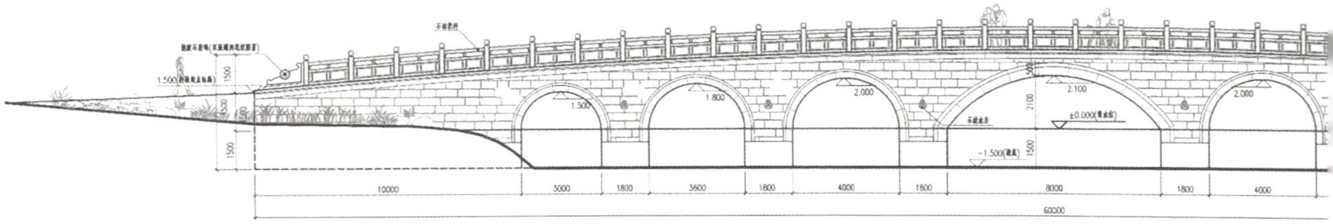

Introduction:

The 42 bridges in this exposition fall into 12 patterns according to the size and dimension, the styles across each age of Chinese and western, becoming a bridge art exposition. The romantic Dufu Bridge, classical Fuxi Bridge, European-style Bridge, single-arch bridge, three-arch bridges, seven-arch bridges…just like beautiful rainbows dotted with this green event. These bridges are not only the traffic link water with land, but also peculiar landscapes in the garden.

Bridge floor of European-style Bridge is paved with granite, both side are decorated with Travertino Romano, full of European style; winding trestle accompany with coquettish green plants and the building play with each other in the ripple of bluish waves, expressing a commanding view of Southeast Asia style; seven-arch bridge between Yefeng Waterfront and Chang' an Valley of Flower is the largest span, and a highlight in the bridges as well, it is 8 meters wide, 60 meters long, the middle arch is the biggest one, others on both sides diminish gradually and rhythmically, reflects the sense of rhythm in symmetry, blends with the natural and cultural landscape as a whole; on the wooden Lianli Bridge that adjacent to Huanxisha Stage, twinned love locker hanged on the ropes, as if pairs of sweet lovers cling to each other.

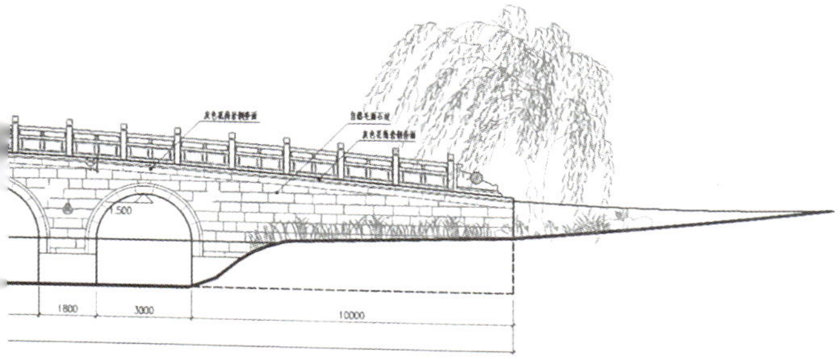

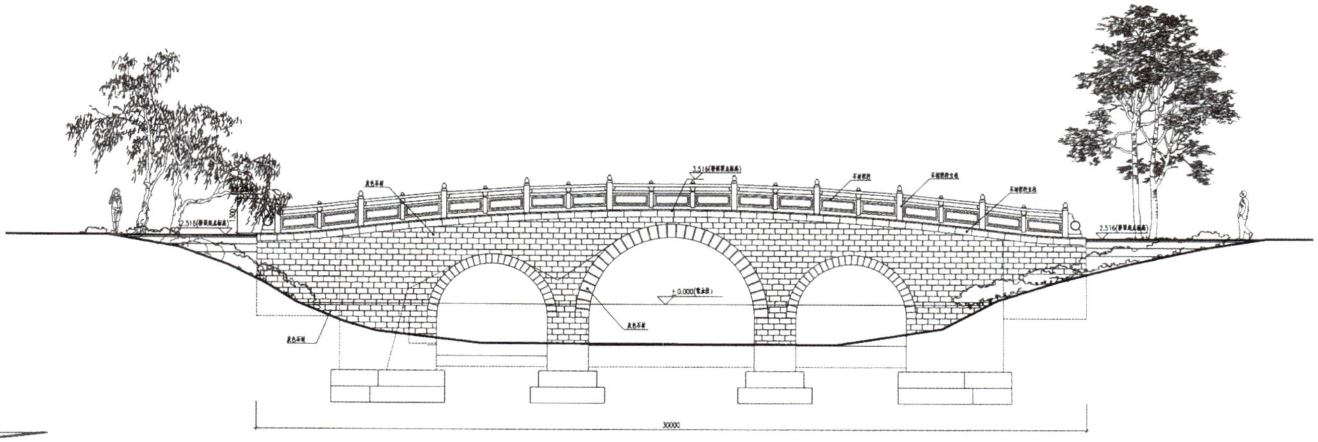

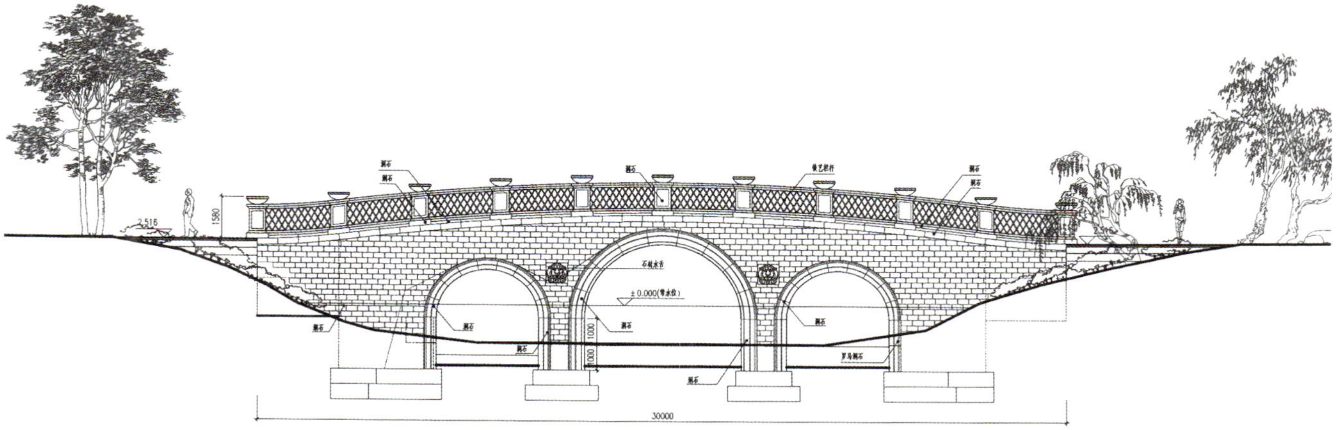

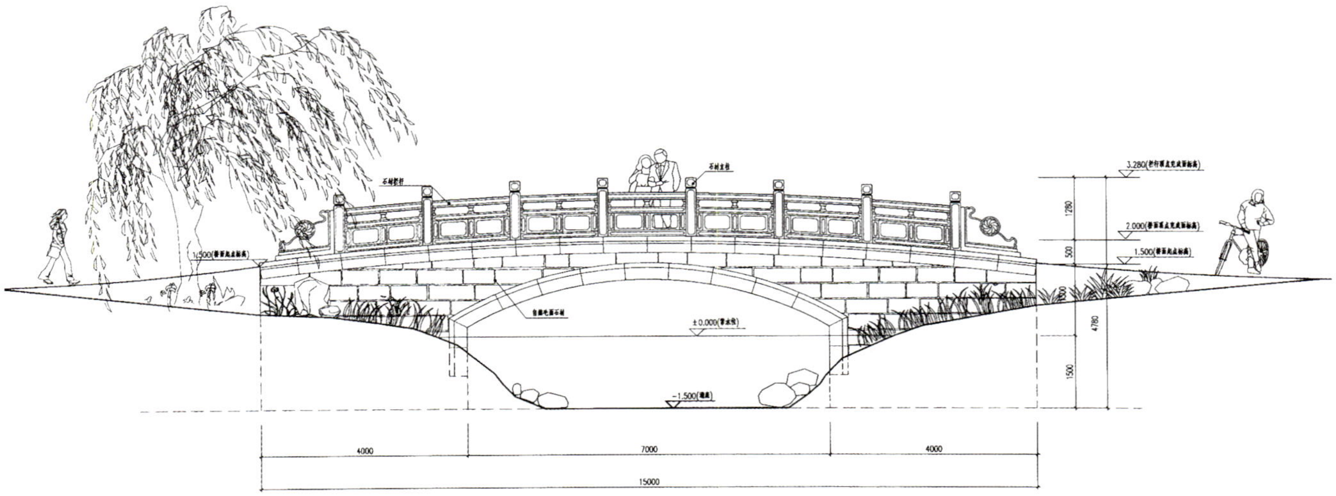

展园介绍：

西安世园会的42座桥，以尺寸规格分为12种类型，风格跨越了中西各个时代，可谓锦上添花的“桥”艺博览。浪漫的连理桥、诗意的杜甫桥、古典的祓禊桥、欧式风格的欧式桥、不同孔数的单孔桥、三孔桥、七孔桥……犹如一道道绚丽的彩虹，点缀着这场绿色盛会。在西安世园会上，桥不仅是沟通水陆的交通纽带，更是园林庭院中别具一格的景观。

欧式桥桥面以花岗岩铺装构成，桥身两侧是罗马洞石装饰面层，极具欧式风情；曲折的栈桥，伴随婀娜多姿的绿色植物与建筑物在碧波的荡漾中交相辉映，东南亚风情一览无余；椰风水岸与长安花谷之间的七孔桥是园区内跨度最大的桥梁，也是园区桥梁的一大亮点，桥宽8米，长60m，七孔桥洞中间最大，两边有节奏地依次排开，逐渐变小，体现了对称中的韵律感和节奏感，与园区自然景观和人文景观融为一体；在浣溪沙舞台附近的木质“连理桥”上，一条条绳索上挂着成双成对的同心锁，仿佛一对对甜蜜恋人，相依相偎……

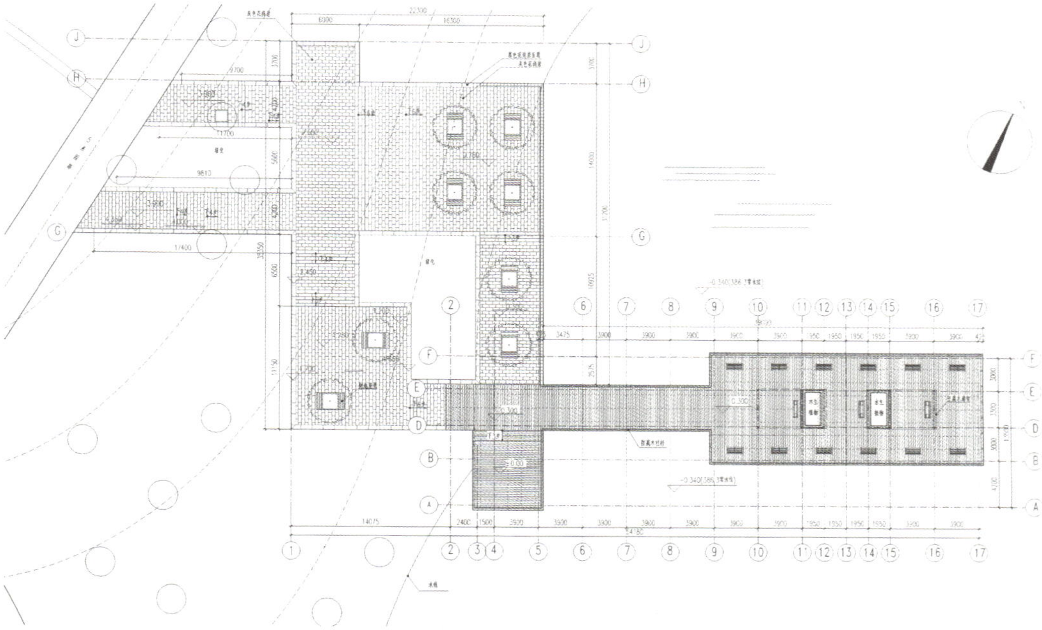

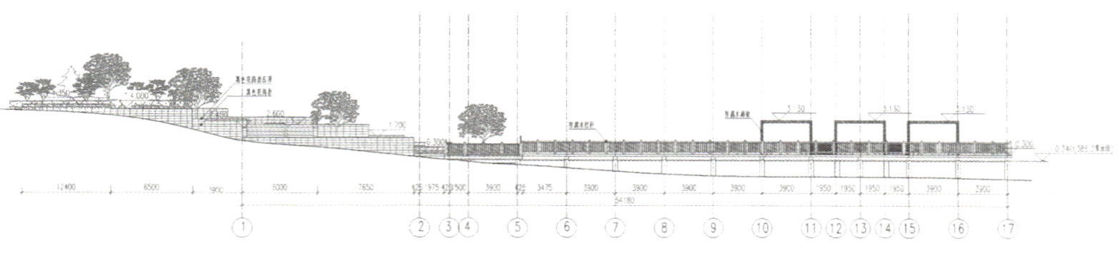

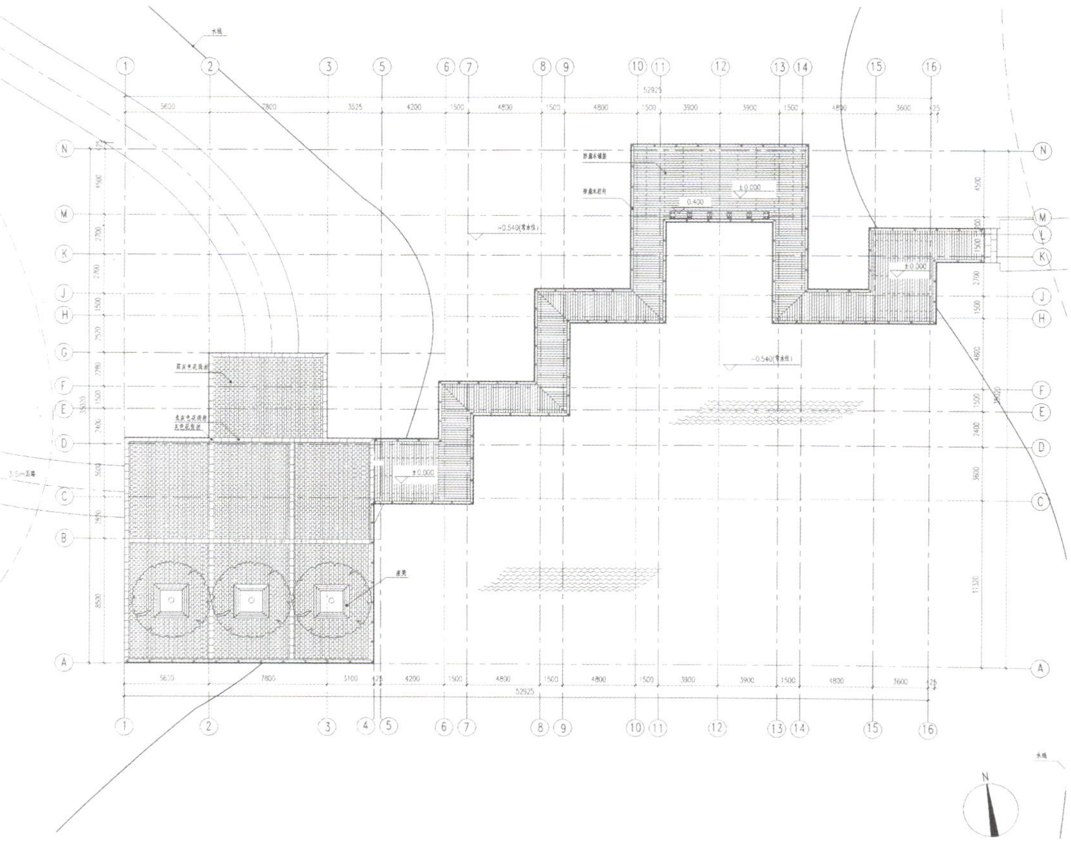

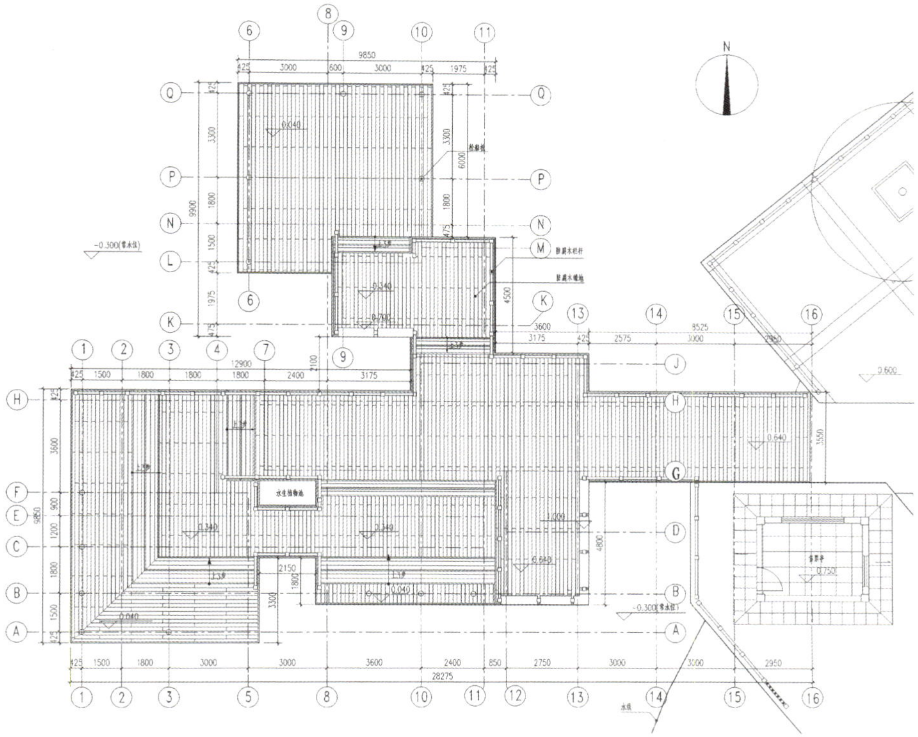

N
9850
12900
28275
-0.300(常水位)
水生植物池

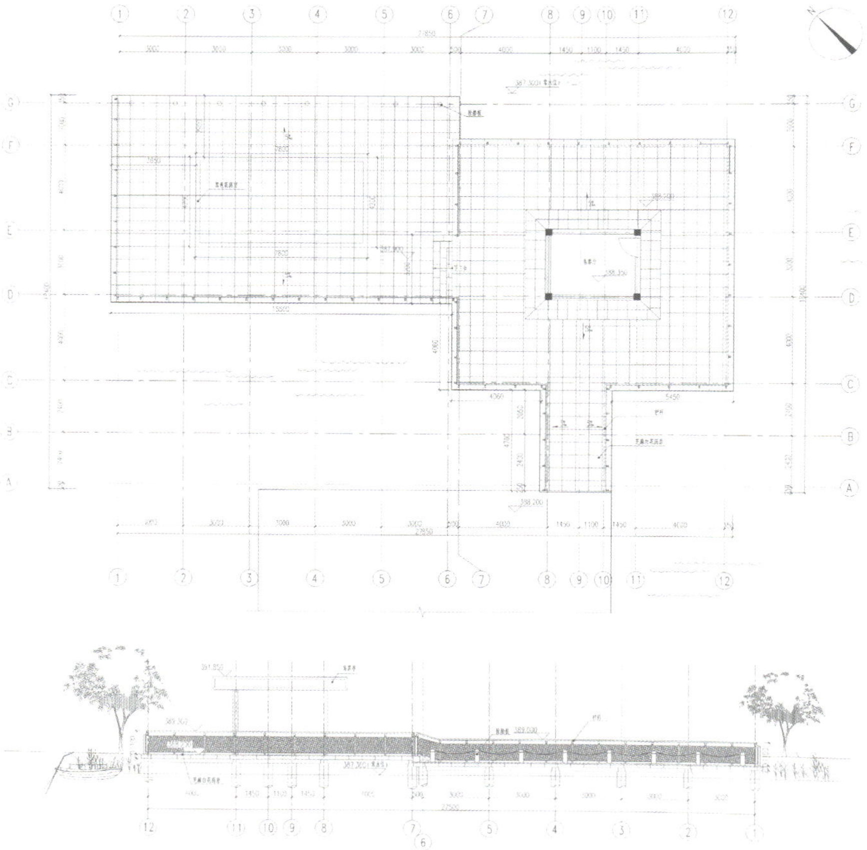

Small Garden Ornaments
世园小品

Varied creative and delicate ornaments have shown the creativity of the design and made the International Horticultural Exposition 2011 Xi' an splendid.

许多创新、别致的小品展现了精品创造的魅力，使整个世园会更增光辉。

Introduction:

There are many characteristic ornaments in the exhibition park, such as the mascot, finger-shaped trees, frog-like sculpture carved on tree, crane-like bamboo canopy, etc. All of them have shown the second theme of the exposition: creation and delicateness. They also combine the whole exhibition park together to provide landscapes everywhere. Ten rules of green lifestyle are penetrated into the ornament design: energy saving and emission reduction, low-carbon travel; cherishing food and green eating; producing according to demand, rational consumption; saving water, cyclic utilization; advocating organic, reducing pollution; planting trees and flowers, beautifying the life.

展园介绍：

西安世园会中还有很多有创意的特色小品，例如吉祥物的摆设、手指型的树木、蛙形绿色树雕、鹤状的竹席棚盖等，无一不展示了世园会中创新、别致的第二主题，另外小品还使整个世园会统一起来，让人感觉路路有景看，处处是创造。其中 10 条绿色生活方式——节能减排、低碳出行；珍惜粮食、绿色饮食；按需定量、理性消费；惜水节水、循环利用；提倡有机，减少污染；植树种花、美化生活等内容也贯穿了世园的小品设计中。

天人长安 创意自然
EXPO XI'AN
花开富贵
开园大吉
Eternal peace & harmony between nature & mankind, nurturing the future earth

休闲体能乐园
体能休闲乐园
温馨提示
注意安全
文明游园
本项目仅限5–15岁少年、儿童游玩

陕西省林业厅

Sand Sculpture

沙雕

Sand sculpture exhibition area shows more than 10 artistic sculptures in Sino-foreign theme, such as Chinese Dragon, Plain of Bailu, Gulliver' s Travel, The Mermaid and etc, reflect the theme of the World Horticultural Exposition.

沙雕展示区展示了华夏龙脉、白鹿原、格列佛游记、美人鱼等10多座中外题材、具艺术风格的沙雕作品，体现了世园会创意自然的主题。

Introduction:

Sand sculpture exhibition area in Xi' an World Horticultural Exposition shows more than 10 artistic sculptures in Sino-foreign theme, such as Chinese Dragon, Plain of Bailu, Gulliver' s Travel, The Mermaid and etc,. So far, it is the largest and longest sand sculpture exhibition area in western China. In the Xi' an Exposition Garden, 9 sand sculpture artist from various places of China carved 6 works in different themes with 7000 m^3sand. These sand sculptures take sand and water as materials, no chemical blinders, reflect the theme "Eternal peace & harmony between nature & mankind, nurturing the future earth" of this exposition. It is a public artistic space and a paradise for self-entertainment.

展区介绍：

西安世园会园区沙雕展示区有华夏龙脉、白鹿原、格列佛游记、美人鱼等10多座中外题材、具艺术风格的沙雕作品，是至今我国西部内陆地区规模最大、历时最长的沙雕展示区。在西安世园会园区，9位来自全国各地的沙雕艺术家用7000 m^2 的沙子，雕刻出6座不同主题的作品。这些沙雕以沙和水为材料，不使用化学粘合剂，体现了世园会"天人长安·创意自然——城市与自然和谐共生"的主题，其既是公共艺术空间，又是自娱自乐的乐园。

2011
西安世界园艺博览会

華夏
龍脈
2011西安世园会 塑沙区
2011西安世园会 塑沙区
2011西安世园会 塑沙区

華夏
龍脈

Qingyuanchun Stage

沁园春舞台

Qingyuanchun stage is a performance venue in lotus shape, there are four performances of Water Moon Flower Dance every day.

沁园春舞台是一个荷花状的演出场地，每天有四场水月花舞的演出。

Introduction:

Qingyuanchun stage is a performance venue in lotus shape, there are four performances of Water Moon Flower Dance every day. There will be more than 5000 excellent performances during the 178 days of the World Horticultural Exposition, at least 28 performances every day on average. The contents of these performance are abundant, including traditional Guzheng, cross talk, Taiji, modern cartoon show, dazzle dance, fancy car parade and fancy boat parade. Show places scattered here and there in the garden, you can enjoy the show whether on land or in water, whether in the daytime of at night.

展园介绍：

沁园春舞台是一个荷花状的演出场地，每天有四场水月花舞的演出。而在为期178天的世园会中，园区内一共会举行5000多场精彩绝伦的表演，平均每天最少表演28场。这些演出的内容丰富多样，有古老的传统表演形式古筝、相声、太极，也有现代气息浓重的卡通人偶表演、青春炫舞，还有花车巡游、花船巡游等。演出地点分布在园区各处，无论在陆地还是在水面，无论是白天还是夜晚都能欣赏。

Supporting Facilities

配套设施

To ensure the success of International Horticultural Exposition 2011 Xi' an, there are many supporting facilities to provide different services, such as rest room, store, toilet, signboard and nameplate.

为了西安世园会有效有序地开展，世园会上开设了很多服务性的配套设施，如休息间、商店、卫生间、指引的标识等设施。

Rest Room

The courtyards of the Expo are all built in beautiful landscape to speak the same theme: modesty and elegance. There are five to six pavilions in different sizes in the greening leisure zone. The non-material cultural heritages shows of old Shanxi and the original Hatha Yoga of India are in these pavilions; others are held in the wooden pavilions.

休息间设施

世园的庭院，可以说全都建设在美景之中，而且都透着一个共同的特征，那就是质朴庄重、典雅清逸。世园会的绿色休闲体验区也有大大小小五六个亭子，大的上百平米，小的有几十平米。经常在休息的庭院里上演的节目是老陕的非遗展演、原汁原味的印度哈他瑜伽等，其大多都选择在木亭中举行。

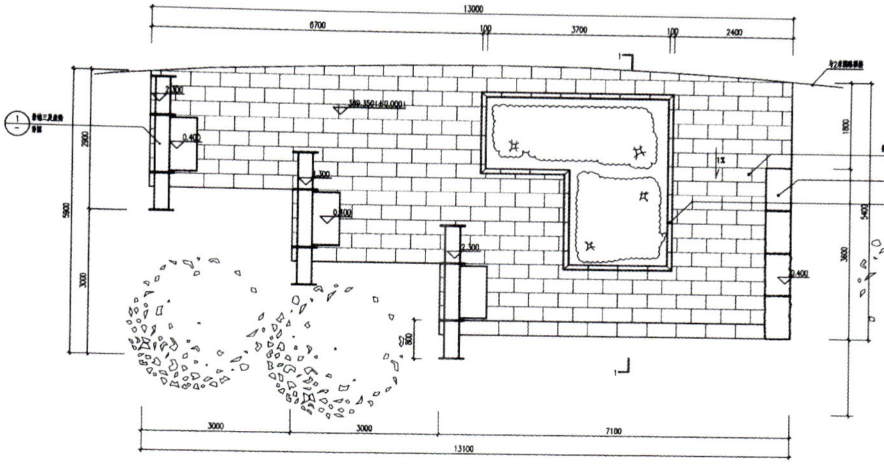

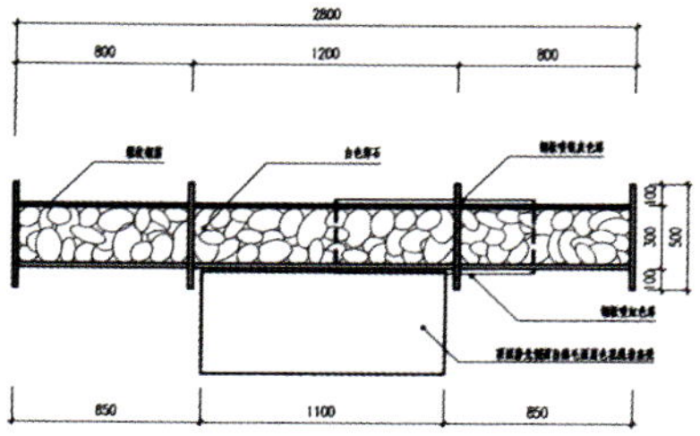

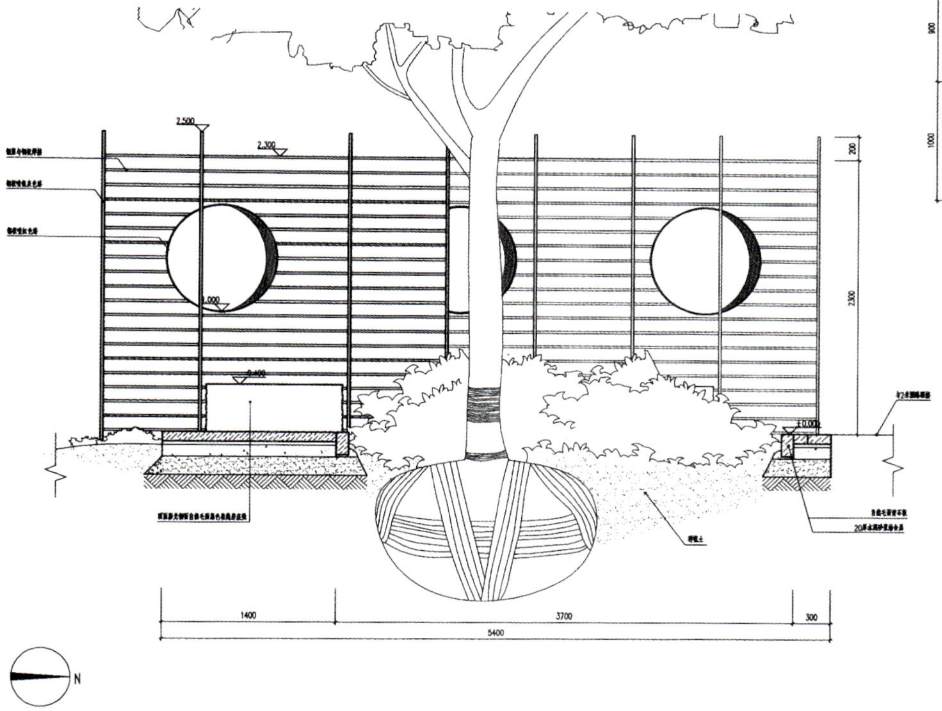

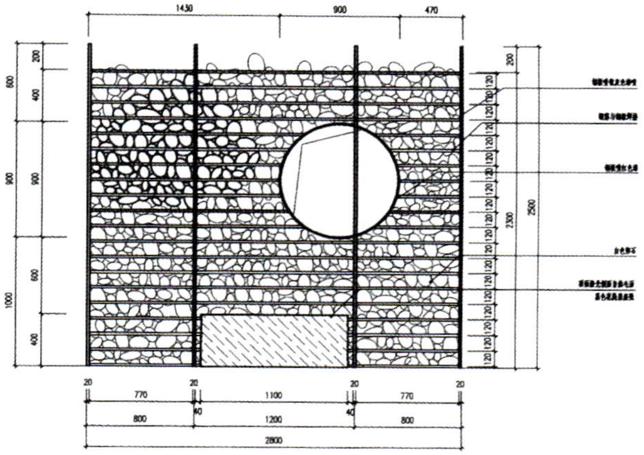

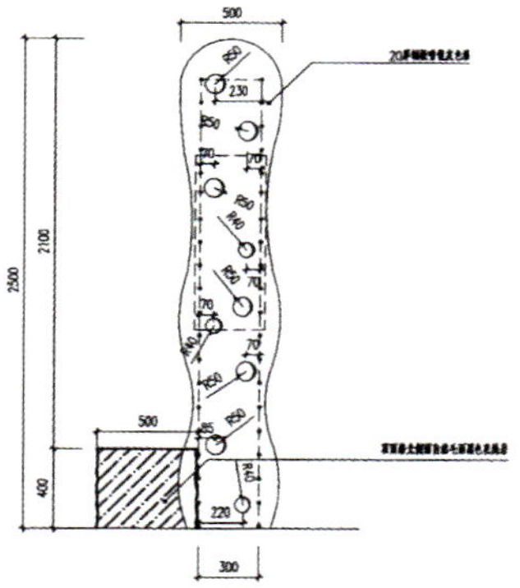

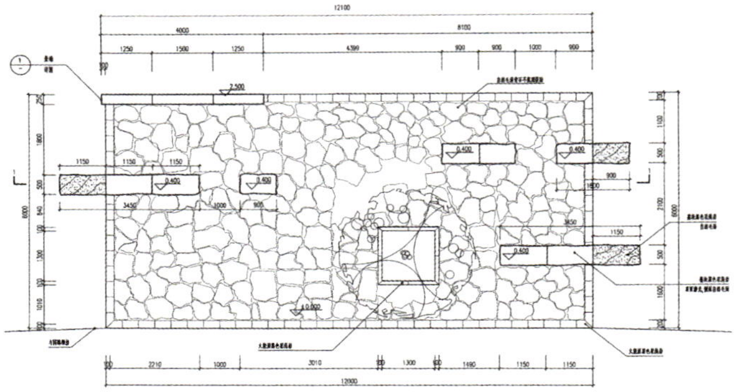
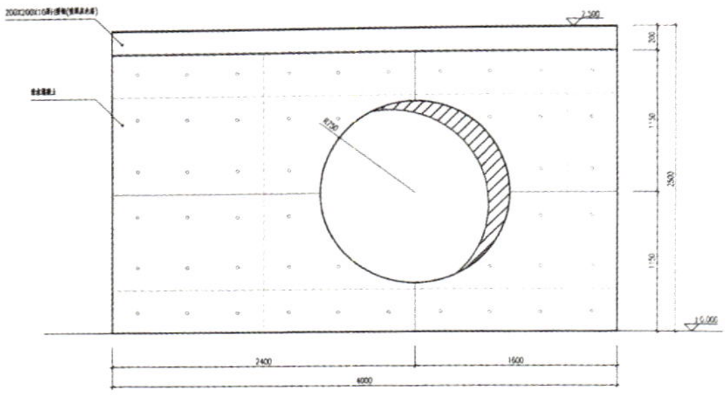
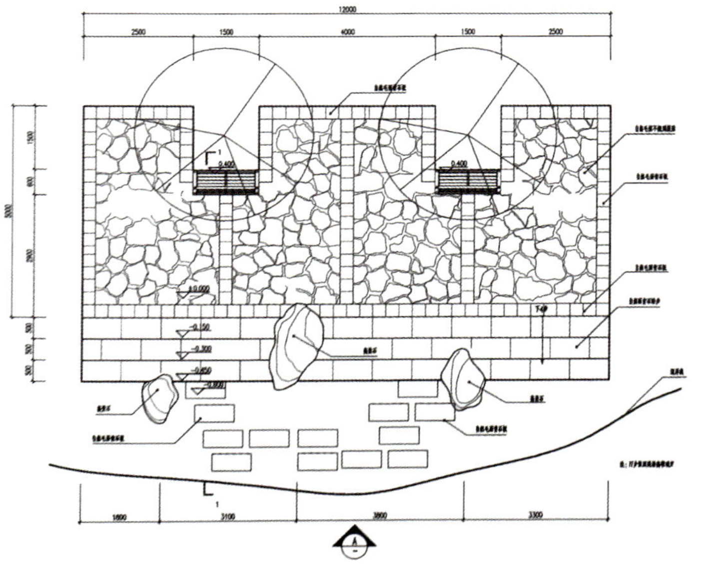
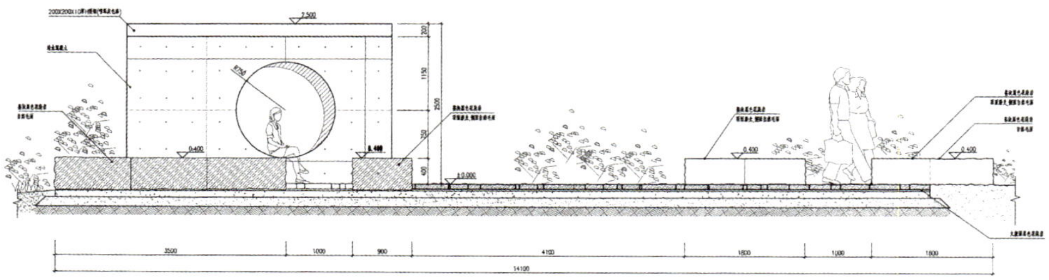
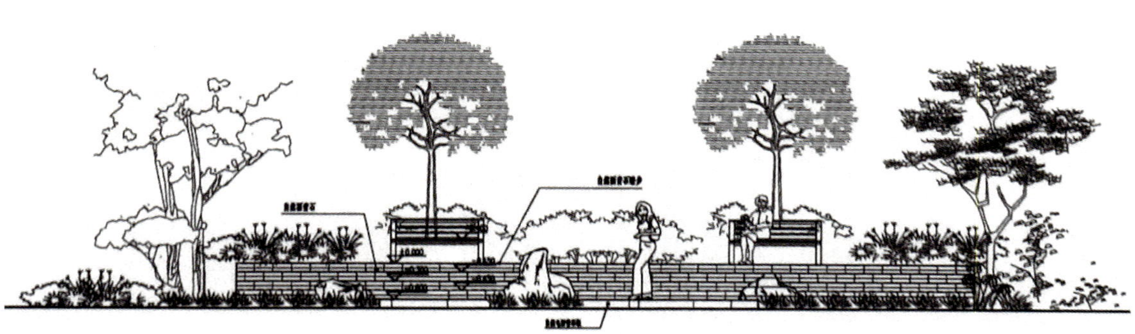

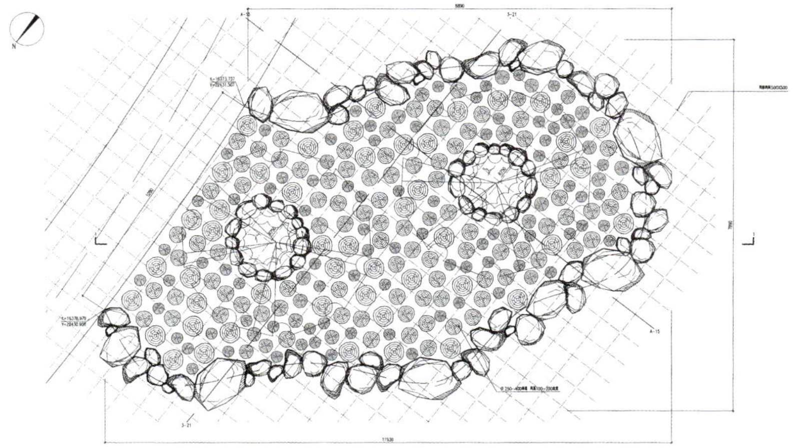

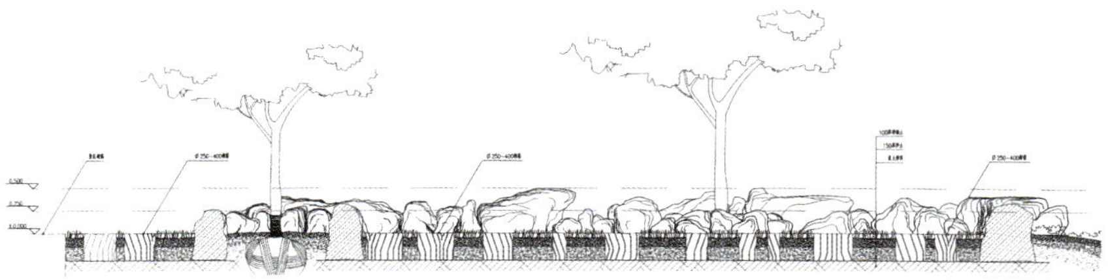

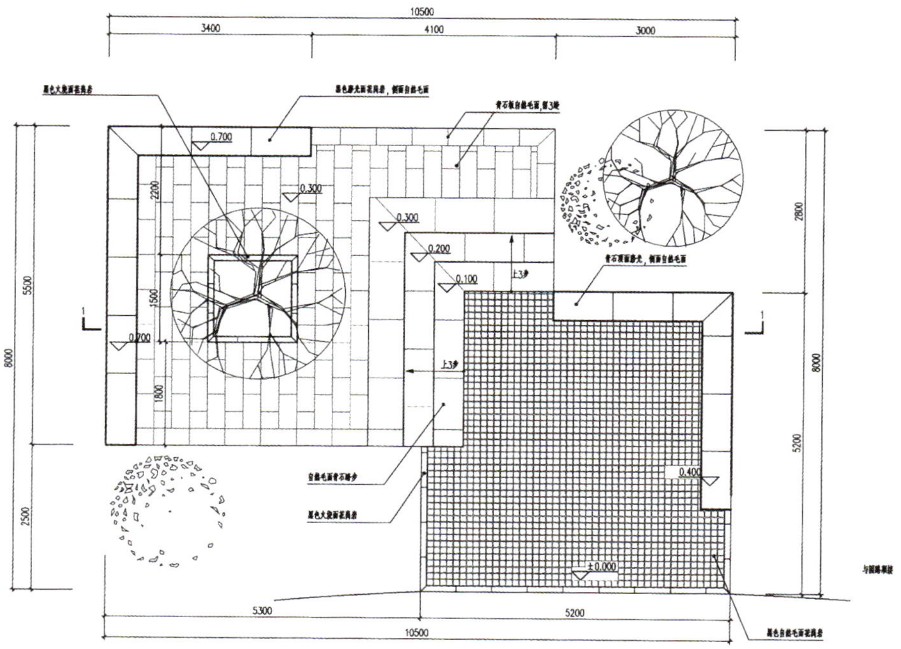
10500
3400
4100
3000
0.700
0.300
0.300
0.200
0.100
0.700
0.400
±0.000
上3步
上3步
2200
1500
1800
5500
8000
2500
2800
8000
5200
5300
5200
10500

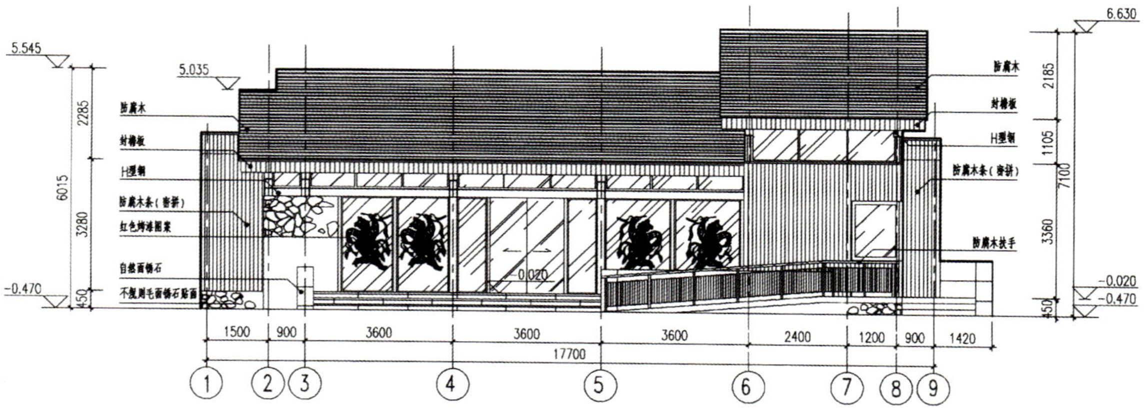
5.545
5.035
6.630
-0.470
-0.020
0.020
2285
6015
3280
450
2185
1105
7100
3360
防腐木
封檐板
H型钢
防腐木条（密拼）
红色烤漆图案
自然面饰石
不规则毛面饰石贴面
防腐木扶手
1500
900
3600
3600
3600
2400
1200
900
1420
17700
1
2
3
4
5
6
7
8
9

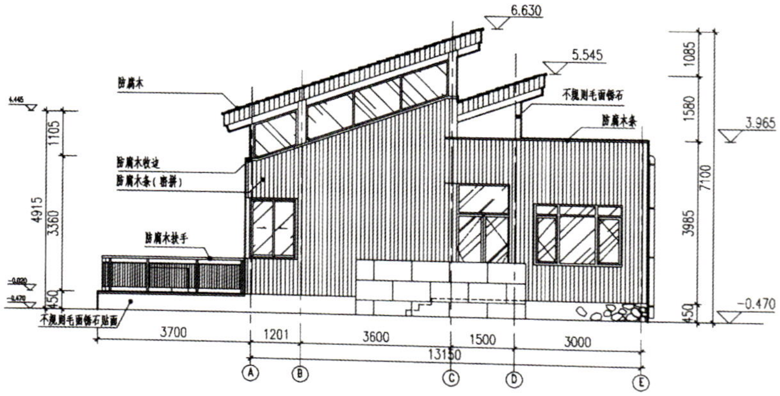
6.630
5.545
3.965
-0.470
1105
4915
3360
450
1085
1580
7100
3985
防腐木
防腐木收边
防腐木条（密拼）
防腐木扶手
不规则毛面饰石贴面
不规则毛面饰石
防腐木条
3700
1201
3600
1500
3000
13150
A
B
C
D
E

Store

There are 17 stores inside the park, most of which are gathering in three places: Romance by Ba River, Coconut Wind and River Bank, and European Avenue. Others are small stores selling souvenirs. In addition, the commodities in the stores and cold drink shops are suggested with unified retail prices.

商店设施

商业设施有17处，其中3处规模较为集中，分布在灞上人家、椰风水岸以及欧陆风情内，其他为小型购物点，主要售卖纪念品。此外，世园会还对商店和冷饮店销售的物品将实行统一的建议零售价。

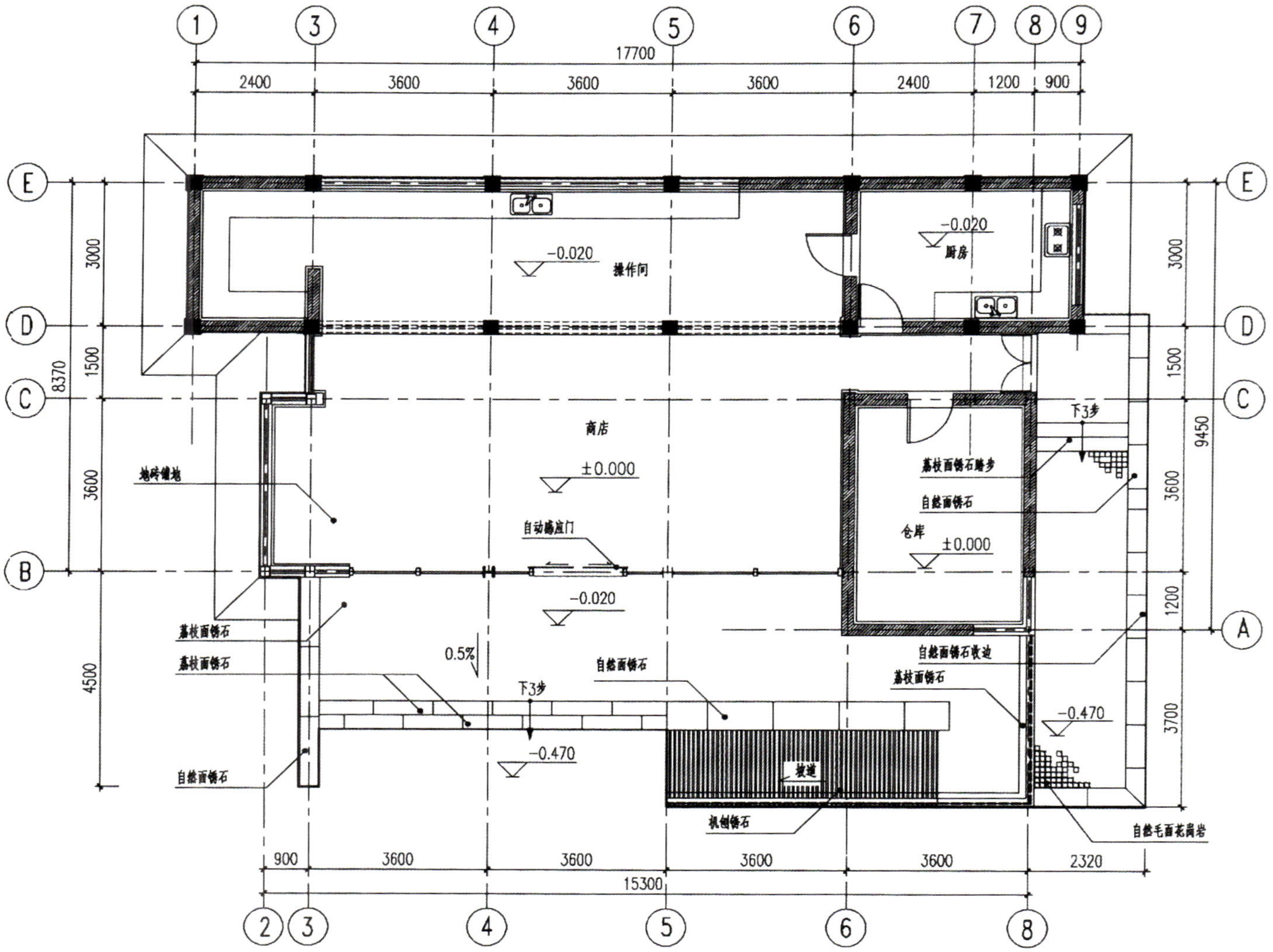

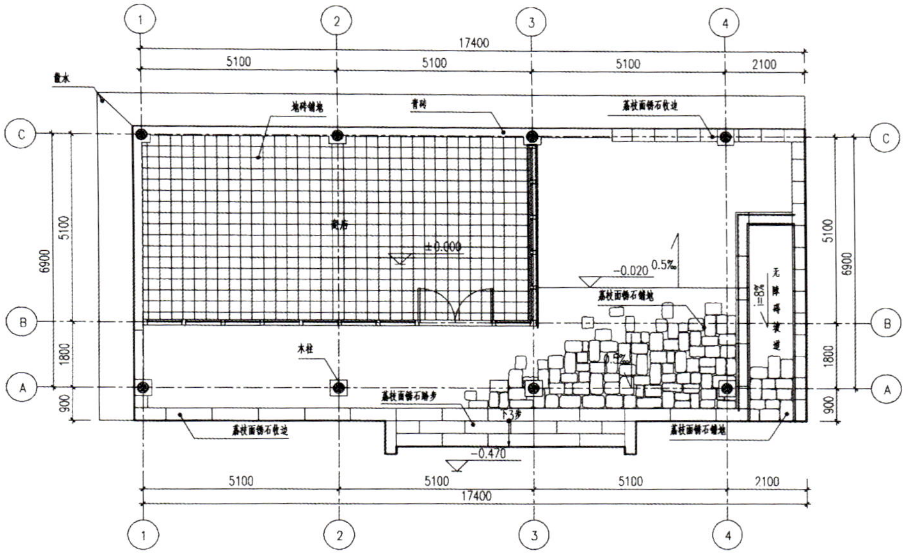
17400
5100
5100
5100
2100
6900
5100
1800
900
±0.000
-0.020
-0.470
青砖
木柱

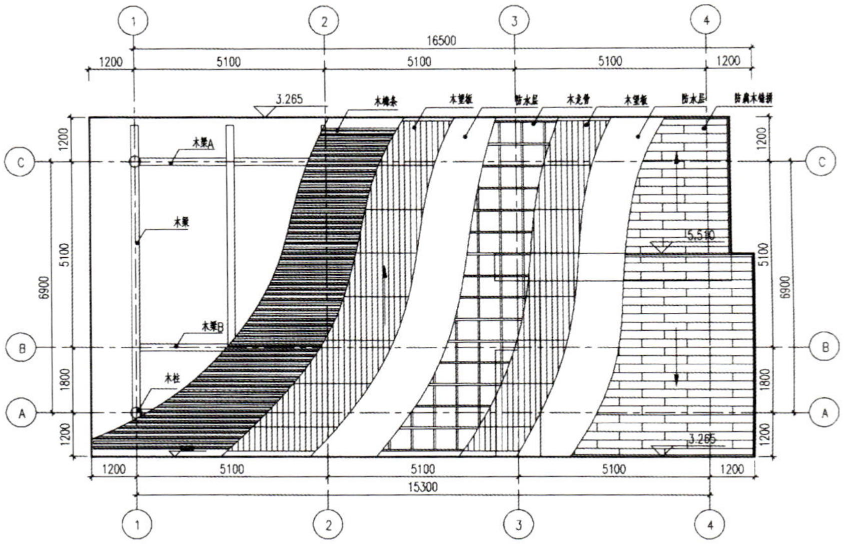
16500
1200
5100
5100
5100
1200
15300
6900
3.265
5.510

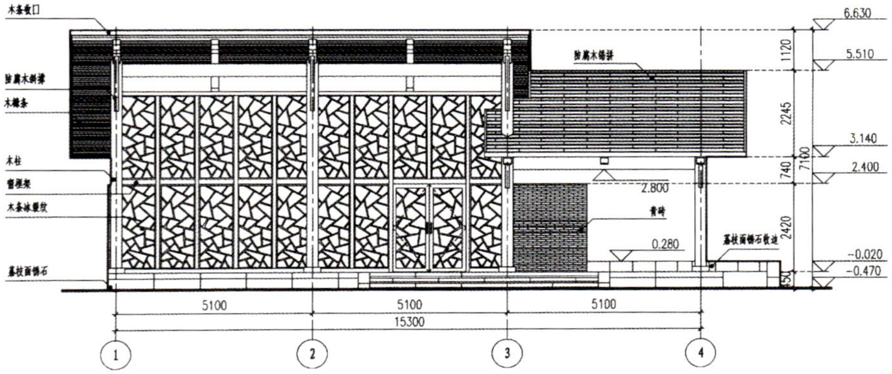
6.630
5.510
3.140
2.400
2.800
0.280
-0.020
-0.470
1120
2245
740
2420
7100
5100
5100
5100
15300
青砖
木柱

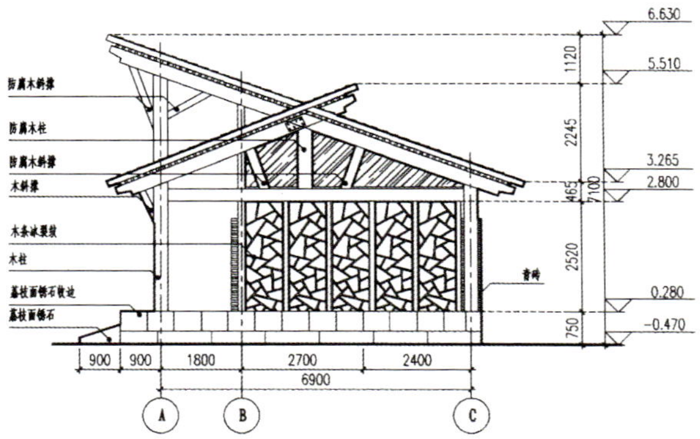
6.630
5.510
3.265
2.800
0.280
-0.470
1120
2245
465
2520
750
7100
900
900
1800
2700
2400
6900
木柱
青砖

EXPO XI'AN
入口

Toilet

For the toilet, in accordance with the relevant national standards, and to consider the situation of extreme visitors flow rate, the exposition sets about 36 toilets with barrier-free access and disabled only, 1200 toilet cubicles in total, among which, 400 for male, 800 for female. The proportion of men' s and women' s is between 1:2 and1:3. The layout is set according to the 100-300m service radius.

Adhering to the concept of environmental protection, water for flushing toilets is recycled water, relative to tap water, sewage and industrial waste water. Once the waste water is collected for batch treatment, it meets the standard for toilet flushing, so as to save water.

卫生设施

卫生间方面，按照相关国家标准，考虑极端日人流状况，世园上设置了约 36 处卫生间，共 1 200 个厕位，其中男厕蹲位约 400 个，女厕蹲位约 800 个，且每处厕所都设置了无障碍通道和残疾人专用蹲位。男女蹲位比例设置为 1:2—1:3 之间。世园会园区内厕所布局按照服务半径 100 - 300 米设置。

园区厕所采用中水冲厕，秉承了绿色环保理念。所谓中水，是相对于上水自来水、下水生活污水和工业废水而言的。西安世园会将废（污）水集中处理后，达到规定标准回用于厕所冲洗，从而达到节水的效果。

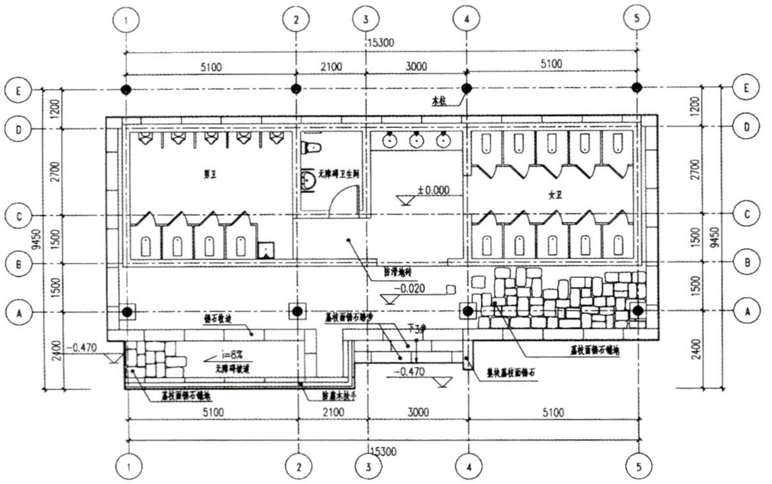
15300
5100
2100
3000
5100
9450
±0.000
-0.020
-0.470
i=8%

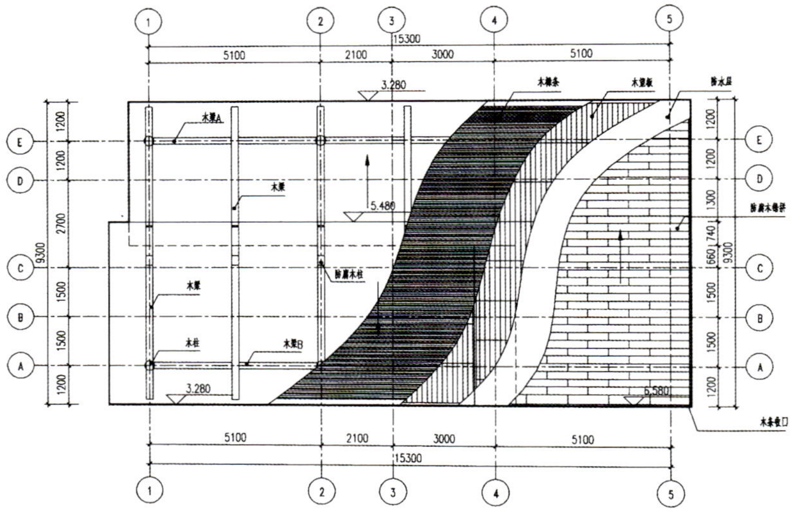
15300
5100
2100
3000
5100
9300
3.280
5.480
6.580

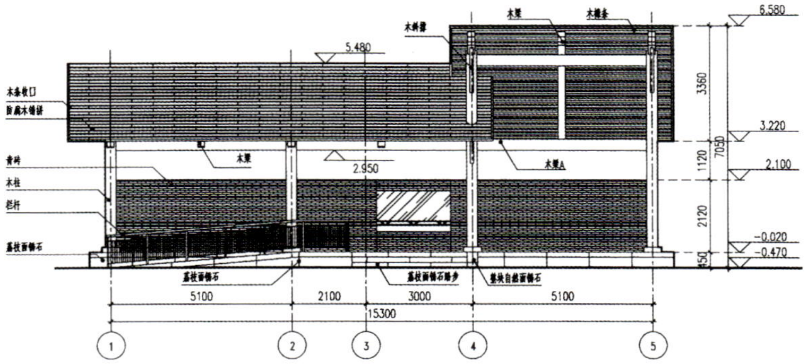
6.580
5.480
3.220
2.950
2.100
-0.020
-0.470
5100
2100
3000
5100
15300

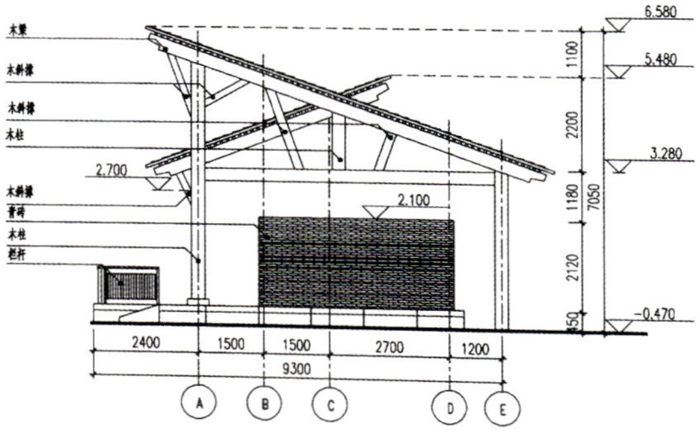
6.580
5.480
3.280
2.700
2.100
-0.470
2400
1500
1500
2700
1200
9300

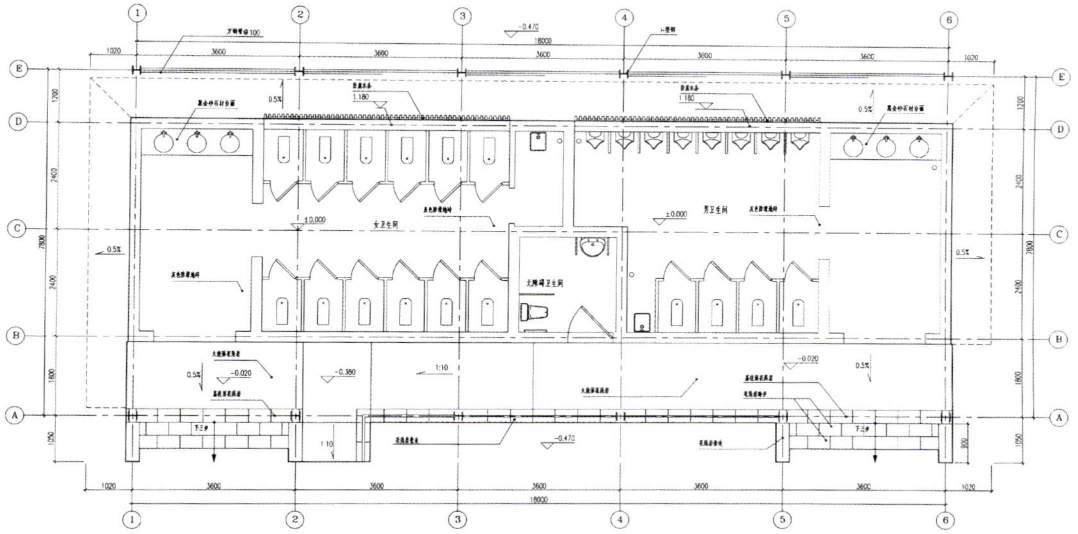

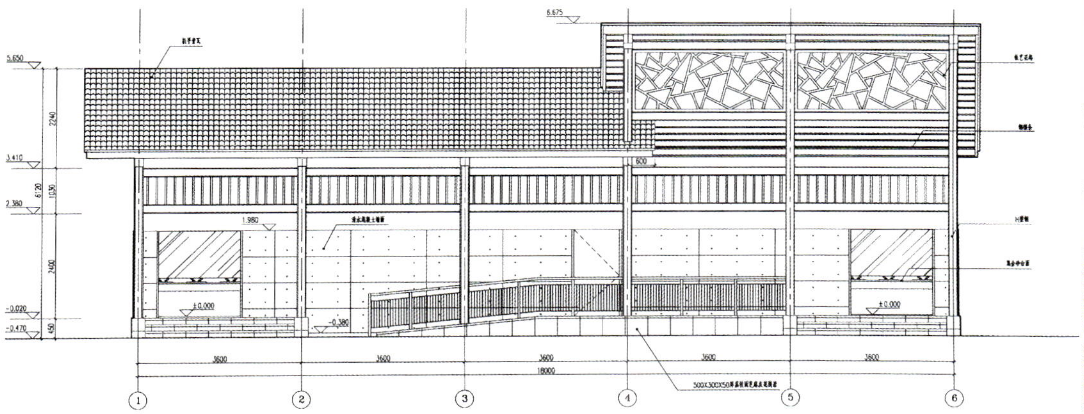

Signboard and Nameplate

The signboards and nameplates in the park are arranged reasonably and scientifically to improve the density. To solve the problem of inconspicuous route guidance, big medias have published the park guides and maps extensively; 200,000 sketch maps are posted on the bus stops, 1,500 sketch maps are posted on the buses of 14 lines. At the entrances and exits as well as the parking places, striking kindly reminders are set.

Inside the park, it has provided the visitors with guide maps containing massive information to show the information of the traffic system, exhibition gardens, service and activities. There is also a set of professional sign system, including the signboards and nameplates of the exhibition halls, exhibition gardens and landscape spots.

When entering into the park, there are introduction signboards beside rare flowers and trees. Important scenic spots are also attached with brief introduction. The visitors can enjoy the beautiful sceneries of the park as well as get some knowledge about gardening and horticulture.

There are also 81 workers and 286 volunteers for tour guide. Meanwhile, alarm signboards are set on many lamp posts along the roads for easy alarming and quick alarm dispatching.

标识设施

园内现有标识标牌进行合理科学规划，提高园内标识布设密度；针对市区到园区路线指引不明显的问题，在各大媒体大篇幅刊登世园会导览图；在全市公交站牌张贴示意图 20 万份，在 14 条世园公交专线张贴示意图 1500 份，在园区各大出入口和停车场设置了醒目的温馨提示。

在世园会园区内，为所有世园会参观者提供包含大量信息的导览地图，全方位展示了西安世园会的交通信息、馆园信息、服务信息、活动信息等，此外还配有一套专业的标识系统，包括展馆、展园及景观的名称标识、指引标识等。

如今进入园区，珍稀花卉、树木旁有介绍标示，重要景点前有简要的介绍说明，许多游客在参观完美丽的世园风景后还会增长不少的园林、园艺知识。

园区还分布了 81 个岗位的导览工作人员及 286 个岗位的导览志愿者。同时园中设置很多路边路灯杆上报警标识牌，方便市民报警及警方处警的同时也方便了 120 救护和 119 处警。

自然馆
Greenhouse
长安塔
Chang'an Tower
盆景园
Bonsai Garden
长安园
Chang'an Garden

国际展园
VIP酒店

浣溪沙舞台
Silk Stage
公共洗手间
Public Bathroom
企业园
Enterprise Gardens
魔石乐园
Devil Stone Paradise
塑沙区
Model Zone
国际竞赛馆
International competition pavilion
灞上人家
Bahe River Side
秦岭园
Qinling Park
人文山水诗意长安园
Poetic Landscape in Chang'an Garden

园区入口
Entrance
出租车场
Taxis Circle the Car Park
1出口 EXIT
公交车停车场
BUS Park
5出口 EXIT 6出口 EXIT
7出口 EXIT

游客服务中心
问询中心
Information Center
物品租赁处
Articles Rental Office
物品寄存处
Checkroom
走失儿童中心
Center For Lost Children
失物招领中心
Center For Lost Articles

温馨提示
观光车线路共设四个售票点，广运门站、锦绣门站、灞上人家站、得宝门站，请广大游客朋友按图示选择售票点购票，谢谢合作！
WARM SUGGESTION
There are four ticket windows for sightseeing shuttle, which are Guangyun Entrance ,Jinxiu Entrence,Bahe river Side and Debao Entrance .Please choose nearby ticket window following the sign board.Thanks for your cooperation!

International Horticultural Exposition

世界园艺博览会

Category of International Horticultural Exposition

The are two categories for the International Horticultural Exposition: the Universal Exposition and the Specialized Exposition. For the Universal Exposition, the participating countries build their own pavilions in the area provided for free by the host country and display their own products or technologies; for the Specialized Exposition, the participating countries present their specialized products in the area provided by the host country and are in charge of the decoration indoor and outdoor and the exhibition setting. In accordance with the rules of international organization, the Universal Exposition will be divided into the General Exposition and Special Exposition; while the Specialized Exposition falls into four categories: A1, A2, B1 and B2.

Given the official approval by AIPH, the international horticultural industry organization, International Horticultural Exposition is divided into four categories: A1, A2, B1 and B2.

A1: Large International Horticultural Exhibitions. This exhibition is allowed to be organized no more than once a year. Its duration may not be less than three months and no more than six months with the minimum area of 500, 000 m^2. The application must be handed in 6–12 years before the opening date of the exhibition and at least 10 countries must agree to participate. The exhibition must showcase all aspects of horticulture.

A2: International Horticultural Exhibitions. These exhibitions can not be organized more than twice a year. When two exhibitions are held on the same continent, there must be an interval of at least three months between the two opening dates. Their duration may be no less than eight days and no more than 20 days, and at least six different countries must participate.

B1: Permanent International Horticultural Exhibitions. These exhibitions are allowed to be organized only once per year. Their duration may be no less than three months and no more than six months.

B2: National Specialized Exhibitions

International Garden Festival

International Horticultural Exposition is the highest–level and specialized international exhibition, which is also called International Horticultural Festival. It is a big exhibition for gardening and horticultural works as well as exotic flowers and herbs all around the world that enhances the communication between countries. It is also the largest grade–A1 international horticultural exposition integrating cultural and scientific achievements together. The exposition usually lasts six months from midsummer to mid–autumn. In history, the International Horticultural Exposition was mostly hosted by the developed European countries and the U.S.A.

Since the first International Horticultural Expo, the scale of the expo has been continually expanding: the area is changed from thousands of square meters to 5000,000 square meters, the participant countries have increased from a dozen to a hundred, the visitors grew to 60,000,000, and the economic income reaches to 0.4 billion U.S. dollars.

Since the first International Horticultural Exposition held in Rotterdam, the Netherlands in 1960, there have been 21 International Horticultural Expos till 1999. The following are the times and places for the International Horticultural Expositions.

List of International Horticultural Exposition

Year	Host Country	Hose City	Exposition Name
1960	Netherlands	Rotterdam	Floriade
1963	Germany	Hamburg	IGA
1964	Austria	Vienna	International Horticultural Exposition
1969	France	Paris	International Horticultural Exposition
1972	Netherlands	Amsterdam (Amstel Park)	Florida
1973	Germany	Hamburg	Hamburg
1974	Austria	Vienna	International Horticultural Exposition
1976	Canada	Quebec	The International Horticultural Exposition
1980	Canada	Montreal	International Horticultural Exposition
1982	Netherlands	Amsterdam (Amstel Park)	Floriade
1983	Germany	Munich	IGA
1984	Britain	Liverpool	International Garden Festival
1990	Japan	Osaka	International Gardens and Greenery Exposition
1992	Netherlands	Amsterdam (Zoetermeer)	Floriade
1993	Germany	Stuttgart	IGA
1994	France	Saint Denis	The International Horticultural Exposition
1995	Germany	Cottbus	Federal Gardening Exhibition
1996	Italy	Genoa	The International Horticultural Exposition
1997	Belgian	Liege	The International Horticultural Exposition
1997	Canada	Quebec	The International Horticultural Exposition
1999	China	Kunming	International Garden Festival
2000	Japan	Awaji	Florart 2000
2002	Netherlands	Amsterdam (Haarlemermeer)	Floridae
2003	Germany	Rostock	IGA
2004	Japan	Hamamatsu	Pacific Flora
2005	Germany	Munich	BUGA
2006	Thailand	Chiang Mai	Royal Flora Ratchaphruek
2006	China	Shenyang	International Horticultural Exposition
2010	Taiwan	Taipei	Taipei International Garden and Horticulture Exposition
2011	China	Xi' an	The International Horticultural Exposition 2011 Xi' an
2012	Netherlands	Venlo	Floridae

世园会的类别

世界园艺的博览会分为综合性和专业性两大类。综合性世界园艺博览会是由参展国政府出资，在东道国无偿提供的场地上建造自己独立的展览馆，展示本国的产品或技术；专业性世界园艺博览会是参展国在东道国为其准备的场地中，自己负责室内外装饰及展品设置，展出某类专业性产品。按照国际组织的规则评定，综合性世界园艺博览会分为一般世界园艺博览会和特殊博览会两种；专业性博览会分为 A1、A2、B1、B2 四个级别。

世界园艺博览会是由国际园艺花卉行业组织——国际园艺生产者协会（AIPH）批准举办的国际性园艺展会，分为 A1、A2、B1、B2 四大类。

A1 类：大型国际园艺展览会。这类展览会举办每年不超过 1 个。A1 类展览会时间最短 3 个月，最长 6 个月，最小面积 50 万 m^2。在展览会开幕日期前 6 ~ 12 年提出申请，至少有 10 个不同国家的参展者参加。此类展览会必须包含园艺业的所有领域。

A2 类：国际园艺展览会。这类展览会每年最多举办两个，当两个展会在同一个洲内举办时，它们的开幕日期至少要相隔 3 个月，展期最少 8 天，最多 20 天。至少有 6 个不同的国家参展。

B1 类：长期国际性园艺展览会。这类展会每年度只能举办一届。展期最少 3 个月，最多 6 个月。

B2 类：国内专业展示会。

世界园艺节

世界园艺博览会是最高级别的专业性国际博览会，也叫世界园艺节。它是世界各国园林园艺精品、奇花异草的大联展，是以增进各国的相互交流，集文化成就与科技成果于一体的规模最大的 A1 级世界园艺博览会。会期通常为 6 个月，自晚春起，经盛夏至中秋。历史上，世界园艺博览会的举办地大都是经济比较发达的欧洲国家和美国。

自首届世界园艺博览会以来，世界园艺博览会的举办规模呈不断扩大的趋势。由占地面积几万平方米，参展国家十几个，逐渐发展到占地近 500 万 m^2，参展国家近百个，观众人数达 6000 多万人次，经济收益最高达 4 亿多美元。

自 1960 年在荷兰鹿特丹举办首次国际园艺博览会以来，至 1999 年，共举办过 21 次世界园艺博览会。以下是世界园艺博览会举行的时间及地点：

历届世界园艺博览会一览表

举辨时间	举办国	举办城市	博览会名称
1960 年	荷兰	鹿特丹	国际园艺博览会
1963 年	德国	汉堡	汉堡国际园艺博览会
1964 年	奥地利	维也纳	奥地利世界园艺博览会
1969 年	法国	巴黎	巴黎国际花草博览会
1972 年	荷兰	阿姆斯特丹(Amstelpark 公园)	芙萝莉雅蝶园艺博览会
1973 年	德国	汉堡	汉堡国际园艺博览会
1974 年	奥地利	维也纳	维也纳国际园艺博览会
1976 年	加拿大	魁北克	魁北克国际园艺博览会
1980 年	加拿大	蒙特利尔	蒙特利尔园艺博览会
1982 年	荷兰	阿姆斯特丹 (Gaasperlaspark 公园）	阿姆斯特丹国际艺博览会（国际园林节）
1983 年	德国	慕尼黑	慕尼黑国际园艺博览会
1984 年	英国	利物浦	国际园林节
1990 年	日本	大阪	大阪万国花卉博览会（花与绿博览会）
1992 年	荷兰	阿姆斯特丹（路特米尔 Zoetermeer)	海牙国际园艺博览会
1993 年	德国	斯图加特	斯图加特园艺博览会
1994 年	法国	圣・丹尼斯	圣・丹尼斯国际园艺博览会
1995 年	德国	哥特布斯	哥特布斯国际园艺博览会
1996 年	意大利	热亚那	热亚那国际园艺博览会
1997 年	比利时	利戈	利戈国际园艺博览会
1997 年	加拿大	魁北克	97 国际花卉博览会
1999 年	中国	昆明	昆明世界园艺博览会
2000 年	日本	兵库县淡路岛	日本淡路花卉博览会
2002 年	荷兰	阿姆斯特丹（哈伦姆米尔 Haarlemermeer)	芙萝莉雅蝶园艺博览会
2003 年	德国	罗斯托克	罗斯托克国际园艺博览会
2004 年	日本	静冈（滨名湖畔）	日本滨名湖国际园艺博览会
2005 年	德国	慕尼黑	慕尼黑联邦园艺展 (BUGA)
2006 年	泰国	清迈	清迈世界园艺博览会
2006 年	中国	沈阳	沈阳世界园艺博览会
2010 年	台湾	台北	台北国际花卉博览会
2011 年	中国	西安	西安世界园艺博览会
2012 年	荷兰	芬罗	芙萝莉雅蝶园艺博览会